Digital Sustainability

Digital Sustainability: Navigating Entrepreneurship in the Information Age explores the intersection of technology and sustainability, offering a panoramic view of innovative strategies and solutions for building a more environmentally conscious and socially responsible future.

From exploring the transformative potential of blockchain technology in sustainable supply chains to harnessing the power of Artificial Intelligence (AI) and machine learning for environmental monitoring and conservation, each chapter presents cutting-edge insights and practical applications. The book highlights the ethical implications of entrepreneurship and data privacy, focusing on the potential of AI and machine learning for sustainable resource utilization and decision-making processes. Delving into areas such as renewable energy integration, data privacy, cybersecurity, IoT entrepreneurship, smart cities, and beyond, this book equips entrepreneurs, policymakers, and researchers with the knowledge and tools needed to drive meaningful change in the digital era.

With a rich tapestry of case studies, future perspectives, and actionable insights, this book offers a roadmap for entrepreneurs, engineers, business professionals, and those interested in technology and sustainability, focusing on redefining business models, fostering innovation, and creating a more connected, sustainable world.

Digital Sustainability

Navigating Entrepreneurship in the Information Age

Edited by

Pankaj Bhambri and Ilona Pawełoszek

CRC Press
Taylor & Francis Group
Boca Raton London New York

CRC Press is an imprint of the
Taylor & Francis Group, an **informa** business

Designed cover image: Shutterstock

First edition published 2025
by CRC Press
2385 NW Executive Center Drive, Suite 320, Boca Raton FL 33431

and by CRC Press
4 Park Square, Milton Park, Abingdon, Oxon, OX14 4RN

CRC Press is an imprint of Taylor & Francis Group, LLC

Library of Congress Cataloging-in-Publication Data

Names: Bhambri, Pankaj, editor. | Pawełoszek-Korek, Ilona, editor.
Title: Digital sustainability : navigating entrepreneurship in the information age /
edited by Pankaj Bhambri and Ilona Pawełoszek.
Description: First edition. | Boca Raton, FL : CRC Press, 2025. |
Includes bibliographical references and index.
Identifiers: LCCN 2024026366 (print) | LCCN 2024026367 (ebook) | ISBN 9781032775555
(hardback) | ISBN 9781032776552 (paperback) | ISBN 9781003484226 (ebook)
Subjects: LCSH: Technological innovations--Economic aspects. |
Technological innovations--Environmental aspects. | Social entrepreneurship.
Classification: LCC HC79.T4 D537 2025 (print) | LCC HC79.T4 (ebook) |
DDC 338/.064--dc23/eng/20240808
LC record available at https://lccn.loc.gov/2024026366
LC ebook record available at https://lccn.loc.gov/2024026367

ISBN: 978-1-032-77555-5 (hbk)
ISBN: 978-1-032-77655-2 (pbk)
ISBN: 978-1-003-48422-6 (ebk)

DOI: 10.1201/9781003484226

Typeset in Times New Roman
by Deanta Global Publishing Services, Chennai, India

Contents

About the Editors

Dr. Pankaj Bhambri is affiliated with the Department of Information Technology at Guru Nanak Dev Engineering College in Ludhiana. Additionally, he fulfills the role of the Convener for his Departmental Board of Studies. He possesses nearly two decades of teaching experience. His research work has been published in esteemed worldwide and national journals, as well as conference proceedings. Dr. Bhambri has garnered extensive experience in the realm of academic publishing, having served as an editor for a multitude of books in collaboration with esteemed publishing houses such as CRC Press, Elsevier, Scrivener, and Bentham Science. In addition to his editorial roles, he has demonstrated his scholarly prowess by authoring numerous books and contributing chapters to distinguished publishers within the academic community. Dr. Bhambri has been honored with several prestigious accolades, including the ISTE Best Teacher Award in 2023 and 2022, the I2OR National Award in 2020, the Green ThinkerZ Top 100 International Distinguished Educators Award in 2020, the I2OR Outstanding Educator Award in 2019, the SAA Distinguished Alumni Award in 2012, the CIPS Rashtriya Rattan Award in 2008, the LCHC Best Teacher Award in 2007, and numerous other commendations from various government and non-profit organizations. He has provided guidance and oversight for numerous research projects and dissertations at the postgraduate and Ph.D. levels. He successfully organized a diverse range of educational programmes, securing financial backing from esteemed institutions such as the AICTE, the TEQIP, among others. Dr. Bhambri's areas of interest encompass machine learning, bioinformatics, wireless sensor networks, and network security.

Asoc. Prof. Ilona Pawełoszek is an academic, teacher, and researcher hailing from Poland. She holds a Ph.D. in Economics with a specialization in Management and Quality Sciences. Her commitment to academic excellence is further reflected in her completion of pedagogical studies focused on training teachers in technical subjects. She graduated from the Faculty of Management at the Czestochowa University of Technology, specializing in "development management and consulting. Since 1999, Dr. Pawełoszek has been a dedicated member of the Faculty of Management at the Częstochowa University of Technology, where she has played various roles in research and education. She fulfilled many responsibilities, such as being an organizational and scientific committee member for more than 20 international conferences, a member of editor boards of three scientific journals, and a reviewer of many scientific articles and conference proceedings. She was a supervisor of two Ph.D. students and an expert in research conducted in cooperation with companies.

Currently, she serves as an assistant professor at the Department of Management Information Systems, a position that underscores her expertise in the field. With an extensive educational background and a prolific research career, she has made significant contributions to the fields of management, business, information technology, and Artificial Intelligence. She is the author of more than 100 scientific publications, including three monographs. Dr. Pawełoszek's research interests revolve around cutting-edge technologies that support managerial decisions. She is a proponent of using semantic web technology, Big Data analysis, and data mining techniques to enhance decision-making processes. Her work has led to several dozen publications, including two original monographs. Her commitment to research has earned her membership in both the Scientific Association of Business Informatics (NTIE) and the Polish Association for Production Management (PTZP). Dr. Pawełoszek's commitment to sharing her knowledge extends beyond her academic work. She has delivered invited lectures and speeches at various international events and institutions. Her topics have ranged from geographic information systems to the integration of Artificial Intelligence and management, showcasing her diverse expertise.

Contributors

Kurbanov Khayrillo Abdurasulovich
Tashkent State University of Economics
Tashkent, Uzbekistan

M. Shuaib Ahmed
C. Abdul Hakeem College of Engineering and Technology
Ranipet, Tamil Nadu, India

Paula Bajdor
Czestochowa University of Technology
Czestochowa, Poland

Aruna Bajpai
Banasthali Vidyapith
Jaipur, Rajasthan, India

Amber Batwara
The LNM Institute of Information Technology
Jaipur, Rajasthan, India

Pankaj Bhambri
Guru Nanak Dev Engineering College
Ludhiana, Punjab, India

K.S. Chakradhar
Mohan Babu University (Erstwhile Sree Vidyanikethan Engineering College)
Tirupati, Andhra Pradesh, India

Harpreet Kaur Channi
Chandigarh University
Mohali, Punjab, India

Grzegprz Chmielarz
Czestochowa University of Technology
Czestochowa, Poland

Iwona Chomiak-Orsa
Wroclaw University of Economics and Business
Wroclaw, Poland

Dariusz Dudek
Czestochowa University of Technology
Czestochowa, Poland

Soumik Gangopadhyay
Institute of Engineering & Management
Kolkata, West Bengal, India

Karlibayeva Gulshat
Tashkent State University of Economics
Tashkent, Uzbekistan

Yogesh Kumar Gupta
Banasthali Vidyapith
Jaipur, Rajasthan, India

Magdalena Jurczyk-Bunkowska
Opole University of Technology
Luboszycka, Opole, Poland

A.B. Feroz Khan
Syed Hameedha Arts and Science College, Keelakarai
Tamilnadu, India

Karlibayeva Raya Khodjabaevna
Tashkent State University of Economics
Tashkent, Uzbekistan

Suresh Kumar
Geeta University
Panipat, Haryana, India

Tomasz Lis
Czestochowa University of Technology
Czestochowa, Poland

Aarchie Maggu
I.K.Gujral Punjab Technical University
Jalandhar, Punjab, India

Mohit Makkar
The LNM Institute of Information Technology
Jaipur, Rajasthan, India

M. Muhammed Mustafa
Sri Krishna College of Engineering and Technology
Coimbatore, Tamil Nadu, India

K. Neelima
Mohan Babu University (Erstwhile Sree Vidyanikethan Engineering College)
Tirupati, Andhra Pradesh, India

Karolina Nurzyńska
Silesian University of Technology
Silesia, Poland

Ilona Pawełoszek
Czestochowa University of Technology
Czestochowa, Poland

Aleksandra Ptak
Czestochowa University of Technology
Czestochowa, Poland

Rachna Rana
Ludhiana Group of Colleges
Ludhiana, Punjab, India

Ritu
I.K.Gujral Punjab Technical University
Jalandhar, Punjab, India

Suchana Roy
Institute of Engineering & Management
Kolkata, West Bengal, India

Ramandeep Sandhu
Lovely Professional University
Jalandhar, Punjab, India

J. Sankari
Anna Adarsh College for Women
Chennai, Tamil Nadu, India

Piyal Sarkar
Barabisha High School
Alipurduar, West Bengal, India

Satyam
Birla Institute of Technology & Science
Pilani, Rajasthan, India

Neelam Sharma
I.K.Gujral Punjab Technical University
Jalandhar, Punjab, India

Sonia Sharma
Lovely Professional University
Phagwara, Punjab, India

Vikram Sharma
The LNM Institute of Information Technology
Jaipur, Rajasthan, India

Klaudia Smoląg
Czestochowa University of Technology
Czestochowa, Poland

Marta Starostka-Patyk
Czestochowa University of Technology
Czestochowa, Poland

Cezary Stępniak
Czestochowa University of Technology
Czestochowa, Poland

Julian Szymański
Gdansk University of Technology
Gdansk, Poland

Bebesh Tripathy
Chandigarh University
Mohali, Punjab, India

Tomasz Turek
Czestochowa University of Technology
Czestochowa, Poland

Paweł Weichbroth
Gdansk University of Technology
Gdansk, Poland

Jędrzej Wieczorkowski
SGH Warsaw School of Economics
Niepodległości, Warsaw, Poland

Leszek Ziora
Czestochowa University of Technology
Czestochowa, Poland

Preface

In an era where technological advancements intertwine with the urgent need for environmental stewardship, this edited volume delves into the intersection of digital innovation and sustainable entrepreneurship. The journey through this book is a testament to the transformative power of technology when harnessed for the greater good. From blockchain's potential to revolutionize supply chains, to Artificial Intelligence (AI)-driven solutions for environmental monitoring, each chapter offers insights into how digital tools can propel sustainability forward.

In the opening sections, we lay the groundwork with an exploration of digital sustainability and entrepreneurship, setting the stage for an enlightening discourse. We then dive deep into specific domains, examining the role of renewable energy in data centers, the ethical implications of AI in entrepreneurship, and the emergence of smart cities driven by Internet of Things (IoT) innovation. Cybersecurity, an ever-pressing concern in our interconnected world, takes center stage as we explore its critical role in sustainable business models. From safeguarding supply chains to protecting digital infrastructure, the importance of robust cybersecurity measures cannot be overstated.

As we conclude our journey, we reflect on key takeaways and gaze into the future, envisioning a landscape where digital sustainability is not just a concept, but a way of life. Through case studies, future perspectives, and a compelling call to action, we invite you to join us in driving the digital sustainability agenda forward.

This book is not just a collection of chapters; it's a manifesto for change, a blueprint for entrepreneurs and innovators to leverage digital technologies in the service of a more sustainable future. We hope it inspires you, challenges you, and ultimately empowers you to make a difference in the world of entrepreneurship and beyond.

Thank you for embarking on this journey with us.

1 Elevating Workforce Strategies

Designing a SMART Recruiting System with ML Models

Aarchi Maggu and Pankaj Bhambri

1.1 INTRODUCTION

In recent times, internet employment portals have been inundated with a large number of resumes that come in various formats, posing a challenge in extracting crucial information from them [1]. Human resource managers face challenges in effectively screening resumes to identify the most suitable candidates [2].

Many firms resolve this issue by transitioning to e-recruitment platforms and comprehending the distinct criteria that recruiters have for online and offline resumes [3]. These platforms help to reduce the time, effort, and cost of manually filtering out the resumes. These systems address different challenges faced during screening candidate resumes. Online job recruitment has simplified the job search process, as they extract important information from resumes and match it with job descriptions to find the best-fit resumes [4,5]. Many recruiters focus on the approach of finding the best possible ways to fill a particular vacancy but often neglect the run-time complexities i.e. they match every resume with all the job descriptions available instead of comparing only those resumes that match the job category [6,7].

In recent years, online recruitment in the IT industry has faced the following challenges:

- Finding suitable candidates from the pack: It becomes difficult for human resource managers to find suitable resumes that match the given job category.
- Understanding diverse formats of resumes: The resumes that come for a job posting are not standard. They are in different styles and formats. Because of this, certain resumes are not easily accessed by e-recruiting platforms, thus a deserving candidate can lose various job opportunities by being overlooked.
- Evaluating the candidate's resume: The next task is to determine whether the specific resume has a resemblance to the given job description.

DOI: 10.1201/9781003484226-1

To overcome the above-mentioned challenges, we will outline a model named SMART Recruiting System, which uses machine learning (ML) models. The resumes are classified according to a provided job category using different classifier models, and then keywords from job descriptions and resumes are extracted using term frequency–inverse document frequency (TF-IDF). Then the similarity between resume and job description is found out using a combination of cosine similarity, Jaccard similarity, Sorensen similarity, and overlap similarity.

The main objectives of the project are:

- To convert unstructured resumes to a structured format.
- To classify the resumes according to job categories.
- Use of a content-based filtering recommendation model, which recommends resumes that best match the job descriptions.
- Topic modeling of resumes is done to provide additional information about the resumes.

1.2 RELATED WORK

In recent years, there has been a significant rise in the number of online job portals that offer recruiting services to job seekers, many of whom apply to the described position [8]. Job recruiters face challenges in selecting the most suitable profile/resume from a pool of resumes. The method of aligning a candidate's resume with a job description is analogous to a recommender system, where the candidate's profile is suggested for a certain position [9]. A recommender system is defined as an important class of machine learning algorithms that offers relevant suggestions to users. In recent years, some of the biggest brands that we engage with daily are built using recommendation engines such as Netflix, Amazon, and Google, and is also used in various domains i.e. book recommendations [10], music recommendations [11], and news recommendations [12].

The authors [13] proposed an approach that there should be a good match between the preferences of the candidates for the job and the recruiters who are in search of the best applicants. To ensure the effectiveness of their strategy, they implemented two separate and exclusive recommendation systems, which were then integrated into their systems to enhance the compatibility between candidates and job opportunities. The authors [14] introduced a method that utilizes collaborative filtering to identify the most suitable candidate for a job. The recruitment process employed two methodologies, namely, matching as well as collaborative filtering. However, the scoring system presented the issue of limited variation in scores. The issue was resolved by the use of collaborative filtering, which enhanced the distribution of scores and facilitated the identification of the most suitable option for the organization. The authors [15] employed an item-based collaborative filtering algorithm to provide employment recommendations. The method was optimized by merging job descriptions and resume material, where the job preference prediction formula was enhanced by including historical delivery weight derived from job descriptions alongside comparable user weight derived from resume content. They utilized four recommendation algorithms for their calculations: Collaborative-Filtering Recommendation, Content-Based Recommendation, Knowledge-Based Recommendation, and Hybrid Recommendation. In their study, the authors [8] introduced a hybrid technique that utilized an ontology-based method to match candidates' profiles with job descriptions. This methodology involved the use of ranking algorithms and a selection-based approach to determine the candidates' rankings. Nevertheless, the primary drawback of this strategy is the substantial expense (in terms of run-time complexity) associated with the matching process [7]. In an attempt to address the limitation identified in the aforementioned paper, [16] sought to minimize the run-time complexity by comparing the resumes of candidates with the job vacancies they are interested in and have applied for. The methodology employed involved utilizing machine learning techniques and natural language processing techniques to ascertain the scores and subsequently rank the candidates. This enabled the selection of the most qualified individuals from a vast pool of applicants for the job.

Hence, we present a novel model that diverges from previous systems by not solely relying on the content of individuals' resumes to recommend job opportunities, which results in modest levels of accuracy. To enhance this, we have suggested a method that is comprised of currentchapter. This chapter consists of the classification of the resumes alongwith recommendations to the hiring committee.

1.3 METHODOLOGY

The major goal of this chapter is to filter out the resumes from the pack of resumes. To achieve this, the system is divided into two parts:

1. Classification
2. Recommendation

1.3.1 Design Approach

Function-oriented design is an approach that is used in which the decomposition of the model takes place into a set of interacting modules or units. Each module or unit has a clearly defined function. Hence, the proposed model is designed from a functional viewpoint.

1.3.2 Detail Design

1.3.2.1 Dataset Collection

In the proposed model, the resumes' dataset is selected from Kaggle, which contains 330 resumes. The dataset of job descriptions is taken from Kaggle and indeed.com.

1.3.2.2 Tokenization

Tokenization separates pieces of text into smaller modules called tokens. Here, tokens are considered as subwords, words, or characters. Therefore, tokenization can be classified into three words that are subwords, words, or characters tokenization. It is performed using the Natural Language Toolkit from Python (NLTK) library. The steps of tokenization are:

a) *Import tokenize module from NLTK library*
b) *Remove the useless punctuations from text*
c) *Print the output of tokenized data*

1.3.2.3 Pre-Processing

Pre-processing refers to the transformation of data into a format that is easily comprehensible by a computer.

First, tokenization using the NLTK library is performed on resumes. Further, the pre-processing steps are done on tokenized resumes which are lemmatization, stop word removal, and stemming.

Stop words: Stop words are phrases in English that do not contribute meaning to a sentence. They can be disregarded without altering the significance of the phrase. For instance, words like "he," "the," and "have" can serve as examples. The corpus already contains such terms, which are imported using the NLTK package.

Stemming and lemmatization: Stemming is the procedure that generates morphological variations of the fundamental word. Stemming codes are commonly known as stemming or stemming algorithms. NLTK has a PorterStemmer class which helps us to implement PorterStemmer algorithms for the word that we want to stem. This PorterStemmer class knows several suffixes and regular word forms with the help of which the input word can be transformed into an output stem. The resulting stem is mostly a shorter word that has the same root meaning. Both stemming and lemmatization yield the root form of the inflected words. The distinction between their outcomes lies in the fact that the stem cannot function as a valid word, whereas the lemma is a legitimate word in the language. The WordNetLemmatizer library is employed for lemmatization.

Redundancy removal: The redundancy of the text can be removed by using a set that reduces the repetitive words and gives the single word that is needed. Further, the target words were extracted as Proper Noun and Relevant Noun words that contain words related to tech and job description using spaCy tags.

1.3.2.4 TF-IDF

The term frequency–inverse document frequency (TF-IDF) method is the predominant approach for computing word frequencies. Frequency is the criterion used to compute a final score for each word. TF-IDF is a scoring method that highlights words that are more interesting and unique within a phrase, rather than across different phrases, without going into the details of the calculations. The TF-IDF vectorizer is capable of acquiring a set of words, enabling the encoding of additional ones, applying inverse frequency weightings, and tokenizing text.

Term frequency (TF) refers to the frequency with which a word or chunk appears throughout a document, divided by the total number of words in the document.

TF(t) = (Number of times term "t" appears in a document) / (Total number of terms in the document)

Term frequency is how common a word is, and inverse document frequency (IDF) is how unique or rare a word is.

IDF(t) = log_e(Total number of documents / Number of documents with term "t" in it)

According to the above-computed equation, the more the TF-IDF score, the more the relevance of that term in the document. In the proposed model, the resumes and job descriptions are modeled in the vector space. This can be accomplished by compiling a dictionary of words found in the papers and then converting them. Each phrase corresponds to a vector space dimension. The TF-IDF matrix was generated for the resumes and the job description.

1.3.2.5 Classification

Five different models are used for classification:

1. **Bernoulli naive Bayes:** Bernoulli naive Bayes is primarily employed for analyzing discrete data. It operates based on the Bernoulli distribution. Bernoulli naive Bayes is characterized by its ability to handle binary characteristics, which can take on values such as yes or no, true or false, 0 or 1, success or failure, and similar options. When the feature values are binary, the appropriate classifier to use is the Bernoulli Naive Bayes classifier. The formulae for Bernoulli naive Bayes is given in equation (1).

$$p(x) = P[X = x] = \begin{cases} q = 1 - p & x = 0 \\ p & x = 1 \end{cases} \tag{1}$$

where

p = probability of success

q = probability of failure

For a random variable "X" in the Bernoulli distribution,

where x is either 0 or 1

2. **Multinomial naive Bayes:** The Multinomial naive Bayes algorithm is a commonly employed probabilistic learning method in the field of natural language processing (NLP). The foundation of this method lies in Bayes' theorem, which forecasts the classification of a text as either a newspaper story or an email. The system computes the likelihood of each tag for a given text and thereafter outputs the tag with the highest probability. The formulae for Bernoulli naive Bayes are in Equation (2). This formula will help us to find the probability of tags in text.

$$P\left(\frac{A}{B}\right) = P(A) * P(B/A)/P(B) \tag{2}$$

where

P(B) = prior probability of B
P(A) = prior probability of class A
P(B|A) = occurrence of predictor B given class A probability

3. **Support vector machine (SVM):** This algorithm is a type of supervised algorithm for machine learning that can be applied to solve regression or classification problems. Nevertheless, it is frequently employed in situations involving classification. The radial basis function (RBF) kernel is stated as

$$K\left(xi, xj\right) = \exp\left(-\gamma \left\|\mathrm{xi} - \mathrm{xj}\right\| 2\right) \tag{3}$$

The RBF is also known as the Gaussian function. It contains parameter γ. If the value of γ is small, then it will make the model behave like SVM.

4. **Decision tree:** A decision tree is constructed by sequentially asking a series of questions based on the dataset of 330 resumes that we have collected. Every time an answer is obtained, a question that is in the queue is asked up until a conclusion about the class label is recorded. The decision tree, which is a hierarchical structure consisting of nodes and their respective directed edges towards them, is used to organize the chain of questions and their possible answers.
5. **Random forest:** Classification and regression problems can be solved by a machine learning algorithm i.e. random forest. Also, it makes use of the ensemble learning technique in which many classifiers that produce complex solutions are combined and used.

In the proposed work, the model is trained on a training dataset that consists of two folders in which selected and rejected resumes for a particular job are present. First on training data, cleaning and lemmatization are done.

After that, a function named training_list is created which will read the resumes present in the training dataset and store it in a list. From the resume list, the data is labeled and then shuffled and split into testing and training sets. Features are extracted from the training and testing sets and corresponding feature matrices using TF-IDF [17–20]. The accuracy using different classifier models was taken out. The classifier models employed include Bernoulli naive Bayes, support vector machine, multinomial naive Bayes, decision tree, and random forest. The random forest classifier yielded the highest accuracy among the options. The confusion matrix was printed using random forest as shown in Figure 1.1.

Once the training model is prepared, it should be saved using the pickle module. The pickle module is responsible for implementing binary protocols that allow for the serialization and de-serialization of a Python object structure. The process of "pickling" refers to the conversion of a Python object hierarchy into a byte stream.

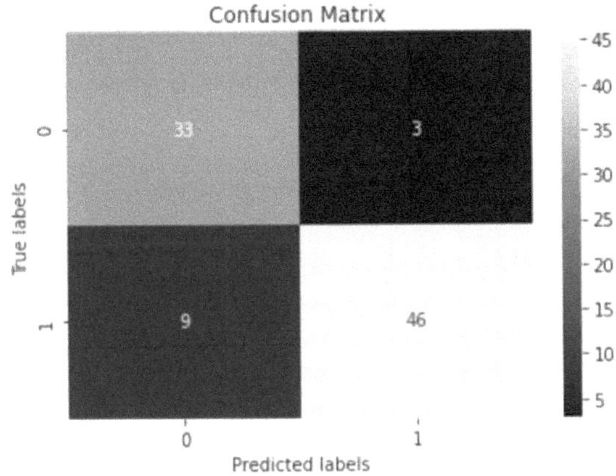

FIGURE 1.1 Confusion matrix using random forest.

By saving the completed model as a pickle, you can avoid the need to train the model each time the program is executed, resulting in significant time savings. After saving your model as a pickle file, you may then load it to make predictions. The proposed model is stored as resume_classifier.pkl and resume_vector.pkl. Once the models are saved, they may be loaded using the pickle module to forecast the class for new, unseen resumes.

Now prediction of new unseen resumes will be done. This model can handle several file types such as doc, docx, odt, pdf, pptx, and odp zip files (zip files will be extracted into the zip extract folder). Pre-processing of each resume will be done. Based on the training model, the results of new, unseen resumes will be printed and the resumes selected for a particular job will be stored in the output folder.

Training algorithm

Step 1: Collect resumes from indeed.com.
Step 2: Store resumes belonging to software developer posts in the selected resume folder and store the rest of the resumes in the rejected resume folder.
Step 3: Read the resumes from the folder.
Step 4: Pre-process the resumes.
Step 5: Tokenize and lemmatize the text of the resumes.
Step 6: Store the text of all resumes in resume_list.
Step 7: Remove punctuations and other unnecessary characters.
Step 8: Label the selected resumes as 1 and the rejected resumes as 0.
Step 9: Shuffle and split the data into a training set and a testing set.
Step 10: Extract words as features from the training and testing sets and make corresponding feature matrices.
Step 11: Find accuracies from other classifier models.
Step 12: Train the model and save it in a pickle file.
Step 13: Evaluate the model using test data.
Step 14: Exit.

TABLE 1.1
Accuracies Obtained from Other Classifier Models

Classifier	Accuracy
Bernoulli Naive Bayes [13]	0.75
Multinomial Naive Bayes [13]	0.71
Support Vector Machine [14]	0.80
Decision Tree [15]	0.75
Random Forest [16]	0.91

1.3.2.6 Results

Based on Table 1.1, including the outcomes of different classifier models, it can be observed that the random forest approach achieves the highest level of accuracy in comparison to the other algorithms. The Bernoulli naive Bayes classifier achieves an accuracy of 75%, the multinomial naive Bayes classifier achieves an accuracy of 71%, the support vector machine classifier achieves an accuracy of 80%, the decision tree classifier achieves an accuracy of 75%, and the random forest classifier achieves an accuracy of 91%, as depicted in Figure 1.2. Therefore, random forest algorithm is chosen for further evaluation of the proposed model and the confusion matrix is shown in Figure 1.4.

Further, for testing purposes, the pickle files were opened, and then scores for resumes put in the testing folder were taken out and according to the scores, the resumes were differentiated into selected and rejected resumes. Then the selected resumes get stored in an output folder, which will be further used for recommendation.

1.3.2.7 Resume Recommendation Model

The recommendation model is specifically developed to analyze job descriptions and resumes, and then provide a list of resumes that closely match the given job description. This is done by giving scores to each resume and rating them based on their suitability for the position.

This is achieved by combining the four similarity algorithms i.e. Cosine similarity, Jaccard similarity, Sorensen–Dice, and Overlap similarity.

1.3.2.8 Content-Based Recommender

This content-based recommendation system matches the various job descriptions supplied by the recruiting team with the content of the resumes. It then recommends the top resumes that have the most similarity to the job descriptions for the recruiters. In addition, the resumes are assigned ratings and subsequently rated to facilitate the selection of the most suitable candidate for the desired job by the recruiting team. The model utilizes the cleaned resumes and job descriptions to calculate the aggregated outcome of the cosine similarity, Jaccard similarity, Sorensen–Dice similarity, and overlap similarity between the job descriptions and the resumes.

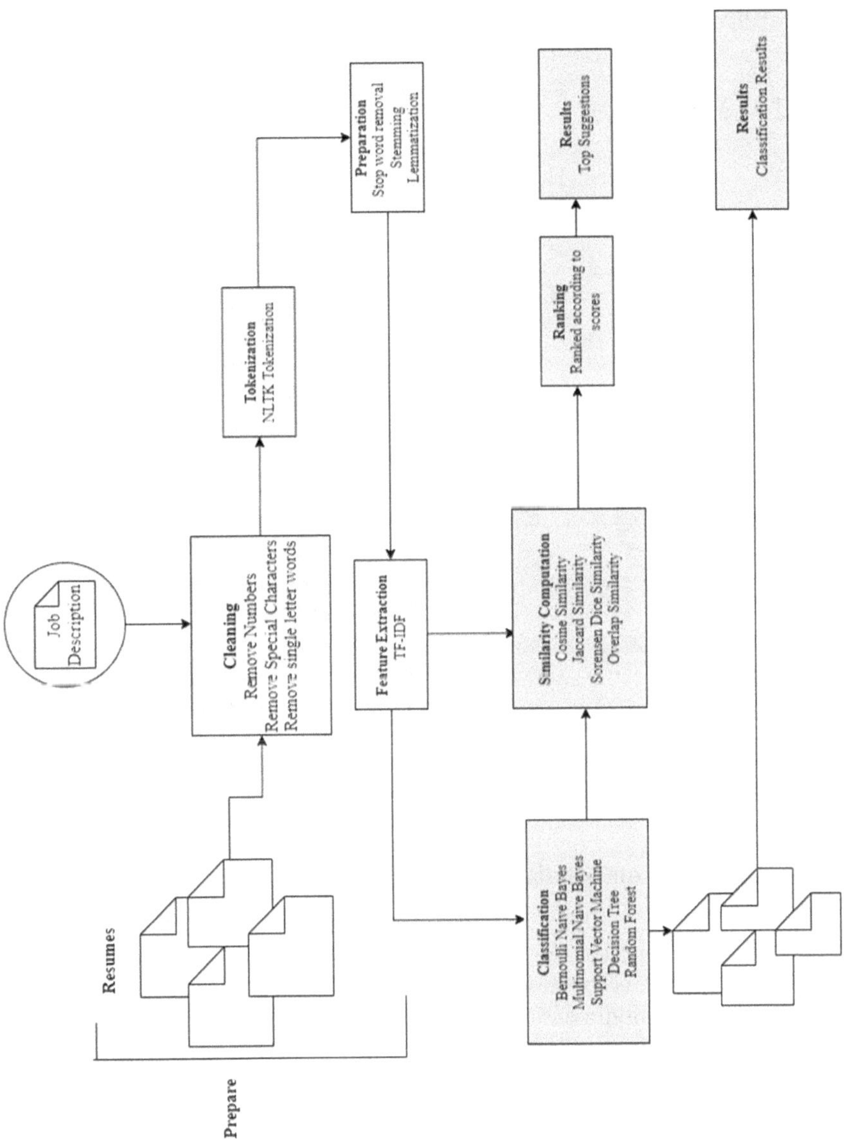

FIGURE 1.2 Detailed design of the proposed model.

FIGURE 1.3 Stemming.

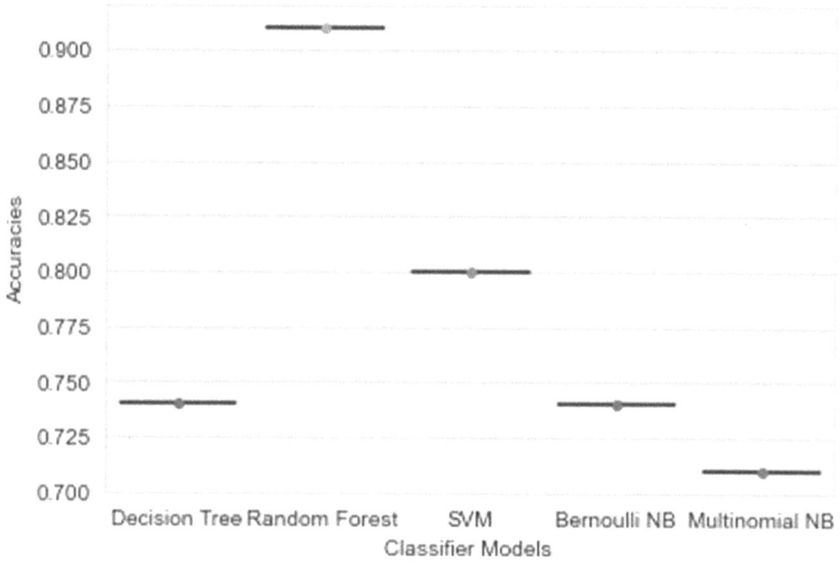

FIGURE 1.4 Accuracies of different classifier models for software developer post.

1.3.2.9 Similarity Computations

1.3.2.9.1 Consine Similarity

The most prevalent criterion for measuring the degree of similarity between texts, regardless of their size, is commonly utilized. In mathematical terms, the projection of two vectors in a multidimensional space is a measure of the cosine of the angle between them. For an illustration, look at the charts in Figures 1.5, 1.6, and 1.7.

The jobs, i.e. the machine learning engineer and the data scientist, are very similar to each other, so much so that θ is close to 0° (zero) and cos(θ) is close to 1 (close to 1 means more similar items).

The jobs, i.e. bartender and data scientist, are not similar to each other in a way that θ is close to 90° (ninety degrees) and cos(θ) is close to 0 (close to 0 means fewer similar items).

The underlying concept in the aforementioned case is that a cosine value close to 1 indicates similarity between items, while a value close to 0 suggests dissimilarity. Conversely, a cosine value of –1 signifies similarity between items that are opposite to that of one another. The formulae of cosine similarity are given in Equation (4).

$$(P,Q)=\frac{|P\cap Q|}{\sqrt{|P|.|Q|}}=\frac{|P\cap Q|}{\sqrt{(|P\cap Q|+|P-Q|)(|P\cap Q|+|Q-P|)}} \tag{4}$$

Where P and Q are empty sets.

1.3.2.9.2 *Jaccard Similarity*

A criterion where two sets of data are available and it is used to measure the similarity between them. The argument can take place that the computation of the size of two given sets is required to measure the similarity. But in this case, the number of similar elements solely cannot inform us how relatively it can be equated to the set's size.

The Jaccard similarity recommends that to assess similarity, we need to calculate the ratio between the intersection with the union of both sets of data. The formulae of Jaccard similarity is given in Equation (5).

$$J(P,Q)=\frac{|P\cap Q|}{|P\cup Q|}=\frac{|P\cap Q|}{|P\cap Q|+|P-Q|+|Q-P|} \tag{5}$$

Where P and Q are non-empty.

1.3.2.9.3 *Sorensen–Dice*

The Sorensen–Dice distance, a statistical metric, is employed to quantify the similarity between two provided sets. The term is mathematically defined as the product of two and the cardinality of the intersection of sets P and Q, divided by the total of the cardinalities of sets P and Q. The Sorensen–Dice formula is provided in equation (6).

$$\mathrm{D}(\mathrm{P},\mathrm{Q})=\frac{2.|P\cap Q|}{|P|+|Q|} \tag{6}$$

Where P and Q are nonempty sets.

Similar to Jaccard similarity, the similarity values in this range span from zero to one. However, in contrast to Jaccard similarity, this dissimilarity measure does not qualify as a metric because it fails to meet the criteria of triangular inequality.

1.3.2.9.4 *Overlap Similarity*

The overlap similarity quantifies the degree of overlap between two given sets. It is defined as the cardinality of the intersection between two sets and the cardinality of the smaller of the two sets. The formulae for overlap similarity are given in Equation (7).

$$O(A,B) = \frac{|A \cap B|}{\min(|A|,|B|)} \tag{7}$$

This method involves comparing the employer's job description with the content of resumes in the database and then suggesting the top resumes that have the closest match to the employer. This model integrates the sanitized resume data with the job description. The Likeness Score of resumes is determined by employing four similarity measures: cosine similarity, Jaccard similarity, Sorensen–Dice similarity, and overlap similarity. These measures are calculated between every set of resume vectors and job descriptions, both of which are represented in vector space. The resulting similarities are then combined to obtain the overall likeness score. At last, the final average score of all the similarities is considered for ranking the resumes according to specified job descriptions.

1.3.2.10 Topic Modeling Using LDA

Classification and Categorization of text in a document takes place using latent Dirichlet allocation (LDA). It also finds the words per topic. Hence, it is a technique of topic modeling. This is represented using Dirichlet distributions and processes.

Latent Dirichlet allocation operates under two assumptions:

1. Documents consist of a combination of subjects.
2. Topics consist of a combination of words.

The probability distribution is used to generate tokens, which are words, on specific themes. In statistical terminology, documents can be referred to as the probability distribution of themes, whereas topics represent the probability distribution of tokens, which are essentially words.

The suggested methodology utilizes LDA for topic modeling, which enables the identification of the specific class or unit to which a resume belongs. Here, the topics are divided into several useful topics and create a cluster of matching topic resumes as shown in Figures 1.8 and 1.9.

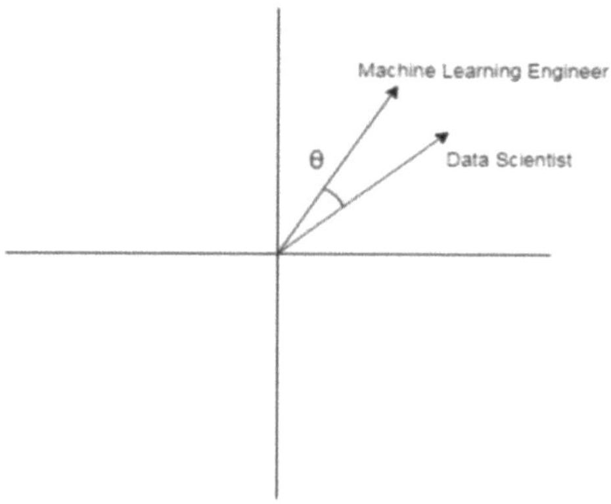

FIGURE 1.5 Similarity between ML engineer and data scientist.

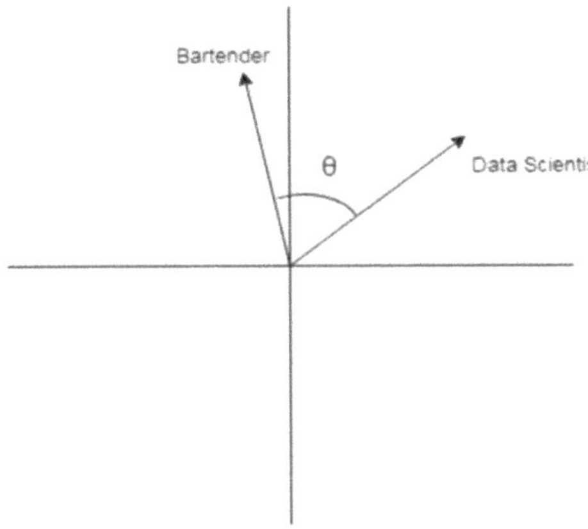

FIGURE 1.6 Similarity between bartender and data scientist.

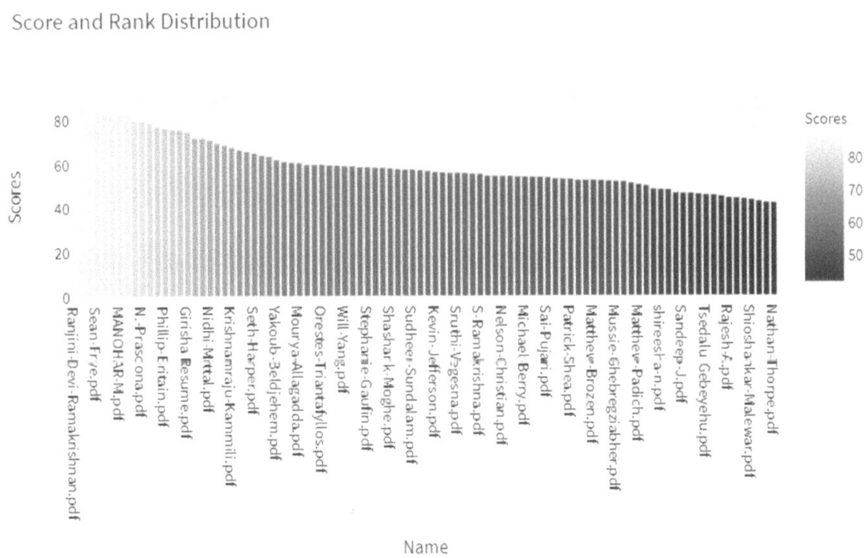

FIGURE 1.7 The score and rank distribution of the resumes.

1.4 MAJOR CONTRIBUTIONS

1. The proposed system converts unstructured resumes to structured resumes.
2. The proposed system automates the process of classification of resumes and applicant ranking based on job descriptions provided, which reduces the run-time complexities.
3. The proposed system produces ranking decisions that are somewhat highly compatible with those of skillful humans.

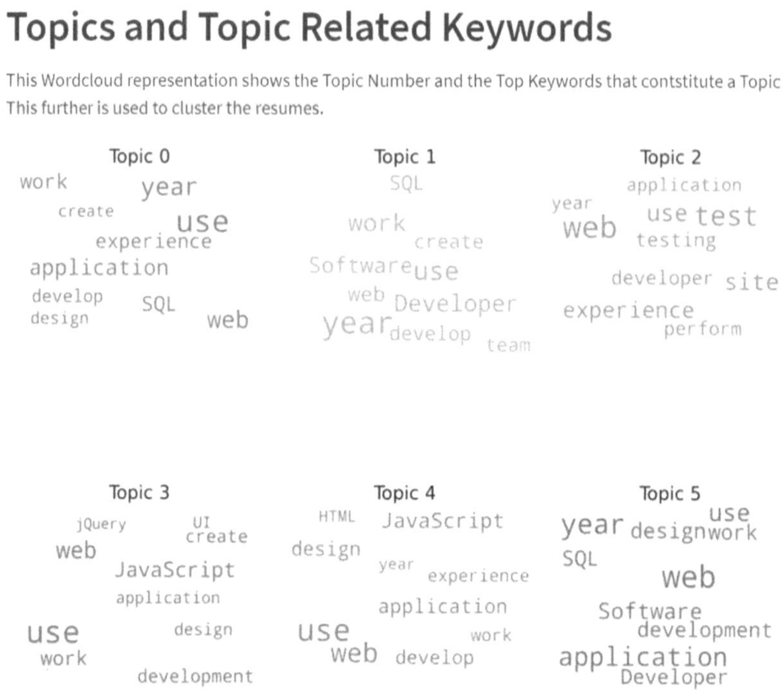

FIGURE 1.8 Representation of topic numbers and topic keywords.

4. Topic modeling is performed using LDA to extract supplementary information from resumes, specifically to determine the category or class to which each resume belongs, as determined by the LDA model.

1.5 CONCLUSION

1. The organization receives a diverse array of applicants for each job posting. Choosing the most qualified candidate's curriculum vitae from a large pool of applicants is a fatiguing undertaking for any organization in the present times. The manual classification of resumes is a highly time-consuming operation that results in a significant waste of resources.
2. To address this problem, we have developed a machine learning model that suggests the most suitable CV to the HR department based on the job description. The proposed system is divided into two phases:first, classify the resumes according to the particular job post; and second, recommend the resume of the applicant that is very similar to the job post and which matches its job description highly.
3. This approach works efficiently as can be seen through the results, but this does not guarantee the job to the applicant because the final decision will be taken when the personal interviews will take place. This model just takes

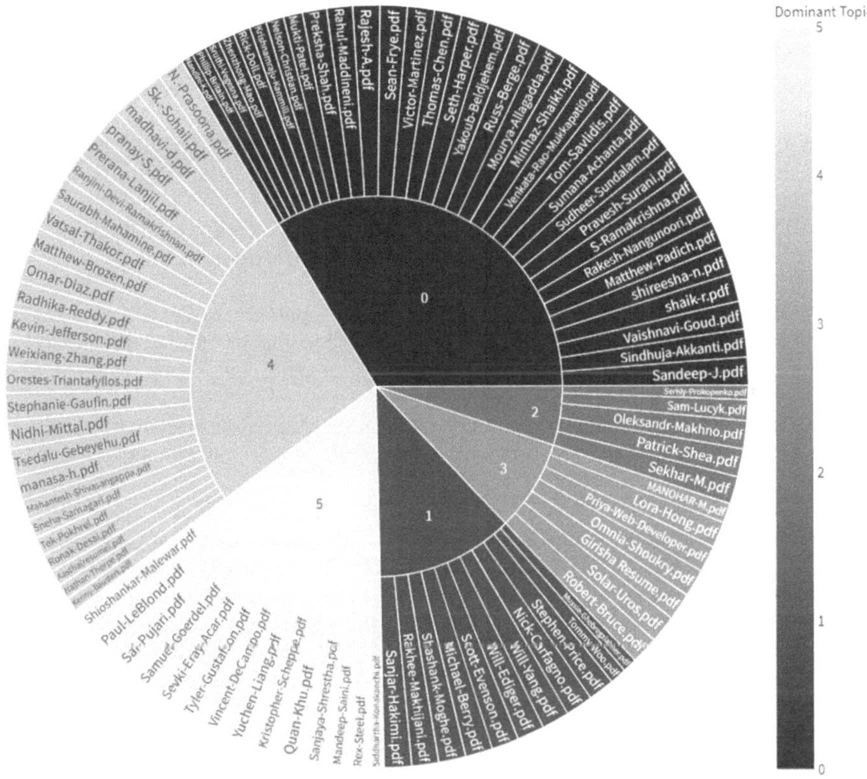

FIGURE 1.9 Topic distribution graph.

the applicant one step closer to fulfilling their dreams and also helps the recruiters so they can find the relevant candidate to whom they can interview further and select them according to their company's criteria. In this way, it reduces the workload of the hiring committee as they can get the best candidates' resumes through this model.

1.6 FUTURE SCOPE

In the future, the model's performance can be improved by utilizing advanced deep learning models like recurrent neural networks, convolutional neural networks, long-short term memory, and other similar models. The machine learning models will remember the candidates that were selected in the past and, therefore, it will rank them according to their past experiences and, hence, the model will work efficiently. Furthermore, the model can be enhanced by taking feedback from the HR professionals and making improvements in the model iteratively. Thus, machine learning has a wide range of potentiality in this area and, therefore, the researchers can explore ways to make the recruitment process even better.

REFERENCES

1. V. Mittal, P. Mehta, D. Relan, and G. Gabrani, "Methodology for resume parsing and job domain prediction," *Journal of Statistics and Management Systems*, vol. 23, no. 7, pp. 1265–1274, Oct. 2020. doi: 10.1080/09720510.2020.1799583.
2. V. Lai, K. J. Shim, R. J. Oentaryo, P. K. Prasetyo, C. V. E.-P. Lim, and D. Lo, "CareerMapper: An automated resume evaluation tool," in *2016 IEEE International Conference on Big Data (BIG DATA)*, 2016, pp. 4005–4007.
3. E. Furtmueller, C. Wilderom, and M. Tate, "Managing recruitment and selection in the digital age: e-HRM and resumes," *Human System Management*, vol. 30, no. 4, pp. 243–259, 2011. doi: 10.3233/HSM-2011-0753.
4. A. Zaroor, M. Maree, and M. Sabha, "A hybrid approach to conceptual classification and ranking of resumes and their corresponding job posts," in *Intelligent Decision Technologies 2017, KES-IDT 2017, PT I*, 2018, vol. 72, pp. 107–119. doi: 10.1007/978-3-319-59421-7_10.
5. S. Guo, F. Alamudun, and T. Hammond, "ResuMatcher: A personalized resume-job matching system," *Expert System with Applications*, vol. 60, pp. 169–182, Oct. 2016. doi: 10.1016/j.eswa.2016.04.013.
6. A. B. Kmail, M. Maree, and M. Belkhatir, "MatchingSem: Online recruitment system based on multiple semantic resources," in *2015 12th International Conference on Fuzzy Systems and Knowledge Discovery (FSKD)*, 2015, pp. 2654–2659.
7. A. Zaroor, M. Maree, and M. Sabha, "JRC: A job post and resume classification system for online recruitment," in *2017 IEEE 29th International Conference on Tools with Artificial Intelligence (ICTAI 2017)*, 2017, pp. 780–787. doi: 10.1109/ICTAI.2017.00123.
8. "ICIAfS 2018 Author Index," in *2018 IEEE International Conference on Information and Automation for Sustainability (ICIAfS)*, 2018, pp. 1–13. doi: 10.1109/ICIAFS.2018.8913378.
9. P. K. Roy, S. S. Chowdhary, and R. Bhatia, "A machine learning approach for automation of resume recommendation system," in *International Conference on Computational Intelligence and Data Science*, 2020, vol. 167, pp. 2318–2327. doi: 10.1016/j.procs.2020.03.284.
10. N. Kurmashov, K. Latuta, and A. Nussipbekov, "Online book recommendation system," in *2015 Twelve International Conference on Electronics Computer and Computation (ICECCO)*, 2015, pp. 1–4. doi: 10.1109/ICECCO.2015.7416895.
11. Y. Song, S. Dixon, and M. Pearce, "A survey of music recommendation systems and future perspectives," Dec. 2012.
12. A. Darvishy, H. Ibrahim, F. Sidi, and A. Mustapha, "HYPNER: A hybrid approach for personalized news recommendation," *IEEE ACCESS*, vol. 8, pp. 46877–46894, 2020. doi: 10.1109/ACCESS.2020.2978505.
13. R. Rana, P. Bhambri, and Y. Chhabra, "Evolution and the future of industrial engineering with the IoT and AI," in *Integration of AI-Based Manufacturing and Industrial Engineering Systems with the Internet of Things* (pp. 19–37). CRC Press, 2024.
14. G. Singh and P. Bhambri, "Simulation analysis of AODV and DSDV routing protocols for secure and reliable service in Mobile Adhoc Networks (MANETs)," in *Integration of AI-Based Manufacturing and Industrial Engineering Systems with the Internet of Things* (pp. 205–216). CRC Press, 2023.
15. P. Bhambri and A. Khang, "Biosensor applications and principles of agricultural and aquacultural sectors," in A. Khang (Ed.), *Agriculture and Aquaculture Applications of Biosensors and Bioelectronics* (pp. 1–17). IGI Global, 2024. doi: 10.4018/979-8-3693-2069-3.ch001.

16. S. M. Shanmuga and P. Bhambri, "Pulmonary and lungs nodule classification using deep learning," in P. Bhambri, S. Rani, & M. Fahim (Eds.), *Computational Intelligence and Blockchain in Biomedical and Health Informatics* (pp. 320–332). CRC Press, Taylor & Francis Group, 2024. doi: 10.1201/9781003459347.
17. M. Singh, P. Singh, P. Bhambri, and R. Sachdeva, "A comparative study: Security algorithms," in U.G.C. Sponsored National Seminar (Abstract Book) (p. 26). Presented at the "Network Security and Its Implementations" seminar at Doaba College, Jalandhar held on March 1, 2005.
18. P. Bhambri, V. K. Sinha, I. S. Dhanoa, and J. Kaur, "Genome DNA sequence matching using HBM algorithm," *International Journal of Control and Automation*, vol. 12, pp. 531–539, 2019. http://sersc.org/journals/index.php/IJCA/article/view/2686/1887.
19. N. Tondon and P. Bhambri, "Novel approach for drug discovery," *International Journal of Research in Engineering and Applied Sciences*, vol. 7, no. 6, pp. 28–46, 2017. www.euroasiapub.org.
20. R. Rana, Y. Chhabra, and P. Bhambri, "A review on development and challenges in wireless sensor network," Paper presented at the International Multidisciplinary Academic Research Conference at CT University, Ludhiana (pp. 184-188) held during September 6–7, 2019.

2 Introduction to Digital Sustainability and Entrepreneurship

Pankaj Bhambri and Ilona Pawełoszek

2.1 INTRODUCTION

Digital sustainability refers to the responsible and efficient use of digital technologies to minimize their environmental impact and ensure their long-term viability. As our dependence on digital infrastructure and technologies continues to grow, concerns about the environmental footprint associated with their production, usage, and disposal have become increasingly prominent. The lifecycle of digital products, from manufacturing and energy consumption during operation to electronic waste at the end of life, contributes significantly to carbon emissions and resource depletion. The urgency to address these issues arises from the escalating global demand for digital services and the rapid evolution of technology, necessitating a comprehensive approach to balance technological advancement with environmental stewardship.

In the broader context, digital sustainability intersects with broader environmental and social sustainability goals. The digital sector has a dual role as both a contributor to environmental challenges and a potential solution through innovations that enhance energy efficiency, promote circular economy principles, and address societal challenges. Recognizing the need for a sustainable digital future, stakeholders across industries are increasingly embracing eco-friendly practices, such as optimizing data center efficiency, adopting renewable energy sources, and designing products with extended lifecycles. Digital sustainability is thus a critical aspect of overall sustainability efforts, fostering a balance between technological progress and environmental responsibility to ensure a resilient and sustainable digital ecosystem for future generations.

2.1.1 Rationale for Exploring Digital Sustainability and Entrepreneurship

The exploration of digital sustainability within the realm of entrepreneurship is motivated by a confluence of environmental, economic, and societal imperatives. In recent years, the escalating threat of climate change has heightened awareness of the ecological impact associated with digital technologies, prompting businesses to reconsider their practices. Entrepreneurs, driven by a sense of responsibility and

 DOI: 10.1201/9781003484226-2

recognizing the long-term risks, are increasingly seeking sustainable solutions to minimize the carbon footprint of their digital ventures (Sharfman & Fernando, 2008). This shift aligns with the growing consumer demand for environmentally conscious products and services, creating a market incentive for entrepreneurs to adopt sustainable practices (Panda et al., 2019). Moreover, as governments and regulatory bodies worldwide tighten environmental standards, entrepreneurs who integrate digital sustainability into their business models not only contribute to global efforts to combat climate change but also position themselves favorably in the evolving regulatory landscape (Ghobadian et al., 2009).

Furthermore, the rationale for exploring digital sustainability in entrepreneurship extends beyond mere environmental considerations. Embracing sustainable practices can foster innovation and resilience within businesses, creating opportunities for cost savings, efficiency gains, and improved brand reputation. Entrepreneurs are recognizing that digital sustainability is not only an ethical imperative but also a strategic advantage in a competitive marketplace (Chesbrough, 2003). As the world transitions toward a more sustainable future, entrepreneurs who proactively embed environmental considerations into their digital ventures are better positioned to thrive, meeting the expectations of environmentally conscious consumers while contributing to the overall well-being of the planet (Rattan et al., 2005). In essence, the exploration of digital sustainability in entrepreneurship is driven by recognition of the interconnectedness between environmental responsibility, economic viability, and societal expectations in the modern business landscape.

2.1.2 Scope and Objectives of the Chapter

The scope and objectives of this chapter are designed to provide readers with a comprehensive foundation for understanding the intricate relationship between digital sustainability and entrepreneurial endeavors in the contemporary Information Age (Shanmuga & Bhambri, 2024). The chapter begins by establishing a solid groundwork through a thorough exploration of the background and context surrounding digital sustainability. By delving into the evolution of the global landscape, it aims to trace the paradigm shifts brought about by the Information Age, examining the transformative impact on traditional business models and the emergence of the interconnected digital ecosystem.

Moving forward, the chapter seeks to define the concept of digital sustainability, offering insights into its ecological considerations as well as the social and economic dimensions inherent in the digital space. The intention is to foster a nuanced understanding of the multifaceted challenges and opportunities that digital sustainability presents to entrepreneurs. As the chapter progresses, it strategically explores the symbiotic relationship between sustainability and entrepreneurship, emphasizing the role of digital entrepreneurs as agents of change. Through the lens of case studies, the chapter endeavors to highlight instances of successful integration of sustainability in entrepreneurial ventures, providing real-world examples that underscore the practical application of sustainable practices (Rana & Bhambri, 2024). Overall, the scope of the chapter is expansive, aiming to lay the groundwork for subsequent discussions on the significance of sustainable practices in the digital ecosystem, the long-term benefits of sustainable entrepreneurship, and the keywords that will be in focus throughout the book. Through this approach, the chapter seeks to set the stage for a comprehensive exploration of the dynamic intersection between digital sustainability and entrepreneurship in the information age.

2.2 THE EVOLUTION OF THE GLOBAL LANDSCAPE

Over the past few decades, the global landscape has undergone a profound transformation, fueled by rapid advancements in technology and communication. This evolution has not only interconnected economies and societies but has also given rise to unprecedented challenges and opportunities (Shrivastava, 1995). The emergence of digital technologies has revolutionized the way businesses operate, breaking down traditional barriers and fostering a borderless entrepreneurial ecosystem. In this dynamic environment, the concept of digital sustainability has gained prominence, emphasizing the need for responsible and ethical practices in the digital realm. As the world becomes increasingly interconnected, entrepreneurs are presented with a unique set of challenges and responsibilities, requiring them to navigate the evolving landscape with an acute awareness of the environmental, social, and economic impacts of their ventures.

The intersection of digital sustainability and entrepreneurship is a pivotal focal point in this evolving global scenario. Entrepreneurs, armed with innovative digital solutions, have the potential to drive positive change by addressing pressing global issues such as climate change, resource depletion, and social inequality. Digital sustainability in entrepreneurship goes beyond just adopting eco-friendly practices; it encompasses a holistic approach that integrates environmental stewardship, social responsibility, and economic viability. As the global landscape continues to evolve, the success and impact of entrepreneurial ventures will be intrinsically tied to their ability to harness the power of digital technologies responsibly, fostering a sustainable future for generations to come.

2.2.1 The Information Age Paradigm Shift

Characterized by the rapid digitization, interconnectedness, and accessibility of information, this era has redefined the dynamics of communication, commerce, and culture (Stubbs & Cocklin, 2008). The proliferation of the internet, mobile technologies, and advanced computing has facilitated real-time exchange of information, fostering a globalized society where borders are transcended, and knowledge flows seamlessly across nations. This interconnectedness has not only accelerated the pace of innovation but has also democratized access to information, empowering individuals and communities worldwide (Mair & Martí, 2006). Consequently, the Information Age has brought about a fundamental reconfiguration of power structures, with information becoming a potent currency shaping economies, politics, and social dynamics on a global scale.

The digitization of industries has led to the rise of new economic models, with technology-driven innovation becoming a key driver of growth. The evolving nature of work, characterized by remote collaboration and the gig economy, reflects the adaptability demanded by this era. Simultaneously, the Information Age has also given rise to new challenges, such as concerns about privacy, cybersecurity, and the digital divide (Jenkins & Yakovleva, 2006). Societal structures are undergoing a metamorphosis as traditional norms are questioned and diverse voices find platforms for expression. Navigating this evolving landscape requires a dynamic approach that embraces technological advancements while addressing the ethical, social, and economic implications of the Information Age (Singh et al., 2004). In essence, this paradigm shift represents a pivotal moment in human history, shaping the trajectory of societies and economies as they navigate the complexities and opportunities of an interconnected, information-driven world.

2.2.2 Impact on Traditional Business Models

The evolution of the global landscape for digital sustainability and entrepreneurship is reshaping traditional business models by emphasizing innovation, sustainability, global connectivity, and adaptability. To thrive in this evolving landscape, businesses must embrace digital transformation, consider environmental and social impacts, and be open to adopting new technologies and collaborative approaches. Several key factors contribute to this transformation:

- Digital transformation: The advent of digital technologies has driven a significant shift in business operations. Traditional models are being replaced or adapted to incorporate digital tools and platforms. Companies are leveraging data analytics, Artificial Intelligence, and automation to streamline processes, enhance efficiency, and stay competitive.
- Global connectivity: Digital sustainability and entrepreneurship have enabled businesses to reach a global audience more easily. With the rise of e-commerce, online marketplaces, and digital communication tools, traditional businesses are increasingly challenged to expand their reach and compete on an international scale.
- Consumer behavior and expectations: Digital sustainability emphasizes eco-friendly practices and responsible resource use. Consumers are becoming more conscious of the environmental impact of products and services, leading traditional businesses to reevaluate their practices. Entrepreneurs, meanwhile, are capitalizing on sustainability trends to create innovative, eco-friendly solutions.
- Entrepreneurial innovation: Digital entrepreneurship has given rise to innovative business models that challenge traditional norms. Startups are leveraging technology to create disruptive solutions, often addressing sustainability challenges (Singh et al., 2005). This creates new market dynamics and forces established businesses to adapt or risk becoming obsolete.

- Sharing economy and collaborative platforms: The sharing economy, facilitated by digital platforms, has disrupted traditional business models in industries such as transportation (Uber, Lyft) and accommodation (Airbnb). Entrepreneurs are leveraging these collaborative platforms to create new business models that prioritize resource sharing, sustainability, and efficiency.
- Blockchain and decentralization: Technologies like blockchain are impacting traditional business models by introducing decentralized and transparent systems. This has implications for supply chain management, financial transactions, and data security. Entrepreneurs are exploring blockchain applications to create more transparent and sustainable business processes.
- Corporate social responsibility (CSR): Digital sustainability has heightened the importance of CSR for businesses. Consumers increasingly expect companies to engage in socially responsible practices. Traditional businesses are adjusting their strategies to incorporate sustainable initiatives, while digital entrepreneurs often build their ventures with a strong emphasis on ethical and sustainable practices.
- Evolving regulatory landscape: The global landscape for digital sustainability and entrepreneurship is influenced by evolving regulations (Jain & Bhambri, 2005). Governments and international bodies are introducing policies that encourage or mandate sustainable practices, affecting both traditional businesses and startups.
- Rapid technological advancements: The fast-paced evolution of technology requires businesses, both traditional and entrepreneurial, to stay agile and adapt to new tools and trends. This constant innovation reshapes industries and demands flexibility in business models.

2.2.3 Emergence of the Digital Ecosystem

In recent years, technological advancements and the widespread adoption of digital tools have given rise to interconnected networks that span industries, economies, and societies. This interconnectedness forms the basis of the digital ecosystem, where various components, such as software, hardware, data, and people, interact and collaborate seamlessly (Elkington, 1997). This evolution has not only facilitated innovation and efficiency but has also become a catalyst for digital sustainability by enabling businesses to streamline processes, reduce resource consumption, and contribute to environmental conservation.

In the realm of entrepreneurship, the digital ecosystem has opened up unprecedented opportunities. Startups and small businesses can leverage digital platforms to reach a global audience, access diverse markets, and establish cost-effective operations (Loorbach, 2007). The accessibility of online resources, cloud computing, and digital marketing tools has lowered entry barriers, allowing entrepreneurs to create and scale their ventures with greater speed and agility. Moreover, the digital ecosystem fosters collaboration and knowledge-sharing, creating a conducive environment for innovation and the development of sustainable business models.

As digital technologies continue to evolve, the global landscape for digital sustainability and entrepreneurship is likely to see further advancements, emphasizing the need for businesses to adapt to the dynamic and interconnected nature of the digital ecosystem.

2.3 UNDERSTANDING DIGITAL SUSTAINABILITY

Digital sustainability refers to the responsible and ethical use of digital technologies to minimize their environmental impact, promote social well-being, and ensure long-term viability. In the contemporary era, as societies become increasingly digitized, the concept of digital sustainability has gained prominence (Cohen & Winn, 2007). It encompasses various dimensions, including environmental conservation, ethical considerations, and the equitable distribution of benefits derived from digital advancements.

From an environmental perspective, digital sustainability involves reducing the carbon footprint and energy consumption associated with digital technologies. This may include optimizing data centers, adopting energy-efficient hardware and software solutions, and promoting responsible e-waste management. Ethical considerations in digital sustainability involve addressing issues like data privacy, security, and algorithmic fairness to ensure that technology serves the greater good without compromising individual rights. Furthermore, fostering digital inclusivity and bridging the digital divide is crucial to ensuring that the benefits of technological advancements reach all segments of society, contributing to a more equitable and socially sustainable digital future (Pearce & Doh, 2012). As organizations and individuals embrace the principles of digital sustainability, they contribute to building a resilient and responsible digital ecosystem for the benefit of current and future generations.

2.3.1 Ecological Considerations in the Digital Space

By considering the ecological factors, stakeholders in the digital space can work towards minimizing the environmental impact and promoting digital sustainability. This involves a collaborative effort from industry players, policymakers, and users to create a more sustainable and environmentally friendly digital ecosystem. Here are some key ecological considerations in the digital space:

- Energy consumption: The operation of data centers, where digital information is stored and processed, requires a significant amount of energy. Using energy-efficient hardware, optimizing server configurations, and implementing renewable energy sources can help reduce the environmental impact. Designing and manufacturing energy-efficient devices, such as smartphones, laptops, and servers, can contribute to lower overall energy consumption in the digital ecosystem.
- E-Waste management: The disposal of electronic waste (e-waste) poses environmental challenges. Promoting responsible e-waste recycling practices, such as proper disposal and recycling of electronic devices, helps

minimize the environmental impact of hazardous materials found in electronic components.

- Product lifespan and design – planned obsolescence: Designing products with planned obsolescence can contribute to unnecessary electronic waste. Encouraging longer product lifespans through durable design, modular components, and upgradability can help reduce the environmental impact of electronic devices.
- Sustainable sourcing of materials – supply chain practices: Consideration should be given to the environmental impact of sourcing materials for manufacturing digital devices. Sustainable and responsibly sourced materials can contribute to reducing the ecological footprint of the digital industry.
- Carbon footprint of internet services: Cloud services and data storage contribute to the carbon footprint of the digital space. Employing energy-efficient data centers and adopting green cloud solutions can help reduce the environmental impact (Waddock & Graves, 1997). The infrastructure that supports the internet, including networks and data transmission, should be optimized for energy efficiency to minimize the overall carbon footprint.
- Digital literacy and awareness: Promoting digital literacy and awareness among users can lead to more responsible and sustainable digital practices. Educating users about the environmental impact of digital activities, such as excessive data storage and energy consumption, can encourage eco-friendly behavior.
- Regulatory measures and standards: Governments and international bodies can implement and enforce regulations that incentivize environmentally friendly practices in the digital industry. Setting standards for energy efficiency and responsible e-waste management can drive positive change.
- Green innovation: Encouraging research and development in green technologies, such as energy-efficient hardware, sustainable manufacturing processes, and eco-friendly materials, can contribute to the overall sustainability of the digital space.

2.3.2 Social and Economic Dimensions of Digital Sustainability

Digital sustainability extends beyond ecological considerations and encompasses social and economic dimensions that play a pivotal role in fostering a balanced and equitable digital ecosystem. On the social front, digital sustainability involves ensuring accessibility and inclusivity, bridging the digital divide, and safeguarding user privacy. Access to digital technologies should be widespread, and efforts should be made to close gaps in digital literacy and internet access, especially in marginalized communities (Bhambri & Gupta, 2005). Additionally, protecting user privacy and data rights is crucial to building trust and maintaining ethical standards in the digital realm. Responsible use of technology should prioritize the well-being of individuals and communities, fostering a digital environment that respects diverse perspectives and promotes social cohesion.

Economically, digital sustainability involves creating a resilient and fair digital economy. This includes addressing issues of economic inequality, promoting innovation, and supporting small businesses. Policies should be designed to prevent digital monopolies, encourage fair competition, and ensure that the benefits of digital advancements are distributed equitably (Porter & Kramer, 2011). Moreover, investing in digital education and skills development can empower individuals to participate in the digital economy, fostering economic growth and reducing disparities. By integrating social and economic considerations into the framework of digital sustainability, stakeholders can work towards a future where technology serves as a catalyst for positive societal and economic change.

2.4 THE INTERCONNECTED THEMES: SUSTAINABILITY AND ENTREPRENEURSHIP

Entrepreneurs, driven by a commitment to sustainable practices, not only seek to build economically viable ventures but also aim to make a positive impact on the planet and society. Sustainable entrepreneurship involves the integration of eco-friendly business models, ethical supply chains, and social responsibility, demonstrating a holistic approach that goes beyond profit generation (Hockerts, 2017). Successful ventures in this space often emphasize resource efficiency, circular economy principles, and community engagement, showcasing the transformative power of entrepreneurship in creating environmentally conscious solutions. As sustainability becomes an integral part of business strategy, entrepreneurs play a pivotal role in driving positive change, demonstrating that profitability and responsible business practices can be mutually reinforcing forces in building a resilient and sustainable future.

2.4.1 Exploring the Symbiotic Relationship

Entrepreneurship, driven by innovation and vision, has the power to be a catalyst for sustainable practices. Sustainable entrepreneurship involves developing business models that not only seek financial success but also prioritize environmental and social considerations. Entrepreneurs are increasingly recognizing the importance of integrating sustainable practices into their ventures, addressing issues such as resource efficiency, ethical sourcing, and community engagement (Carayannis & Campbell, 2010). In turn, embracing sustainability can enhance the long-term viability and resilience of entrepreneurial endeavors, contributing to reputation building, stakeholder trust, and access to emerging markets with a growing demand for eco-friendly products and services.

On the other side of the spectrum, sustainability can inspire and shape entrepreneurial initiatives. The pressing global challenges related to climate change, resource depletion, and social inequality provide fertile ground for entrepreneurs to identify innovative solutions (Schaltegger & Wagner, 2006). Entrepreneurs who embed sustainability into their business models not only contribute to environmental conservation but also tap into a burgeoning market driven by conscientious consumers. The pursuit of sustainability can also lead to cost savings through energy efficiency and waste reduction, aligning economic success with environmental responsibility. Ultimately, the symbiosis between sustainability and entrepreneurship reflects a transformative force, where business innovation becomes a driving force for positive change, fostering a more harmonious relationship between commerce and the planet (Hitt et al., 2017).

2.4.2 Digital Entrepreneurs as Agents of Change

Digital entrepreneurs serve as potent agents of change, leveraging technology to drive innovation and address societal challenges. With a dynamic and forward-thinking approach, these entrepreneurs harness digital tools to create disruptive solutions, fostering economic growth, and transforming traditional industries (Short et al., 2009). Their initiatives often focus on environmental sustainability, social impact, and inclusivity, utilizing digital platforms to bridge gaps and connect communities (Baumann et al., 2002). Digital entrepreneurs play a pivotal role in shaping the future by challenging the status quo, introducing novel business models, and promoting digital sustainability. Through their endeavors, these change agents not only revolutionize industries but also contribute to building a more interconnected, resilient, and socially conscious global landscape.

2.4.3 Case Studies: Successful Integration of Sustainability in Entrepreneurial Ventures

The following case studies illustrate how companies can integrate sustainability into their core strategies, achieving both environmental and business success.

- Tesla, Inc.: Tesla, led by Elon Musk, is known for its electric vehicles and sustainable energy solutions. Tesla's core business revolves around sustainable transportation, with a focus on electric cars and renewable energy products like solar panels and energy storage. By promoting clean energy and electric vehicles, Tesla has contributed to reducing carbon emissions and advancing sustainable technology adoption in the automotive industry.
- Patagonia: Patagonia is a renowned outdoor clothing and gear company. Patagonia has a strong commitment to environmental sustainability, including using recycled materials, fair labor practices, and encouraging customers to buy only what they need through initiatives like "Worn Wear." Patagonia's sustainability efforts have strengthened its brand and attracted environmentally conscious consumers, demonstrating that a company can thrive by aligning its business with eco-friendly values.
- Unilever: Unilever is a multinational consumer goods company. Unilever has a Sustainable Living Plan with ambitious goals to improve health and well-being, reduce environmental impact, and enhance livelihoods across its value chain. By integrating sustainability into its business model, Unilever has not only reduced its environmental footprint but also appealed to consumers who prioritize ethical and sustainable products.

- Etsy: Etsy is an e-commerce platform focused on handmade and vintage items. Etsy promotes sustainability by supporting small businesses, artisans, and sellers who often prioritize eco-friendly and ethical practices. Etsy's platform aligns with sustainability values, offering consumers a marketplace where they can find unique, handmade products with lower environmental impact compared to mass-produced alternatives.
- Interface, Inc.: Interface is a global modular flooring company. Interface has been a pioneer in sustainable business practices, aiming to eliminate its environmental impact by 2020. This includes using recycled materials and adopting a closed-loop manufacturing process. Interface's sustainability initiatives have demonstrated that a large manufacturing company can successfully reduce its environmental footprint while maintaining profitability.

2.5 THE SIGNIFICANCE OF SUSTAINABLE PRACTICES IN THE DIGITAL ECOSYSTEM

As the world increasingly relies on digital technologies, addressing the ecological footprint of data centers, electronic waste, and energy consumption becomes imperative. Sustainable practices in the digital realm involve optimizing energy efficiency, adopting renewable energy sources, and minimizing electronic waste through recycling and responsible disposal (Christensen, 1997). By integrating such practices, businesses and individuals not only contribute to global environmental preservation but also position themselves to thrive in a future where sustainability is a key determinant of success. Embracing eco-friendly approaches in the digital ecosystem is essential for fostering a harmonious balance between technological advancement and ecological well-being.

2.5.1 Building Resilient Businesses

As technology rapidly evolves, companies face diverse challenges such as climate-related risks, supply chain disruptions, and changing consumer expectations (Spence & Rutherfoord, 2001). Embracing sustainability not only mitigates environmental impact but also enhances operational adaptability and long-term viability. By incorporating eco-friendly measures, businesses can reduce resource dependencies, optimize energy consumption, and foster innovation. Moreover, sustainable practices fortify brand reputation, attracting environmentally conscious consumers and investors (Potluri et al., 2019). In the digital realm, where agility and adaptability are paramount, businesses that integrate sustainability demonstrate a forward-thinking approach, positioning them to navigate uncertainties and thrive in an ever-changing landscape.

2.5.2 Mitigating Risks and Embracing Opportunities

As organizations increasingly rely on digital technologies, they face risks associated with environmental impact, data security, and societal concerns. Sustainable practices, such as energy-efficient data centers and responsible data management, can mitigate environmental risks (Hart, 1995). By adopting eco-friendly technologies and reducing carbon footprints, companies contribute to a more sustainable digital ecosystem. Simultaneously, addressing cybersecurity threats through robust practices ensures the protection of sensitive information, thereby mitigating reputational and financial risks.

Moreover, embracing sustainability opens doors to numerous opportunities in the digital realm. Innovations such as green technologies and circular economy models not only reduce ecological impact but also foster operational efficiency and cost savings (Brown, 2003). Businesses that prioritize sustainability in the digital ecosystem can tap into growing consumer demand for environmentally conscious products and services, enhancing brand reputation and customer loyalty. Collaboration and information sharing within the industry on sustainable best practices can also lead to new business partnerships and industry alliances, creating a positive ripple effect across the entire digital ecosystem. Overall, integrating sustainable practices in the digital landscape not only mitigates risks but also positions organizations to thrive in a rapidly evolving and conscientious market.

2.5.3 Long-Term Benefits of Sustainable Entrepreneurship

Sustainable entrepreneurship in the digital ecosystem brings about a range of long-term benefits, including environmental conservation, resource efficiency, innovation, positive social impact, and regulatory compliance (Rennings, 2000). By integrating sustainable practices into their business models, entrepreneurs cannot only contribute to a healthier planet but also build resilient, future-proofed businesses with positive societal impact. Here are some key long-term benefits:

- Environmental conservation: Sustainable entrepreneurship encourages the use of eco-friendly technologies, energy-efficient practices, and carbon-neutral approaches (Yunus et al., 2010). This helps in minimizing the carbon footprint associated with digital operations, data centers, and electronic devices. Companies adopting sustainable practices in the digital realm are likely to prioritize waste reduction, recycling, and responsible disposal of electronic waste, contributing to a circular economy.
- Resource efficiency and cost savings: Sustainable entrepreneurs often invest in energy-efficient technologies and practices, leading to reduced energy consumption. This not only benefits the environment but also results in cost savings over the long term. By adopting sustainable supply chain practices, entrepreneurs can minimize resource wastage, improve efficiency, and build resilience against disruptions. This, in turn, helps in ensuring the long-term availability of resources.
- Innovation and market position: Sustainable entrepreneurship fosters innovation in products and services. Consumers increasingly value environmentally conscious businesses, and companies that embrace sustainability gain a competitive edge, attracting a growing market of environmentally aware customers. Adapting to sustainable practices in

the digital ecosystem prepares businesses for evolving market trends and regulations. This resilience ensures the company's long-term viability in an ever-changing business landscape.

- Positive social impact: Sustainable entrepreneurship involves considering the interests of a broader range of stakeholders, including employees, local communities, and customers (Kolk & Lenfant, 2010). Building positive relationships with these stakeholders enhances the company's social license to operate and fosters long-term trust. Companies that engage in sustainable entrepreneurship often contribute to community development through initiatives such as education, skill development, and social welfare programs. This creates a positive societal impact and strengthens the company's reputation.
- Regulatory compliance and risk mitigation: Sustainable practices align with evolving environmental regulations and standards (Seelos & Mair, 2005). Entrepreneurs who proactively adopt sustainability measures are better positioned to comply with future regulations, reducing the risk of legal and financial repercussions. Businesses that prioritize sustainability are better equipped to manage reputational risks associated with environmental or social issues. This is crucial for long-term success and maintaining a positive brand image.

2.6 KEYWORDS IN FOCUS

Keywords in focus for this dynamic space include "circular economy," emphasizing the importance of resource efficiency and waste reduction; "green technology," signaling the adoption of environmentally friendly solutions; "social impact," underscoring the role of businesses in addressing societal challenges; "digital inclusion," highlighting the need to bridge the digital divide; and "resilience," recognizing the ability of digital enterprises to adapt and thrive in a rapidly changing landscape (Vachon & Klassen, 2006). By embracing these keywords, entrepreneurs can cultivate ventures that not only drive economic growth but also contribute to a sustainable and inclusive digital future.

2.6.1 Global Landscape

In the realm of digital sustainability, it reflects the imperative for businesses to consider their impact on a global scale, addressing environmental, social, and economic challenges. Entrepreneurs navigating this landscape must leverage technology to create innovative solutions that not only drive economic growth but also contribute to sustainable development goals. The global landscape underscores the need for collaboration, diverse perspectives, and responsible practices, emphasizing the interconnectedness of digital initiatives and their potential to shape a more sustainable and inclusive future. Entrepreneurs, guided by a commitment to global well-being, can harness digital tools to foster positive change, ensuring that their ventures contribute to a more resilient and equitable world.

2.6.2 Digital Ecosystem

Digital ecosystem refers to the strategic integration of digital innovations to address environmental challenges and foster eco-friendly practices (Leitner & Güldenberg, 2010). Entrepreneurs leverage this ecosystem by creating innovative digital solutions, fostering collaboration, and adapting to rapidly evolving technologies. Digital sustainability within the ecosystem involves not only minimizing environmental impact but also promoting economic viability and social responsibility. As entrepreneurs navigate this dynamic digital terrain, embracing a holistic approach that balances innovation, environmental consciousness, and economic growth becomes pivotal for fostering a resilient and sustainable digital future.

2.6.3 Sustainable Practices

Sustainable practices serve as a pivotal keyword in the realms of digital sustainability and entrepreneurship, encapsulating a commitment to environmentally responsible and socially conscious strategies (Smith & Bhatia, 2008). In the digital sphere, this involves optimizing energy consumption, minimizing electronic waste, and promoting eco-friendly technologies. Entrepreneurs embracing sustainable practices not only contribute to global environmental preservation but also enhance their long-term viability by aligning with ethical consumer preferences. From adopting renewable energy sources for data centers to designing products with a life cycle perspective, digital sustainability fosters innovation and resilience. Entrepreneurs leveraging these principles not only mitigate environmental impact but also cultivate a competitive edge in a world increasingly valuing responsible business practices. This synergy between sustainability and entrepreneurship underscores the imperative for a harmonious coexistence between digital advancement and ecological well-being.

2.6.4 Resilient Business

In the digital era, where rapid technological advancements and unforeseen disruptions are inherent, businesses must cultivate resilience to navigate uncertainties successfully. A resilient business not only embraces digital transformation but also strategically leverages technology to fortify its operations against potential disruptions, ensuring sustainable growth. Entrepreneurs, cognizant of the digital landscape's volatility, are increasingly incorporating resilience into their business models, fostering agility and responsiveness. By prioritizing adaptive strategies, diversifying digital assets, and embracing innovative solutions, resilient businesses not only survive but thrive in the ever-evolving landscape of digital sustainability and entrepreneurship.

2.7 CONCLUSION

This chapter has provided a foundational understanding of the intersection between digital sustainability and entrepreneurship within the dynamic landscape of the Information Age. We delved into the background and context, recognizing the transformative paradigm shift that the Information Age has brought about. As traditional business models evolve, the emergence of the digital ecosystem presents both challenges and opportunities for entrepreneurs. By defining digital sustainability and exploring its ecological, social, and economic dimensions, we established a framework for examining how sustainable practices can be integrated into entrepreneurial ventures.

The interconnected themes of sustainability and entrepreneurship were explored, emphasizing the symbiotic relationship between the two. Digital entrepreneurs were identified as crucial agents of change, and real-world case studies illustrated successful integration of sustainability in various entrepreneurial ventures. The significance of sustainable practices in the digital ecosystem was highlighted, emphasizing their role in building resilient businesses, mitigating risks, and embracing long-term opportunities.

2.8 FUTURE DIRECTIONS AND IMPLICATIONS

Examining the implications for future entrepreneurial endeavors, we consider how sustainable practices will continue to shape and redefine business strategies. Moreover, we highlight the burgeoning research and innovation opportunities that lie ahead, emphasizing the critical role of ongoing exploration and adaptation in navigating the dynamic intersection of digital sustainability and entrepreneurship in the Information Age. As we conclude, this chapter sets the stage for continued dialogue and exploration in subsequent studies, urging entrepreneurs to embrace the evolving challenges and opportunities presented by a digitally sustainable future.

2.8.1 Anticipated Trends in Digital Sustainability

As technological advancements continue to reshape industries, we foresee a surge in demand for innovative solutions that seamlessly integrate sustainability into digital business practices. Trends such as the rise of circular economies, increased emphasis on transparency and ethical considerations, and the mainstream adoption of eco-friendly technologies are expected to play pivotal roles. Additionally, the chapter will delve into the evolving regulatory landscape, highlighting the growing importance of compliance with environmental standards and the emergence of new business models that prioritize both profitability and environmental responsibility. Entrepreneurs who proactively embrace these trends are likely to position themselves at the forefront of a sustainable digital future, creating opportunities for growth, resilience, and positive impact on a global scale.

2.8.2 Implications for Future Entrepreneurial Endeavors

As technology continues to advance and the global landscape evolves, entrepreneurs must navigate a dynamic environment where sustainability is not just a choice but a strategic imperative. The chapter explores how integrating sustainable practices into digital entrepreneurship can foster innovation, enhance competitiveness, and contribute to long-term success. By anticipating and adapting to emerging trends in digital sustainability, entrepreneurs can position themselves as leaders in a socially and environmentally conscious business ecosystem. The implications extend beyond individual ventures, influencing industry norms and shaping the broader narrative of responsible and forward-thinking entrepreneurship. As we look ahead, the chapter not only underscores the importance of embracing sustainable business models but also offers insights into the potential rewards and challenges that lie ahead for entrepreneurs in the information age.

2.8.3 Research and Innovation Opportunities

The following research and innovation opportunities aim to provide a roadmap for scholars, practitioners, and policymakers interested in advancing the field of digital sustainability and entrepreneurship. By addressing these areas, the digital landscape can continue to evolve in a way that promotes both environmental responsibility and economic success.

- Smart technologies for sustainability: Investigate the development and integration of smart technologies (e.g., IoT, AI, blockchain) to enhance sustainability practices in digital entrepreneurship. Explore how these technologies can optimize resource usage, reduce energy consumption, and improve overall efficiency in business operations.
- Circular economy in the digital space: Research opportunities in adopting circular economy principles within the digital ecosystem. Examine ways to minimize electronic waste, promote recycling of digital devices, and extend the lifespan of digital products through innovative business models.
- Metrics and measurement tools: Develop comprehensive metrics and measurement tools for assessing the environmental, social, and economic impact of digital entrepreneurial ventures. Explore methodologies for quantifying the sustainability performance of businesses in the digital age.
- Social impact of digital entrepreneurship: Investigate the social implications of digital entrepreneurship, focusing on job creation, community development, and inclusivity. Explore strategies for ensuring that digital entrepreneurship contributes positively to societal well-being.
- Policy and regulatory frameworks: Examine the development of policies and regulatory frameworks that encourage and support sustainable practices in digital entrepreneurship. Investigate the role of governments and international organizations in shaping a conducive environment for sustainable digital businesses.
- Education and awareness initiatives: Explore the effectiveness of educational programs and awareness campaigns in promoting digital sustainability among entrepreneurs and consumers. Develop strategies for integrating sustainability education into business curricula.
- Collaborative platforms for sustainability: Explore the potential of collaborative platforms that facilitate knowledge sharing, resource pooling, and

collective efforts towards sustainable goals in the digital space. Investigate how digital entrepreneurs can collaborate across industries to address common sustainability challenges.

- Innovative financing models: Research alternative financing models that support sustainable entrepreneurship, such as impact investing, green bonds, and other financial instruments. Explore how financial institutions can incentivize and reward businesses adopting sustainable practices.
- Ethical AI and data practices: Investigate ethical considerations in the use of Artificial Intelligence and data analytics in digital entrepreneurship. Develop frameworks for responsible and transparent AI and data practices to ensure ethical business conduct.
- Resilience in the face of technological disruption: Explore strategies for building resilience in digital businesses, considering the rapid pace of technological change. Examine how entrepreneurs can navigate uncertainties and disruptions while maintaining a focus on sustainability.

REFERENCES

Baumann, H., Boons, F., & Bragd, A. (2002). Mapping the green product development field: Engineering, policy and business perspectives. *Journal of Cleaner Production*, 10(5), 409–425.

Bhambri, P., & Gupta, S. (2005, March). A survey & comparison of permutation possibility of fault tolerant multistage interconnection networks. In National Conference on Application of Mathematics in Engg. & Tech. (pp. 13).

Brown, M. T. (2003). The precautionary principle: Making it work in practice. *Environment: Science and Policy for Sustainable Development*, 45(8), 10–21.

Carayannis, E. G., & Campbell, D. F. (2010). Triple Helix, Quadruple Helix and Quintuple Helix and how do knowledge, innovation and the environment relate to each other? A proposed framework for a trans-disciplinary analysis of sustainable development and social ecology. *International Journal of Social Ecology and Sustainable Development (IJSESD)*, 1(1), 41–69.

Chesbrough, H. (2003). *Open Innovation: The New Imperative for Creating and Profiting from Technology*. Harvard Business Press.

Christensen, C. M. (1997). *The Innovator's Dilemma: When New Technologies Cause Great Firms to Fail*. Harvard Business Review Press.

Cohen, B., & Winn, M. I. (2007). Market imperfections, opportunity and sustainable entrepreneurship. *Journal of Business Venturing*, 22(1), 29–49.

Elkington, J. (1997). *Cannibals with Forks: The Triple Bottom Line of 21st Century Business*. Capstone.

Ghobadian, A., O'Regan, N., & Sims, M. (2009). The search for innovation in emerging ecoparadigms: A content and bibliometric analysis. *International Journal of Innovation Management*, 13(3), 397–421.

Hart, S. L. (1995). A natural-resource-based view of the firm. *Academy of Management Review*, 20(4), 986–1014.

Hitt, M. A., Ireland, R. D., & Hoskisson, R. E. (2017). *Strategic Management: Concepts and Cases: Competitiveness and Globalization*. Cengage Learning.

Hockerts, K. (2017). Eco-entrepreneurship: Definition and evidence from the clean energy sector. *Entrepreneurship & Regional Development*, 29(3–4), 196–214.

Jain, V. K., & Bhambri, P. (2005). *Fundamentals of Information Technology & Computer Programming*. KATSONS.

Jenkins, H., & Yakovleva, N. (2006). Corporate social responsibility in the mining industry: Exploring trends in social and environmental disclosure. *Journal of Cleaner Production*, 14(3–4), 271–284.

Kolk, A., & Lenfant, F. (2010). MNC reporting on CSR and conflict in central Africa. *Journal of Business Ethics*, 93(2), 241–255.

Leitner, S., & Güldenberg, S. (2010). Corporate sustainability in the health care sector: A structured review. *International Journal of Environmental Research and Public Health*, 7(6), 2402–2419.

Loorbach, D. (2007). Transition management: New mode of governance for sustainable development. *International Journal of Sustainable Development & World Ecology*, 14(1), 78–91.

Mair, J., & Martí, I. (2006). Social entrepreneurship research: A source of explanation, prediction, and delight. *Journal of World Business*, 41(1), 36–44.

Panda, S. K., Reddy, G. S. M., Goyal, S. B., Thirunavukkarasu, K., Bhambri, P., Rao, M. V., Singh, A. S., Fakih, A. H., Shukla, P. K., Shukla, P. K., & others. (2019). Method for management of scholarship of large number of students based on blockchain. IN Patent App. 201,911,034,937 A.

Pearce, J. A., & Doh, J. P. (2012). The high impact of collaborative social initiatives. *MIT Sloan Management Review*, 53(2), 26–32.

Porter, M. E., & Kramer, M. R. (2011). Creating shared value. *Harvard Business Review*, 89(1/2), 62–77.

Potluri, S., Tiwari, P. K., Bhambri, P., Obulesu, O., Naidu, P. A., Lakshmi, L., Kallam, S., Gupta, S., & Gupta, B. (2019). Method of load distribution balancing for fog cloud computing in IoT environment. IN Patent App. 201,941,044,511.

Rana, R., & Bhambri, P. (2024). Healthcare computational intelligence and blockchain: Real-life applications. In P. Bhambri, S. Rani, & M. Fahim (Eds.), *Computational Intelligence and Blockchain in Biomedical and Health Informatics* (pp. 155–168). CRC Press, Taylor & Francis Group. DOI: 10.1201/9781003459347

Rattan, M., Bhambri, P., & Shaifali. (2005). Information retrieval using soft computing techniques. In National Conference on Bio-informatics Computing (pp. 7).

Rennings, K. (2000). Redefining innovation: Eco-innovation research and the contribution from ecological economics. *Ecological Economics*, 32(2), 319–332.

Schaltegger, S., & Wagner, M. (2006). Managing sustainability performance measurement. *Journal of Business Ethics*, 63(3), 267–284.

Seelos, C., & Mair, J. (2005). Social entrepreneurship: Creating new business models to serve the poor. *Business Horizons*, 48(3), 241–246.

Shanmuga, S. M., & Bhambri, P. (2024). Bone marrow cancer detection from leukocytes using neural networks. In P. Bhambri, S. Rani, & M. Fahim (Eds.), *Computational Intelligence and Blockchain in Biomedical and Health Informatics* (pp. 307–319). CRC Press, Taylor & Francis Group. DOI: 10.1201/9781003459347

Sharfman, M. P., & Fernando, C. S. (2008). Environmental risk management and the cost of capital. *Strategic Management Journal*, 29(6), 569–592.

Short, J. C., Moss, T. W., & Lumpkin, G. T. (2009). Research in social entrepreneurship: Past contributions and future opportunities. *Strategic Entrepreneurship Journal*, 3(2), 161–194.

Shrivastava, P. (1995). The role of corporations in achieving ecological sustainability. *Academy of Management Review*, 20(4), 936–960.

Singh, P., Singh, M., & Bhambri, P. (2004, November). Interoperability: A problem of component reusability. In International Conference on Emerging Technologies in IT Industry (pp. 60).

Singh, P., Singh, M., & Bhambri, P. (2005, January). Embedded systems. In Seminar on Embedded Systems (pp. 10–15).
Smith, N. C., & Bhatia, A. (2008). A process for evaluating environmental management and sustainability degrees. *Journal of Cleaner Production*, 16(5), 537–547.
Spence, L. J., & Rutherfoord, R. (2001). Social and environmental reporting and the corporate ego. *Business Strategy and the Environment*, 10(1), 74–86.
Stubbs, W., & Cocklin, C. (2008). Conceptualizing a "sustainability business model". *Organization & Environment*, 21(2), 103–127.
Vachon, S., & Klassen, R. D. (2006). Extending green practices across the supply chain: The impact of upstream and downstream integration. *International Journal of Operations & Production Management*, 26(7), 795–821.
Waddock, S. A., & Graves, S. B. (1997). The corporate social performance–financial performance link. *Strategic Management Journal*, 18(4), 303–319.
Yunus, M., Moingeon, B., & Lehmann-Ortega, L. (2010). Building social business models: Lessons from the Grameen experience. *Long Range Planning*, 43(2–3), 308–325.

3 Data Privacy and Entrepreneurial Responsibility

Grzegorz Chmielarz

3.1 INTRODUCTION

Undoubtedly, the role of cybersecurity in enterprises has evolved in recent years. It is no longer perceived as "good if we have one", but rather as "we must have one". This represents a modern outlook on the cybersecurity landscape in enterprises. The major problem in the previous decade was that the responsibility for maintaining data privacy frequently almost entirely rested on the shoulders of IT departments. It was not unusual for the managerial personnel to view them as incurring extra costs for the company's operations and providing few, if any, advantages in return. This, fortunately, is a thing of the past as the ongoing onslaught of cybercriminals has necessitated the change of the outlook on the domain of data privacy. What seems to be a prevailing attitude in contemporary enterprises is that cybersecurity requires combined efforts of all the involved stakeholders within a firm, starting at the managerial level, through IT teams, to individual employees. Certainly, one must bear in mind that the size of an enterprise is a crucial factor in assessing and attaining particular cybersecurity priorities. Needless to say, larger enterprises can afford to spend more on broadly understood cybersecurity aspects, whereas the small- and medium-sized ones struggle with limited resources and sometimes do not even employ an IT team or an IT expert. The problem is that contemporary firms, regardless of their size, work with data and it is data that feeds their computer systems and is used to generate value. This means that the value of data as such has increased dramatically. Data and information are frequently referred to as the fourth production factor, whose importance is equal to work, capital and land. Digitalisation of data and moving it from paper-form records to computer databases have increased its potential of utilisation in organisations of all types, not just enterprises. It is enough to look at the pace at which we generate data and how much of it was generated a decade ago to understood how critical the role of data is to all areas of human existence. According to the statistical portal Statista, the amount of total data/information that was created, captured, copied, and consumed globally reached the level of 12.5 zettabytes (over six times more than in 2010). In 2024 it is projected to reach 147 zettabytes (over twelve times more than in 2014) and continue its growth

DOI: 10.1201/9781003484226-3

to 185 zettabytes in 2025. Not surprisingly, a new high in the amount of utilised data was recorded in 2020 during the COVID-19 pandemic, which can be explained by a higher number of people working and learning from home and also making a more extensive use of home entertainment (Statista, 2024). The amount of data that contemporary enterprises deal with imposes a great responsibility for ensuring greater privacy and security. Failure to do so can result in entrepreneurial responsibility for violating the legislation in force, but more importantly can lead to a long-lasting effect of consumer mistrust that may become catastrophic for any enterprise.

3.2 ROAD TO DATA-DRIVEN ENTERPRISE

Several phenomena are responsible for such a development in the area of data creation and consumption. The first one is the new paradigm of working with computers that emerged in the first years of the 21st century, namely cloud computing. According to one of the abundant definitions, "cloud computing is the on-demand availability of computing resources (such as storage and infrastructure), as services over the internet. It eliminates the need for individuals and businesses to self-manage physical resources themselves, and only pay for what they use" (Google Cloud, n.d.). Despite the initial scepticism about the adoption of cloud services, which was primarily driven by the concerns about data security, the cloud computer market has been growing dynamically. In 2022, the value of the global cloud computing market was $446.51 billion and it is expected to surpass $1 trillion by 2028 and reach $1.6 trillion by 2030 (Precedence Research, 2023). Application of cloud computing by enterprises leads to migration of their data repositories to virtual locations. As a result, no investments are needed in expensive computer hardware and software, which in most cases are rarely fully utilised, so that considerable savings can be made on hiring IT specialists to oversee maintaining and controlling the equipment. Also, less space is needed for bulky servers or in-house data centres.

On the negative side, however, cloud computing contributed greatly to the rapid growth of the demand for data storage and processing capabilities, also on the side of enterprises. Consequently, the data centre market has been booming in the recent years with as many as 3107 data centres, located in 81 countries in 2023. The main driving force of the enterprise data centre market are industries that have recognized the flexibility and reliability of this solution. It is expected that the unstoppable development of data-driven technologies will keep this market expanding at a compound annual growth rate (CAGR) that can range between 6.12% and 10.9%, according to different sources, between 2023 and 2030 (Data Centers, 2024).

The second phenomenon, which should be perceived not as one but rather a combination of several intertwined technological trends, is the development of Big Data, Internet of Things, Industry 4.0, and extensive use of Artificial Intelligence. Big Data can be conducive to improving enterprise's business methods as collecting and analysing huge amounts of data allows them to obtain valuable insights. The most measurable value of mining data in BD is predicting future trends and patterns. However, it also means the burden of responsibility for capturing, processing, storing, analysing and sharing this valuable data (Chunling & Yumei, 2023). Big Data is closely related to the concept of the Internet of Things and Industry 4.0. In IoT, smart devices are capable of interconnecting and can mutually communicate and interact utilising the global network. Present smart devices are small but their processing power is significant. They are equipped with advanced embedded systems which provide them with the capabilities of communication, sensing, actuation; and in the case of data their capabilities include its retrieval, collection, storage and processing in real time (Singh & Bhambri, 2024). The technology offers a number of possible applications, functions and services in everyday life and a wide range of domains. These smart devices have become an indispensable part of the development of the concept of Industry 4.0, where the driving forces are data, computational power and connectivity of computer systems, analytics and intelligence, human-machine interaction and digital-to-physical conversion (Lampropoulos et al., 2019). Undoubtedly, enterprises can leverage the potential of IoT and Industry 4.0, yet they are both based on data that is constantly sent, collected, analysed, processed and utilised.

And last but not least is the sky-rocketing application of AI-based solutions in all areas of life. Data feeds the algorithms that are trained in Machine Learning or Deep Learning models. There is huge demand for data that can be used to train algorithms to behave as human-like as possible (Rana & Bhambri, 2024). This bears two basic problems. The first one is the availability of data; not all data is publicly available and can be used at the stage of training algorithms. The problem of illegal or unlawful acquisition of data arises, then. The second problem for contemporary data-driven enterprises is that AI technologies are vulnerable to adversarial attacks. It is believed that security is not a strong point of developing AI systems (Singh et al., 2004). One of the main threats is that a training dataset can be infected with fake data or input attributes or data labels can be changed, leading the way to a misclassification of the input data (AL-Khassawneh, 2023).

Thus, it can be concluded that the more advanced the technologies utilised in modern enterprises become, the more data they need. This dependence on omnipresent data means assuming responsibility for using it appropriately. Enterprises operating in the 21st century have no other choice, they have to work with data. However, entrepreneurs need to bear in mind that ensuring protection of data held by them requires them to introduce internal protective measures that need to be adhered to in the entire organisational eco-system.

3.3 CYBERSECURITY AND CYBERTHREATS – PROTECTING DATA PRIVACY LANDSCAPE

According to one of the many definitions, cybersecurity is the practice of protecting systems, networks, and programs from digital attacks. These cyberattacks are usually aimed at accessing, changing, or destroying sensitive information; extorting money from users via ransomware; or interrupting normal business processes (Cisco, n.d.). Another definition perceives cybersecurity as

> the practice of defending computers, servers, mobile devices, electronic systems, networks, and data from malicious attacks. It is also known as information technology

> security or electronic information security. The term applies in a variety of contexts, from business to mobile computing.
>
> **(Kaspersky, n.d.)**

Cybersecurity can also refer to

> any technology, measure or practice for preventing cyberattacks or mitigating their impact. Cybersecurity aims to protect individuals' and organizations' systems, applications, computing devices, sensitive data and financial assets against simple and annoying computer viruses, sophisticated and costly ransomware attacks, and everything in between.
>
> **(IBM, n.d.)**

As one can observe, the presented definitions focus on two key aspects. The first one is the commonly perceived fact that information and data resources are targeted by criminals, and as such, require protection that can take various forms (Singh et al., 2005). More importantly, however, the definitions prove that cyberattacks that are aimed at contemporary enterprises have become so commonplace that introducing measures to protect possessed data is no longer a matter of choice, but by all means a necessity.

Having stated so, it needs to be stressed that there are no protections that ensure total impregnability against cyberattacks. Being a complex issue, a cybersecurity policy needs to be tailored to individual requirements of a given enterprise, ensuring flexibility and balance between the enterprise's smoothness of operations and achieving its data protection objectives. Thus, a cybersecurity policy should be perceived in terms of an array of technological safeguards, human dedication and system solutions implemented at the managerial level. It is crucial that that the entire team of an enterprise are kept up-to-date with regard to current threats to data privacy as well as methods counteracting them.

It is beyond question that data and information constitute an important asset of an enterprise's competitiveness. Having said so, it should be underscored that entrepreneurial responsibility for data protection can take various forms. The first, and possibly the most obvious shade of responsibility, is the financial one. On the one hand, owing to the fact that both data and information belong to the group of intangible assets, estimating their actual value is, in most cases, impossible. On the other hand, what can be done to determine the worth of data in the possession of an enterprise is assessing the cost of its loss. The value of data held in the enterprise's databases can then be attributed the value of total expenses related to an infringement to data privacy and all the actions that are necessary to restore its proper security level. When looking at the issue from this perspective it is much easier to acknowledge the importance of maintaining data security in contemporary enterprises.

Cybersecurity is an indispensable element of ensuring enterprises' immunity against potential data breaches as it frequently generates tremendous costs (Jain & Bhambri, 2006). As far as the measurable cost of data breaches is concerned, it is worth stressing that according to The Cost of Data Breach Report in 2023, drawn up by IBM Security, the average cost of a data breach reached the level of US $4.45

million, up by US $100,000 against 2022. Compared to 2022, this is an increase by 2.3%, yet, when analysing the past three years, the average total cost of a data breach rose by 15.3%. An important conclusion that can be drawn when analysing the report is that data breaches are not region-specific or country-specific. The phenomenon can be referred to as a global cyber-pandemic that inflicts all the corners of the world.

For example, the list of top five countries and regions where the cost of a data breach is the highest includes: the USA – US $9.48 million; Middle East – US $8.7 million; Canada – US $5.13 million; Germany – US $4.67 million; Japan – US $4.52 million (IBM, 2024). The analysis of data provided by the statistical portal Statista demonstrates clearly that certain industries are under particular attack of cybercriminals and the cost of data breaches there is even higher than the average. A breakdown into the industries where not having good protection of data privacy is especially costly shows that the healthcare industry comes first in this ranking, with the average cost of a data breach of almost US $11 million followed by the financial sector (US $5.9 million). The third and fourth positions are occupied by Pharmaceuticals and Technology – US $4.82 million and US $4.66 million, respectively. The fifth place in the ranking belongs to the Energy sector with an average cost of a data breach totalling US $4.78 million (Statista, 2023).

Fortunately, the recognition of the potential negative effect for enterprises affected by cyber-incidents in the form of data breaches is on the rise among the decision-makers. According to a survey conducted by PricewaterhouseCoopers (PwC), a company that is a leading global provider of professional business services, including accounting, auditing, human resources consulting, and strategy management, 62% of the respondents confirm that such an incident would translate directly into their inability to perform their everyday operations. Needless to say, that this is just a tip of an iceberg. The surveyed executives of global enterprises indicated in the first place the immediate costs that enterprises have to incur once they experience a data breach. They include remediation costs paid to security staff for investigating the reasons behind a breach, finding and sealing gaps in security systems of enterprises, and eventually assessing the extent of the breach. If an enterprise falls victim to a ransomware attack, immediate costs include the cost of paying the ransom for retrieval of the blocked data, costs of restoring security of the affected systems, and frequently also legal costs related to failing to comply with legislation (PWC, 2023).

However large the financial responsibility is, there are also indirect costs of data security incidents, which more often than not arise as a consequence of data privacy infringements. Undoubtedly, for contemporary enterprises that deal with huge amounts of customer data, losing it translates into losing customers' trust towards them. Thus, apart from the financial responsibility, their long-term reputation is at stake. According to the research conducted by Centrify, customers that have experienced their private data being compromised in 65% lost trust in the enterprise that had not protected the data properly. Another study by IDC has found that 80% of customers whose private data is breached will cease to remain loyal and will leave these businesses where security of their data was unsatisfactory. It is evident that big businesses, whose names are widely recognised worldwide, will undoubtedly suffer most as such data breaches will gain coverage in all the media. However, the potential notoriety can easily affect enterprise of smaller sizes. In the era of social media, it is easy to spread dissatisfaction with a brand or company among other customers (Rattan et al., 2005). A survey conducted by Cybermagazine proves that 85% of customers whose data privacy was infringed tell others about the unpleasant experience, with 33.5% of them using social media to do so, while 20% express their dissatisfaction directly on the company's website (Cybermagazine, 2021). It seems that potential consequences of data breaches are not limited to immediate financial expenses but more frequently can involve the whole range of stakeholders from the management of an enterprise through its suppliers, customers, even to shareholders. It is essential to recognise that rebuilding the trust towards an organisation can be a long-term process that in most cases cannot be easily addressed. Affected businesses need time to recover from the outcomes of a data breach incident. However, time is frequently what they lack as they have to simultaneously struggle with limited operational efficiency (human labour engaged in remedial actions), the problem of lost opportunities (normal activity of selling or manufacturing is disrupted), and churn.

3.4 LEGAL ASPECTS OF ENTREPRENEURIAL RESPONSIBILITY

The advent of new technologies resulted in new, frequently automated, methods of collecting data on users of cyberspace. In the last few years one can observe a particularly intense use of personal data that is used in machine learning and by AI solutions. Businesses around the world use personal information without its owners actually knowing how their private data is processed and, more importantly, protected. According to the findings of a Privacy and Trust Report by IAPP, only 29% of global consumers understand how well their private data is protected by enterprises that collect it. Most of the survey respondents were unable to say whether a given company has proper privacy regulations in place, and their knowledge on the types of personal data collected about them is very limited (IAPP, 2023). Yet, this rapid increase in processing of sensitive and personal information by businesses has resulted in growing consumer concerns pertaining to the manner in which private information is used by companies or external entities (Mani & Chouk, 2019). It can be stated that contemporary consumers have become very sensitive about the way their private data is used and managed. They have become much more aware of their rights in this respect. The tool that has enabled them to stand their ground against misuse of private data by enterprises is legislation. As the amount of private data collected by businesses was growing, it became obvious that previous law regulations failed to address the challenges of securely maintaining this data. This has led to introduction of new legal measures that, to a large extent, regulate the problem for both interested parties, consumers and businesses. This means that entrepreneurial responsibility over private data of consumers is also imposed by law. One of the examples of such legislative steps is the General Data Protection Regulation of the European Union that entered into force on 25 May 2018. It is believed to be the toughest privacy and security law in the world. It was drafted and passed by the European Union (EU), but must be complied with by any organisations that target or collect data on EU citizens. This is particularly important for all global enterprises that do business in Europe as they become responsible for violating privacy and security standards imposed by the EU's regulation. As consumers' awareness of their private data processing is growing, it constitutes a serious threat to companies that do not comply with the GDPR and are responsible for data privacy breaches. The regulation foresees harsh fines of 4% of company revenue or a minimum of €20 million if a company is deemed responsible for not maintaining data privacy (GDPR, n.d.).

As companies commonly share personal data with their business partners, they are increasingly losing control over data, which makes them or their partners more vulnerable to violating customer privacy (Schäfer et al., 2023). The efforts to regulate the issues of data privacy in cyberspace have been also undertaken by the US government. In March 2023 a new National Cybersecurity Strategy was released by President Joe Biden. The strategy is based on close collaboration between the public and private sectors perceiving it to be an essential requirement for securing cyberspace. Most importantly, however, the strategy challenges the previous approach of placing too much responsibility for protection of data and information systems on individual users and small organisations. Also, instead of focusing on blaming the business sector for insufficient care of private data, it promises to rebalance the responsibility for ensuring cybersecurity so it is more effective. This is supposed to encourage organisations to consider long-term investments in security, resilience, and promising new technologies. The underlying goal of the strategy is to build a "defensible, resilient digital ecosystem where it is costlier to attack systems than defend them, where sensitive or private information is secure and protected, and where neither incidents nor errors cascade into catastrophic, systemic consequences". The good news for enterprises in this respect is that within the strategy the US government and Congress, industry, civil society, and local territorial governments are going to collaborate so as to strengthen norms of responsible behaviour and implement effective cybersecurity practices across all the sectors of the economy (Whitehouse, 2023).

Entrepreneurial responsibility is also an important element of the new National Cybersecurity Strategy of India. The first cybersecurity policy was introduced in this country in 2013 and presently works are underway by the National Cyber Security Coordinator and the National Security Council Secretariat (NSCS) to integrate India's cybersecurity architecture and policies. What is important from the perspective of Indian entrepreneurs conducting business outside the country is that the draft National Cybersecurity Strategy, 2023, focuses on ensuring compliance with global security. The core of the 2023 strategy is the approach of common but differentiated responsibilities (CBDR). In this way the strategy differentiates responsibilities of individuals, businesses, academia and the government. It is the responsibility of all of the stakeholders using cyberspace and information networks to secure the part that they own. This sets clear requirements towards enterprises with reference to how they are supposed to protect private data in their possession. However, it is still unclear how this whole-of-nation approach will be adopted (Kapoor, 2024).

These are just few examples of how the issues pertaining to data privacy and security are becoming embedded in a national framework of cybersecurity. In fact, all leading countries and regions of the world have already prepared their national cybersecurity strategies. While analysing them, it can be observed that all of them are based on several strategic pillars that include (AlDaajeh et al., 2022):

- The achievement of a strategic vision for a country to become cybersecurity-resilient requires a joint effort of government, industry, and community.
- Both the government and private sector must protect their information resources against cybercrime. To do so they need cybersecurity professionals for whom there is a high demand.
- Investments are indispensable in research and developments of cybersecurity countermeasures to protect a country against emerging sophisticated attacks, also the ones targeting their critical infrastructure.
- It is necessary to improve societies' maturity and awareness of cybersecurity as these are key factors in the fight against cybercrime.

It can be concluded that entrepreneurial responsibility for maintaining data privacy constitutes an essential part of all the legal regulations applicable to enterprises on the global scale. Growing awareness of the global society pertaining to cybercrime and cybersecurity necessitates businesses all over the world to strive even harder to ensure compliance with legislation in force. However, on the positive side, it can be observed that the global trend is to ensure equal engagement of all parties involved in use and protection of private data, rather than put this burden exclusively on the shoulders of entrepreneurs.

3.5 MAINTAINING DATA PRIVACY IN THE ENTERPRISE

Cyberattacks aimed at gaining unauthorised access to data utilised by contemporary enterprises are on the increase. This affects enterprises of all sizes. Yet, the large ones can definitely spend more on hardware and software protections, let alone the cost of hiring IT personnel trained in cybersecurity area. Therefore, the situation is particularly threatening to small- and medium-sized enterprises. Compared with large ones, this group of enterprises has always struggled with limited spending on cybersecurity and IT personnel. As many of them still carry the burden of the pandemic period, the situation is unlikely to improve; they are frequently unable to find enough staff and time to address all the threats. The data of the report by Digital Ocean (Digital Ocean, 2023) shows that the primary concern of global businesses in 2023 was finding it hard to manage security due to lack of time. That was the opinion of 25% of the study participants. Nor surprisingly, though, the next three concerns included data loss or data theft (23%), ransomware attacks (12%), and DDoS attacks (10%). This demonstrates that for many businesses the risk of having their systems attacked remains at a relatively high level. The attacks themselves may take various forms and be driven by a combination of factors, not always purely financial ones. Moreover, as attackers are not bound by time or geographical location, they join forces while preparing their campaigns. They also adapt promptly to use to their advantage, changes of the political, technological or regulatory nature. It needs to be stressed that for attackers it is enough to find only one weakness of an organisation's cyberdefence system. This puts defenders in a disadvantageous position as they have to act on the whole front of countermeasures against cybercrime while for attackers it is enough to make good use of the discovered system's vulnerability (World Economic Forum, 2024).

Nevertheless, the responsibility for ensuring security of their information repositories does lie with enterprises that process data on unprecedented scale. Given the various form of entrepreneurial responsibility for mistreatment of data privacy, all enterprises need to define a data security strategy that will integrate the efforts of all the enterprise employees, not only IT departments, frequently being the only persons perceived as actually in charge of data protection. Certainly, strategies will differ as they need to be tailored to the organisational culture of a given enterprise. Yet, having analysed the guidelines and reports of three leading IT companies that provide solutions for data protection, it can be assumed that there exist several cornerstones of a well-constructed data security strategy. They have been summarised in Table 3.1.

As one can observe in Table 3.1, the good practises of data protection for enterprises provided by leading global IT companies have a lot in common. IBM and Microsoft stress the importance of securing facilities where company's data is

TABLE 3.1
Cornerstones of Cybersecurity for Enterprises

IBM	Microsoft	Kaspersky
Ensure physical security of servers and user devices	Verify thoroughly. Check creditworthiness of users' and devices' prior access to resources is granted	Refrain from clicking on links in emails that have been sent by strangers or come from unfamiliar websites
Ensure access management and controls	Apply Zero Trust principles - use least privilege access	Use strong passwords
Ensure that application security is up-to-date	Keep up-to-date	Update your software and operating system
Do backups	Use extended detection and response (XDR) and antimalware	Use antivirus software
Educate employees	Assume breach	Opening email attachments received from unknown senders should be prohibited
Monitor and control network and endpoint security	Protect your data	Unsecured Wi-Fi networks operating in public places should not be used

Source: Author's elaboration based on IBM (n.d.; Microsoft, 2023; Kaspersky, n.d.)

stored, against unauthorized access, maintaining them in good condition and making sure they have adequate fire suppression measures and climate controls in place. The same companies urge enterprises to treat rights to access data very seriously. The most advisable approach in this area is applying the principle of "least-privilege access" in the entire IT environment of the enterprise. This means that persons that need to access certain resources should be granted access exclusively to them. Not surprisingly, all three analysed companies point out the fact that software that is used in enterprises will only play its defensive role when it is kept up-to-date. Therefore, installing patches and updates is one of the essential practices for entrepreneurs. Data can be lost. The simplest possible piece of advice, then, is to protect the key data, which means knowing where it is and what protective measures have been implemented. As the report by Microsoft underscores, entrepreneurs should assume that their systems are not fully immune and a breach can occur. To minimise the risk of a potential responsibility for data infringement, it is essential that enterprises use antivirus and antimalware software that can automatically detect and block attacks. It is also highly advisable to implement monitoring and controls over the entire IT infrastructure of an enterprise to reduce the risk of a data breach. The thing that cannot be neglected is also performing regular backups of key data that can be quickly restored if a data incident happens.

The rules of data handling suggested by experts from Kaspersky Lab put stress on the human factor and potentially dangerous activities of IT systems users. It is a well-known fact that all protected IT systems are as strong as their weakest link. Unfortunately, this link is frequently people and their carelessness that makes them open emails from unknown senders, clicking links in such emails or use public networks when working with company's data. Therefore, training employees, promoting good security practises and making them use strong passwords are the actions that can successfully support hardware and software measures applied in enterprises.

A large number of enterprises, while formulating and implementing their data security strategy, decide to entrust protection of private data in the hands of experts and recruit cybersecurity experts. However, there is a huge gap on the market that cannot be easily filled. According to the estimations of the World Economic Forum's Global Cybersecurity Outlook 2023 report, the present global demand for cybersecurity experts reaches 3.4 million of people. The same report shows that 59% of business leaders and 64% of cyber leaders ranked talent recruitment and retention as a key challenge for managing cyber resilience in their enterprises. However bad the news seems, enterprises can be supported in their cybersecurity efforts by Artificial Intelligence.

There are four basic stages of ensuring cybersecurity in enterprises: prediction, prevention, detection and reaction. Application of AI in the two first stages is limited. It is impossible to predict and prevent all threats. Prediction and prevention of attacks uses data on what is done after an attack is detected and what steps are implemented as a result. AI can be successfully applied in the remaining two stages – detection and reaction, where it is of utmost importance to detect threats immediately after they arise and define the procedures of handling them. Data analysis and detection of occurrences of out-of-the-pattern nature are mundane and time-consuming for human personnel, yet they can signal threats to the organisational network and cannot be neglected, so they can be done by AI.

Machine learning and deep learning are the technologies that guarantee that parameters of detection are constantly updated. Anomalies are detected with the use of behavioural analysis. Cybersecurity is the domain where AI-powered tools can be applied to a great extent. Unlike humans, it can work with huge datasets and being equipped with pre-defined patterns uses heuristics to predict potentially dangerous content. These features of AI make it an ideal technology to develop secure intra-organisational platforms. Its capabilities will include a possibility of automatic identification and categorisation of an organisation's IT infrastructure and information resources. More importantly, it will be able to determine its critical weak points and incessantly protect the network against attempts of unauthorised access to data and information stored in it (Chmielarz & Pabian, 2020).

SUMMARY

All contemporary businesses possess data that constitute the basis of their operations and, therefore, is valuable from the perspective of a given business organisation. However, the value of private data stored by enterprises has been spotted by cybercriminals who target businesses of all sizes in all geographical locations. The forecasts are not optimistic in this respect and further waves of cybercrime are expected to surge across the global industries. The vulnerability to being a victim of cybercrime should prompt decision-makers of global enterprises to revise and update their strategies of maintaining cybersecurity. Due to legislative measures that have been introduced worldwide, lack of compliance with data handling regulations can result in financial penalties to be paid by entrepreneurs. However, there is a lot more for them at stake than just financial aspects of data breaches. The long-term consequences of losing consumer trust and returning to the pre-breach efficiency of operations can also take their toll.

On the positive side it should be stressed that a modern global trend is collaboration of all the stakeholders, data owners, businesses that process private data, and governments and regulators. In this context entrepreneurial responsibility for protection of private data of customers becomes part of a nationwide regulatory framework. This seems to be a step in the right direction as it is conscious participation of the business sector in the global anti-cybercrime scheme that may prove more effective than the prospect of an imminent fine for data-related incidents, which cannot be entirely eliminated and which entrepreneurs are not always to blame for. One thing is certain, the data-dominated nature of current computer systems of enterprises will require them to adapt to the constantly evolving landscape of cyberthreats. Fortunately, emerging technologies more often than not seem to be a double-edged sword. Artificial Intelligence, although utilised also by cybercriminals, may prove to be a game changer for enterprises as it may reduce the workload of IT teams in large enterprises and constitute the only cyber protection for SMEs with tight budgets for spending on cybersecurity.

REFERENCES

AL-Khassawneh Y. A. (2023). A review of artificial intelligence in security and privacy: Research advances, applications, opportunities, and challenges. *Indonesian Journal of Science and Technology, 8*(1), 79–96.

AlDaajeh S., Saleous H., Alrabaee S., Barka E., Breitinger F. & Choo K. K. R. (2022). The role of national cybersecurity strategies on the improvement of cybersecurity education. *Computers & Security, 119*, 102754.

Chmielarz G. & Pabian A. (2020). AI development and application of AI-based solutions in the area of information and data management in Poland. Proceedings of 16th International Conference on Network and Service Management (red.) Zincir-Heywood Nur, Ulema Mehmet, Sayit Muge, Clayman Stuart, Kim Myung-Sup, Cetinkaya Cihat. IEEE.

Chunling Z. & Yumei X. (2023). Institutional innovation essence and knowledge innovation goal of intellectual property law in the big data era. *Journal of Innovation & Knowledge, 8*(4), 100417.

Cisco. (n.d.). *What is Cybersecurity?* https://www.cisco.com/c/en/us/products/security/what-is-cybersecurity.html

Cybermagazine. (2021). *What Causes the Most Damage, Losing Data or Trust?* https://cybermagazine.com/cyber-security/what-causes-most-damage-losing-data-or-trust

Data Centers. (2024). *Global Data Center Locations.* https://www.datacenters.com/locations

Digital Ocean. (2023). *Small Businesses and Cybersecurity.* https://www.digitalocean.com/reports/cybersecurity-smbs-2023#download

GDPR. (n.d.). *What is GDPR, the EU's New Data Protection Law?* https://gdpr.eu/what-is-gdpr/

Google Cloud. (n.d.). *What is Cloud Computing?* https://cloud.google.com/learn/what-is-cloud-computing

IAPP. (2023). *Privacy and Consumer Trust Report.* https://iapp.org/media/pdf/resource_center/privacy_and_consumer_trust_report_summary.pdf

IBM. (n.d). *What is Cybersecurity?* https://www.ibm.com/topics/cybersecurity

IBM. (2024). *Cost of a Data Breach Report 2023.* https://www.ibm.com/reports/data-breach?

Jain V. K. & Bhambri P. (2006). *Fundamentals of Information Technology & Computer Programming* (4th reprint ed.). Katson Publishers. ISBN 81-88458-59-7.

Kapoor A. S. (2024). *National Cybersecurity Strategy Should Spur Digital Resilience.*https://www.tribuneindia.com/news/comment/national-cybersecurity-strategy-should-spur-digital-resilience-527844

Kaspersky. (n.d.). *What is Cyber Security?.* https://www.kaspersky.com/resource-center/definitions/what-is-cyber-security

Lampropoulos G., Siakas K. & Anastasiadis T. (2019). Internet of Things in the context of Industry 4.0: An overview. *International Journal of Entrepreneurial Knowledge, 7*(1), 4–19.

Mani Z. & Chouk I. (2019). Impact of privacy concerns on resistance to smart services: Does the 'Big Brother effect' matter? *Journal of Marketing Management*, The role of smart technologies in decision making: Developing, supporting and training smart consumers, *35*(15–16), 1460–1479.

Microsoft. (n.d.). *Microsoft Digital Defense Report 2023.* https://www.microsoft.com/en-us/security/security-insider/microsoft-digital-defense-report-2023

Precedence Research. (2023). *Cloud Computing Market.* https://www.precedenceresearch.com/cloud-computing-market

PWC. (2023). *Companies Face a Big Trust Gap on Data Security.* https://www.pwc.com/gx/en/issues/c-suite-insights/the-leadership-agenda/companies-face-a-big-trust-gap-on-data-security.html

Rana R. & Bhambri P. (2024). Artificial Intelligence and Internet of Things (IoT) facilitated digital twin for Industry 4.0 application domains. In S. Rani, P. Bhambri, S. Kumar, P. K. Pareek, & A. A. Elngar (Eds.), *AI-Driven Digital Twin and Industry 4.0: A Conceptual Framework with Applications* (1st ed., pp. 168–181). CRC Press. https://doi.org/10.1201/9781003395416

Rattan M., Bhambri P. & Shaifali. (2005). Information retrieval using soft computing techniques. In National Conference (pp. 58–60) on Bio-informatics Computing. Presented at TIET, Patiala held during March 18–19, 2005.

Schäfer F., Gebauer H., Gröger Ch., Gassmann O. & Wortmann F. (2023). Data-driven business and data privacy: Challenges and measures for product-based companies. *Business Horizons*, *66*(4), 493–504.

Singh P., Singh M. & Bhambri P. (2004). Interoperability: A problem of component reusability. In International Conference (Abstract Book) (p. 60). Presented at the "Emerging Technologies in IT Industry" at PCTE, Ludhiana on November 27, 2004.

Singh P., Singh M., & Bhambri P. (2005). Embedded systems. In U.G.C. Sponsored National Seminar (pp. 10–15). Presented at the "Embedded Systems" seminar at KMV, Jalandhar held during January 24–25, 2005.

Singh M. P. & Bhambri P. (2024). AI-driven digital twin conceptual framework and applications. In S. Rani, P. Bhambri, S. Kumar, P. K. Pareek, & A. A. Elngar (Eds.), *AI-Driven Digital Twin and Industry 4.0: A Conceptual Framework with Applications* (1st ed., pp. 127–141). CRC Press. https://doi.org/10.1201/9781003395416

Statista. (2023). *Average Cost of a Data Breach Worldwide from May 2020 to March 2023, by Industry*. https://www.statista.com/statistics/387861/cost-data-breach-by-industry/

Statista. (2024). *Volume of Data/information Created, Captured, Copied, and Consumed Worldwide from 2010 to 2020, with Forecasts from 2021 to 2025*. https://www.statista.com/statistics/871513/worldwide-data-created/

Whitehouse. (2023). *National Cybersecurity Strategy 2023*. https://www.whitehouse.gov/wp-content/uploads/2023/03/National-Cybersecurity-Strategy-2023.pdf

World Economic Forum. (2024). *Global Cybersecurity Outlook 2024. Insight Report*. https://www3.weforum.org/docs/WEF_Global_Cybersecurity_Outlook_2024.pdf

4 Ethical AI and Machine Learning in Entrepreneurship

Tomasz Lis and Aleksandra Ptak

4.1 INTRODUCTION

Information and knowledge are currently the basic intangible resources of enterprises. Their importance results from their primordial nature in relation to all human activity. Technological development, an important element of which is the development of information and communication technology (ICT), improves information and knowledge management. At the same time, it contributes to increasing the dynamics of the functioning of people and enterprises. This makes information and knowledge management one of the most important areas of market competition (Tunçer, 2015).

One of the most important human limitations in this area is the ability to absorb and analyze. This, naturally, negatively affects the effectiveness of the decision-making process considered from the point of view of speed and optimality. One way to solve the problem is Artificial Intelligence tools. They allow for a significant reduction in operational time while increasing the amount of data and information. Artificial Intelligence is not only about almost immediate search, identification, and analysis of information (Buck & Morrow, 2018); it is also an inference, the key element of which, taking into account the market dynamics, is the ability to self-learn. All this aims to shorten the company's response time as much as possible from the moment information appears. Basing key areas of a company's operation on Artificial Intelligence brings with it many problems and threats, among which ethical threats play an important role.

Nowadays, there is an increase in the popularity of the use of Artificial Intelligence elements. This applies primarily to the area of professional activity, although it should be noted that this technology also enters private life. This is inherently associated with redirecting a number of activities traditionally performed by humans to technical devices, as well as with activating the use of technology – in areas where it has already been involved (Raisch & Krakowski, 2021). Examples of the practical use of Artificial Intelligence tools include: supporting the functioning of courts, diagnoses and proposed methods and stages of treatment, management of resources and supplies, investing and other issues related to financial management, and intelligent means of transport.

DOI: 10.1201/9781003484226-4

Along with the analyzed and indicated benefits, the issue of threats and dangers of using Artificial Intelligence is increasingly being discussed (Bécue et al., 2021). They are related, for example, to the creation of various types of scientific works. Such works are, to varying degrees, in extreme cases entirely, "written" by Artificial Intelligence tools. They do not influence the development of the authors, and at the same time very often contain substantive errors – which contributes to the blurring of scientific ideas and the spread of false information and knowledge. This problem does not only concern this type of solution, but the internet in general. The content contained therein is not always a reliable source of information and knowledge, but as the authors of the work noted, they are very often considered to be such. Another problem with the use of AI is related to its implementation into means of transport, which will result in improved transport processes (Bendiab et al., 2023). However, unattended transport poses a risk of unforeseen disruptions, including those that may result from unauthorized access by third parties.

Artificial Intelligence, which includes machine learning, is a man-made technology. As such, it is subject to all the limitations that are inherent in the essence and characteristics of a man. As an imperfect creation, it has its flaws, including actions to the detriment of others, fallibility, and. actions that cause disruption to others that may be unintentional or intentional. Every development, including technological development, gives people the opportunity to improve and optimize, and at the same time offers the cybercriminals an opportunity to take negative actions consciously (Derave et al., 2022). The scale of this phenomenon is particularly visible since the development of digital technologies, including the internet. This is manifested, among other crimes , by the theft of data and financial resources, impersonating others, and remotely disrupting the functionality of data-storing devices (Rattan et al., 2005). Any technology or new solution may also be burdened with hidden defects that occurred unintentionally. They most often result from errors in assumptions and thinking, as well as ignorance and lack of knowledge. It should be noted that the area of analysis, reasoning, organizing, and planning prior to the development of technology may also be subject to intentional negative actions. Consciously spreading false knowledge, building false assumptions, and disrupting the thought process in any way are just some of the examples of activities that may result in the development of technologies with defects, including those whose nature allows them to be used in a negative way (Chanda & Banerjee, 2022). It can be said that it is the man who decides on the direction (positive/negative) of the use of the technologies he develops, and at the same time, it is assumed that he decides on their purpose (positive/negative) at the development stage.

The aim of this chapter is to present the essence, possibilities, advantages, benefits, but also dangers of using tools based on Artificial Intelligence. The analysis of ethical threats is a vital topic to discuss, whichresults from the authors' belief that the development of digital technologies and Artificial Intelligence gives people great opportunities to improve themselves and their activities. However, taking into account the negative features of selfishness, tendency to aggression, egoism, ignorance, envy, and injustice, AI gives the possibility of negative impact not only on other people, but even on human existence. The authors are aware that the subject matter is extensive and it is not possible to analyze it exhaustively in one work. They state that there is a need to continue research in this direction.

4.2 ARTIFICIAL INTELLIGENCE AS A POTENTIAL SOURCE OF UNETHICAL ACTIVITIES

Not so long ago, Artificial Intelligence was considered in the context of fantasy literature and films. There was often an image of machines that, using their powers and computational capabilities, as well as the ability to repair, dominated the human sphere of life. Machine-human cooperation was based on activities that had a negative impact on other people. These actions were almost always unethical. This has undoubtedly contributed to the development of available solutions in the field of Artificial Intelligence technology in the context of the potential threat and the AI-based tools allowing for bypassing regulations and legal standards, and using technology, to conduct unethical activities.

Despite these concerns, AI-based tools are treated as the future of technology, being used in virtually every dimension of human functioning (Cave & Dihal, 2019). Knowledge of the essence, features, real possibilities of use, disadvantages, and threats of technology are still poor (Bhambri & Gupta, 2005). The authors of this work identify many IT tools with AI that have been in use for a long time. At the same time, they often come into contact with opinions according to which Artificial Intelligence will only be created, and this will happen when "devices" that have their own consciousness and can live independently next to humans are introduced to the market. The most general and one of the first definitions of AI is "the ability of a machine to simulate or imitate human intelligence" (Zalewski, 2020).

Considering that intelligence and independent, conscious action are a feature of living organisms, especially humans, most definitions of the concept of AI contain a reference/comparison to a human being. Other definitions include:

- "Systems that think and behave like humans, that is, think and behave rationally" (Russell & Norvig, 2003).
- "The process of automating certain activities such as decision-making and learning, which are a reflection of human intelligence" (Bellman, 1978).

There are many types/forms of Artificial Intelligence. They differ from each other, among others: a way of teaching, i.e. acquiring the proverbial "intelligence." One of them is machine learning. Its idea is to learn by analyzing large datasets and looking for patterns, dependencies, and relationships in them. This is done due to the implementation of special algorithms. "Machine learning algorithms are fundamentally designed to classify data, find patterns, predict outcomes, and make informed decisions. Algorithms can be used individually or in combination to achieve the highest possible accuracy when data is complex and highly unpredictable" (Uczenie, 2024).

The term intelligence in relation to current AI solutions does not fully reflect its capabilities. These are not yet tools/machines that are able to function completely on their own without the help or cooperation of humans. They do not take into account what distinguishes a human from a machine, i.e. feelings and empathy (Kaplan & Haenlein, 2020). According to the authors of this work, it can be said that they currently reflect the abilities, skills, and approach to solving problems of their authors. Every technology created by man was and is intended to help in his everyday functioning. Facilitating and improving the implementation of activities is not necessarily associated with achieving overall positive effects. Due to human characteristics, technologies can be used to carry out activities aimed at cheating, theft, extortion, etc. What they have in common is a negative emotional impact. In such case, we talk about unethical activities.

Modern Artificial Intelligence tools are created by humans based on, often, very large data resources. As a result of the analyses, dependencies, similarities, and relationships are identified. IT tools, having a specific historical map of the events they research, identify convergences in their "experience and science" (Fjeld et al., 2020). This results in the generation of a set of output information that forms a solution to the problem that the AI is currently dealing with. Tasks carried out in this way do not constitute anything creative, they are not the result of the use of intelligence; they are only a duplication/arrangement of available information into a specific whole. This is how the available solutions work, allowing you to obtain content that answers your questions and "creates" music or images. This method of operation raises widespread discussions about copyright and, ultimately, its ethicality.

According to the authors of this work, these tools were created to present the possibilities offered by Artificial Intelligence. Their naturally fearful behavior, resulting in the generation of ethically "questionable" solutions, is the result of the user's intended activity. In the history of man, there are many technologies known that were created with the intention of supporting his functioning, but which can be used in an unethical and harmful way. An example would be dynamite (Wittenberg, 2022).

Another potential possibility of generating unethical actions by AI is directly related to the principle of its operation. By analyzing datasets that contain racist content that have a negative connotation towards any community, Artificial Intelligence can generate content (in any form: written, audio or image) of a similar nature. They will then be unethical. This is a threat that can only be solved by filtering data resources intended for training. At the same time, any limitation of data resources leads to limitation of knowledge bases, reducing the effectiveness of the learning process, which negatively affects the efficiency of the technology.

Artificial Intelligence increasingly performs tasks related to supporting and optimizing transport (Singh & Bhambri, 2024). This applies to both individual, group, and freight transport. Intelligent means of transport, as well as e.g. AI and information technology supporting drivers, are exposed to unauthorized access by third parties. They may contain errors from the design and creation stages (Bo et al., 2023). This may result in undesirable events, including those that threaten human life and health. In such a case, we can also talk about the potential unethical nature of the technology. The question should be asked whether it is even possible to create a tool or technology that, as a result of its use, would not allow for misuse? Therefore, can we talk about the ethicality of AI? Shouldn't this problem be considered in the context of human ethics, both as the party designing and producing, as well as using ready-made solutions? This is confirmed by Kasparov, who notes: "there is no evidence that machines threaten us. The real danger comes not from killer robots, but from humans – because humans still have the monopoly on evil" (Garri, 2024).

4.3 ETHICS IN THE FUNCTIONING OF ENTERPRISES IN THE ERA OF DEVELOPMENT OF DIGITAL TECHNOLOGIES

Ethics in the functioning of enterprises has been an issue that has been taken into account since the end of World War II. As time passed, it gained in importance. Currently, it is widely commented on and analyzed and has become an important factor in the economy. Increasing importance can be associated with technological development, especially in the area of digital techniques – ICT. This is due to the increase in employee self-awareness, the transition from the producers' market to the customers' market, as well as globalization. These factors, in turn, are a direct result of facilitating access to and dissemination of information, increasing the importance of information and knowledge in the management and general functioning of the entity (Bampton & Cowton, 2002).

According to the online version of the PWN encyclopedia, ethics is

> a set of moral norms recognized at a certain time by a social group as a reference point for assessing and regulating behavior in order to integrate the group around certain values, a synonym of morality; in the philosophical sense – ethics is the science of morality, considered separately in its aspects: normative as the science of morality (so-called normative ethics or proper ethics) and descriptive-explanatory as the science of morality (so-called descriptive ethics or ethology). (...) Ethics in its normative aspect deals with determining what is morally good and what is bad, and – based on the adopted assessments and the duties related to them setting directives, i.e. norms of morally positive behavior and indicating ways of transforming common morality in such a way that adapt it to the adopted moral ideal (...). Ethics in the descriptive aspect deals with the analysis, description and explanation of morality actually accepted (in circulation) in various eras and social environments, indicating the sources, structure, functions of morality as a form of social awareness and detecting the regularities of its development; therefore examines assessments, norms, personal models, ideals, ways of justifying them as data without engaging in their evaluation; in this understanding, ethics includes the sociology of morality, moral psychology, and the semantics of morality (called metaethics in the narrow sense), which studies the language of morality; Descriptive ethics also includes the history of morality and the accompanying ethical doctrines.
>
> **(Etyka, 2024)**

To sum up, it can be said that ethics are the norms of behavior adopted and established among the population in a given area. What they have in common is universal recognition as a model of the proper existence of the individual and the communities created by him. This definition of the concept often allows for significant differences between the ethics of communities depending on their geographical location. The form of the adopted ethical standards was influenced by geographical and environmental conditions, access to technology, and access to other communities (Rana & Bhambri, 2024).

Ethics is inextricably linked with man. There is often talk about appropriate and inappropriate behavior in this regard. In relation to business activity, ethics is represented in the decisions made, which always result in specific activity. Decisions concern the people making them, but also those who constitute the executive element (Bhambri & Singh, 2005). As Kisiel notes, "management ethics covers the standards of behavior of managers that guide them when performing their work. The way the organization treats employees and the attitude of employees towards the organization are important. These areas raise many controversies related to human resources management in general (pay and working conditions, dissemination of confidential information about co-workers, interference in the private sphere or revealing company secrets, etc.)" (Kisiel, 2014). Analyzing the importance of ethics in an organization, the author notes that her interests include the following aspects: social responsibility of the organization, hiring and firing employees, employment conditions, mobbing (Kisiel, 2014). According to the authors of this work, the presented aspects concern the organization as a structure operating within the community and employing workers. Ethics refers to social relations (organization as an element) and those directed from the management down to the organizational structure of the enterprise (Bhambri et al., 2005). There are no relationships taken into account in the directions: from the community (which can be treated as a collective of potential and active customers and stakeholders) to the enterprise, as well as between all elements of the organizational structure.

The issue of ethics in management cannot currently be contemplated without considering every participant of everyday life whose activity, even only in the informal sense, has or may have an impact on the functioning of the organization. This is related, for example, to activities that generate false information that has a negative impact on the company and/or its partners. Walczak-Duraj notes that the topic of ethics is often treated instrumentally in and by companies. It is a tool, an approach that is taken into account as one of many elements intended to achieve goals. At the same time, Walczak-Duraj notes that the paradox of managing contemporary organizations, dressed in the guise of humanistic management, in which the individual is to be treated subjectively, while maintaining all the dignity, is that organizations function quite well in the changing world, and the fate of individuals entangled in management processes usually looks completely different. Employees are treated as objects, are lost, unable to identify with work and a given organization for a long time, confused, knowing only fragments of production processes, often trapped among the "murky" network rules of the organization (Walczak-Duraj, 2017). Walczak-Duraj also states that the idea of ethics in business activity is the integration of ethical principles with the organizational culture of the company. All employees should voluntarily and consciously perform their tasks in a manner consistent with the adopted and applicable principles, whose starting point is morality and ethics.

The relationship between technologies used in business and ethics is often considered in two basic dimensions. In the first one, technology is the element intended to improve and optimize work and increase the level of human safety but also the environment in which the company operates. In the second case, carrying out activities traditionally performed by human resources causes the reduction of full-time positions and termination of jobs, which contributes to an increase in the risk of losing employment (Nath & Sahu, 2020). Technological development, especially in relation to automation, causes an increase in interest in low and high competencies of employees. At the same time, interest in medium competences is decreasing. This may result in a situation where many professions will not be relevant to work needs, which will ultimately lead to their disappearance. Artificial Intelligence can successfully perform the actions of lower-level employees. This applies to most professional activities, especially those related to production, but also increasingly to analysis, reasoning, and even software creation. In the case of limiting the number of employees in many areas of professional activity, according to the Nath and Sahu, it may lead to the creation of a generational professional "hole." The lack of employees gaining experience from scratch will increase the dependence on the digital factor of work.

As Kumięga rightly notes, technological development, including in the area of automation and the introduction of solutions based on Artificial Intelligence, forces a change in the approach to teaching in secondary and higher schools. Pupils and students must have access to new solutions used in practice. It must also be supported by access to theoretical resources of information and knowledge. This, in turn, requires closer cooperation between science and industry. School students at various levels who do not have access to appropriate knowledge resources and do not have the opportunity to "touch" the technology have to learn it all from scratch after coming to work. Kumiga also draws attention to the fact that in such a reality, engineering education programs should include humanistic elements including those related to ethical issues in the new world of work, which is increasingly based on digital solutions. In the dynamically changing reality, it is necessary to pay attention to retraining opportunities. Appreciating the changes caused by technological development, including AI, Kumięga notes that in the near future it will be necessary to take a new look at the issue of ethics in enterprises. According to the author, this is due to the fact that tools are starting to take over the roles traditionally assigned to humans:

> Transhumanists deal with ethical aspects of the relations between people, robots and artificial brains. The subject of their reflection, which is also appropriate for posthumanists, is the problem of future relations between humans and Artificial Intelligences created in the future (which may ultimately exceed human intellectual predispositions). Digital technologies will function in the service of humans and enhance their intellectual capabilities or will gain high independence, typical of humans or exceeding their abilities. Nowadays, a new understanding of objects is required, different from that of previous eras. One of the proposals is Karen Knorri-Cetina's posthumanism, which assumes that currently some objects have taken over the role of people in social relationships, becoming a permanent component of life, while generating a new type of relationship, called "post-social."
>
> **(Kumięga, 2016)**

According to the authors of this work, relationships are the natural element accompanying the coexistence of all individuals who are active in their everyday life. In this approach, the parties to the relationship were always people/employees.

Currently, as a result of technological development, humans activate devices and/or systems that are characterized by a certain autonomy and the ability to perform certain activities only under the supervision of a human factor. In such conditions, it seems justified to include not only people, but also independent "devices" in the relationships. Relationships are an indispensable element of activity in a group/collectivity, the effects of which directly and/or indirectly affect the participants of this group/collectivity.

4.4 ETHICS OF ARTIFICIAL INTELLIGENCE

In his work, Nowakowski notes that, according to research, free access to ICT generates data occupying 2.5 quintillion bits every day. It consists of data relating to every sphere of human activity, regardless of his/her geographical place of existence. This virtually unlimited set is very important for every human activity, the effects of which depend on matching existing needs and expectations. Such an area is undoubtedly the business sphere/economic activity. With access to very large amounts of data, the key to their effective use is to rely on modern technological solutions. These tools should implement all activities related to information and knowledge management, starting from identifying and acquiring, through searching for relationships and dependencies, to reasoning. This indiscriminate and continuous generation of data poses a potential threat. It applies to both data creators and entities that obtain and use them. By generating data, we run the risk of unauthorized persons gaining access to sensitive information, which they can use directly against the creator and/or to one's detriment. The same applies to entities using information and knowledge. Obtaining false, outdated, negative data always results in disruptions in the analysis and conclusion process. It is also harmful because of the analysis and then conclusion tool. Taking into account the amount of data and the time a human would have to devote to processing it, enterprises use IT tools, including those based on Artificial Intelligence. These are "machines" that uncritically analyze data. As a result, there is a possibility of errors in analyses and decisions made on their basis. These negative and often conscious human actions are commonly treated as dishonest and unethical. These factors, according to Nowakowski, determine the need to include ethics in the process of designing and using new digital technologies, especially Artificial Intelligence (Nowakowski, 2022).

Taking into account the fact that Artificial Intelligence is a man-made product, it can be said that it is as ethical in its actions as its designers and producers. (Miernicki & Ng, 2021). This statement is important because the term intelligent does not actually correspond to the accepted definition of this concept. Intelligence considered from the human perspective is related to the ability to recognize semantics (meaning), while from a systems perspective it is based on the ability to analyze and recognize relationships between elements of very large data resources. Therefore, Artificial Intelligence is sensitive to disruptions in databases, as well as to actions taken during its creation. Chojnowski's work states that it is common to associate dangers and threats resulting from the use of AI with literature and science fiction films. Chojnowski emphasizes that at the current level of technology development, the vast majority of threats are pure and unjustified fantasy. It also identifies the three most important groups of challenges related to the use of AI, which are (Chojnowski, 2022):

- Threats related to machine learning – one of the basic elements here is training AI systems. Datasets that are not representative, including those that do not take into account the entire reality described, will not allow for valid and reliable analysis and conclusions in the future. This dimension also applies to situations in which AI reaches conclusions in a way that are incomprehensible even to people co-creating the system. The lack of transparency of causes and motivations is a manifestation of an arbitrary approach to the problem, which is unacceptable in terms of its impact on human functioning. Yet another example is the ability of systems to identify relationships between neutral and actually unrelated data, as well as to identify and link legally protected data – sensitive information about specific people.
- Problems related to communities functioning in a widely digitized world – problems of this nature are related, for example, to the impact on the reduction of employees' self-esteem and their importance in structures, e.g. enterprises. Employees operating automated workstations, relying on decisions generated by IT systems, lose their sense of value and influence on the actions taken, they embody technology and, in their opinion, give it a higher position in the structure. The largest corporations in the world have systems at their disposal that have a large impact on people in their groups and even societies. There is concern based on real possibilities of surveillance and manipulation of opinions. This group also pays attention to the energy consumption required to develop and implement AI tools, which adversely affects the natural environment.
- Issues related to the essence of life features, human nature, differences between humanity and "artificial/machine beings," the status of technological products in societies – this group considers issues of the legal and moral significance and position of AI in human life. It is called abstract, which results from the lack of invention of machines that are intelligent enough to function in societies. There is currently no need to define their rights and obligations. However, this problem must be considered in connection with more and more new information as to the time when such solutions will appear. This is also motivated by the currently known technology of implanting nanotechnology-based implants into living organisms.

In practice, the presented groups of problems should not be considered separately. In fact, they cover a set of issues interconnected by social relations. This is due to the primary role of humans as the causative factor, user and recipient of the effects of using technology. An example is the phenomenon of algorithms omitting data relating to specific and very few social groups. The same occurs in the case of economic events that occur rarely, heterogeneously, the basic common part of which is a human being. This phenomenon is called algorithmic exclusion, which can be defined as a phenomenon which, among other things, through inappropriate selection of data for analysis, causes some groups of people – as a result of such technology – to be better and others, worse. This is because society and culture are difficult to capture in digital data. "It's usually worse for people who are marginalized in society anyway" (Algorytmy, 2024).

The ethics of Artificial Intelligence is often considered in the context of its implementation and use. We talk about ethical use or action. It comes down to the technology production stage, as well as to its end user. This problem is only an element of the relationship between technology and ethical issues. Its basis is the activity of the human using AI for professional and/or private purposes. The existing ethical issues are then closely related to that user. One may use the tool within the permissions and capabilities provided by the creators. The actions are unethical, but this functionality was not intended by the creators. This is where another dimension of the issue of AI ethics comes from. It is related to the stage of developing, designing, and creating AI (Mittelstadt, 2019).

One form of creating Artificial Intelligence is machine learning. Its basis is the preparation of the learning algorithm, then training on initial data, at the next stage the system's performance is assessed (on a different dataset than before), and finally, testing takes place. Ethics as the feature of the system depends mostly on the course of this initial stage. Limited, incomplete data that do not take into account all aspects of the examined reality do not make it possible to prepare an optimal and effective tool. These shortcomings may result from the sheer level of complexity of the phenomenon being studied, but also from the intentional actions of the co-creators (Hamilton & Davison, 2022). Also at this stage, the operational parameters, permissions, and all aspects that the future final user will have to deal with are determined.

The issue of AI ethics is a very important issue in the modern world. Its importance was noticed by the European Union institutions. Already in 2018, the European Commission established the Independent High-Level Expert Group on Artificial Intelligence. Its work produced a document containing "ethical guidelines for trustworthy Artificial Intelligence." The authors identified three basic features that trustworthy AI should have:

- Legality: The technology must ensure compliance with all applicable laws and regulations.
- Ethicality: Ensuring compliance with ethical principles and values.
- Reliability: Ensuring full reliability in the development and production process, which results from the potential occurrence of unintended negative effects of its use.

Experts stated that the ethics of AI require that it be developed, implemented and used while maintaining respect for human autonomy, preventing harm, maintaining fairness, and justification. At the same time, they pointed out that there may be situations in which there are contradictions between the rules, which must always be the subject of appropriate adjustment actions. The document also draws attention to the need to pay special attention to social groups characterized by high social sensitivity: children, disabled people, exposed to the risk of exclusion, exposed to the phenomenon of asymmetry of power or access to information, in particular in connection with employer-employee and enterprise – consumers relations. To sum up, it was noted that, along with the benefits, the use of AI is associated with risks and specific threats. The nature of potential threats refers to the impact on democracy, the rule of law and justice, and even the human mind. By identifying threats and pointing to the requirement for ethical treatment, attention was drawn to the need to introduce control tools and monitor progress in the development and involvement of technological solutions based on AI (Wytyczne, 2024).

The issue of AI ethics was raised in the draft regulation of the European Commission presented in April 2021. The project approved by the European Parliament was created based on the analysis of the risks associated with the use of Artificial Intelligence. According to the provisions, AI systems will be classified into one of four groups. The basis for the division is the size of the estimated risk. For example, systems enabling automatic translations will be placed in the lowest risk group. Solutions with an unacceptable level of risk are to be systems whose operation is in contradiction with the values presented by the EU, e.g. in connection with the violation of fundamental rights. This group includes systems that allow even potential manipulation of people, including those based on subliminal techniques. This also applies to systems that make it possible to exploit the limitations and weaknesses of specific social groups, including children, elderly people, disabled people, etc. According to the provisions of the project, public administration bodies will not be allowed to use point systems for assessing social behavior that would be based on Artificial Intelligence technology. The ban also applies to the use of remote biometric identification, which would take place in real time. With regard to high-risk systems, their use and even trade in the EU would only be permitted if certain requirements are met and compliance with laws and rules is assessed by suppliers before they are placed on the market. After possible introduction, the use of the systems is to be monitored. Moreover, it is pointed out that it is necessary to register high-risk systems in a special EU database. This is intended to contribute to the transparency of the application of solutions, as well as maintaining real control and supervision over them (Piecuch, 2024).

As the report notes "ethical AI is the tangible benefit for companies," approximately nine out of ten managers are aware of situations related to ethical problems as a result of using AI tools. At the same time, approximately 77% of managers are not sure of the ethicality and transparency of the solutions implemented in their companies. Almost 50% of consumers confirmed that ethical problems resulting from the use of Artificial Intelligence tools have occurred in the last two years. Analyzing the level of assessment of the ethicality of AI depending on the place of residence of users, it was found that residents of China have the most negative experiences – 74%. In the case of Europe, this applies to France, 49%. When examining companies using AI solutions, it was found that at least 40% of employees encountered ethical problems resulting from their use during their daily activities. Approximately 44% of employees stated that the existing solutions enable them to undertake potentially unethical activities. At the same time, 42% of employees expressed their opposition to the misuse of personal data and sensitive information by AI. The authors of the report also examined consumers' attitudes towards Artificial Intelligence tools used for business purposes. It was found that 76% of them believe that AI applications should be organized and standardized, which should be handled by governments and other capable independent organizations. Analyzing the issues of carrying out preventive and corrective actions against disruptions, it was found that 64% of the surveyed enterprises introduce long-term strategies aimed at solving problems. At the same time, two out of five managers declared that their companies abandoned the implementation and further use of AI solutions as a result of ethical problems.

Based on the conducted research, the positive impact of ethical activities in the area of using AI on customer trust and satisfaction was also noticed. This was confirmed by approximately 66% of respondents. The most important unethical activities of Artificial Intelligence were also identified, which turned out to be excessive reliance on decisions guided by machines without disclosing information on this subject in the banking sector; collecting and processing personal data of patients without their consent; ambiguous or unclear recommendations regarding diagnosis, care or treatment; inability to obtain an explanation on what basis the decision was made to refuse to grant a loan or pay compensation; inability to undergo an audit when making decisions; use in recruitment, which results in discrimination of candidates, including on the basis of gender, origin, age; and discriminatory policy pricing resulting from decision automation. The most important factors contributing to the ethicality of AI include human control and supervision, as well as the ethicality of employees creating and operating these tools (Raport, 2024).

4.5 CONCLUSIONS

The issue of ethics in relation to Artificial Intelligence technologies, including machine learning, is one of the most current issues, both in terms of scientific research and practical application. This applies to every sphere of its functioning, starting from the design, creation, implementation, and use stage. Two basic directions of analysis can be noticed. One is the area of design and manufacturing, and the other is the area of final application. It should be noted that this applies to both people's private lives and the activities of enterprises.

The basic ethical concerns of AI include copyright, impact on humans, individual and social security, including world peace. While the possibilities of the real threat to world peace are currently infinitesimal, the remaining ones are undoubtedly valid. At the same time, it should be noted that the topic of global security in terms of the use of Artificial Intelligence must be addressed today, which results from the scale of impact and potential consequences; and as history shows, the creation and development of technology very often led to its use in the direction that threatens security. This applies to technologies that have a large and very large impact on humanity. AI is undoubtedly such an area of development.

The ethicality of Artificial Intelligence in the context of copyright is related to the generation of content based on very large data resources, which are the subject of analysis and inference. This corresponds to the essence of current AI solutions, which, in fact, reduce the meaning of the concept of intelligence to inference based on historical data. At the moment, AI solutions are not intelligent in the literal sense of this meaning, considered as the feature of living organisms, primarily humans. They cannot function as self-sufficient elements of life. Therefore, their operation involves specific configuration of available data in order to achieve the expected effect and, importantly, the effect indicated each time by the user. Therefore, AI-generated content is not something creative. According to the authors of this chapter, it is difficult to refer to this aspect of the use of AI from the point of view of the ethicality of the effects of their operation. The current statement here is that in this dimension, Artificial Intelligence is only as ethical as its creators, but primarily its users.

Ethicality in the process of designing and developing AI tools is also an important aspect. It is people who decide on the size, scope, and completeness of the research material on the basis of which Artificial Intelligence learns. Failure to include and/or limit the set of data for training results in the quality of results that may be treated as unethical – racist content that discriminates against groups of people. This is a problem that the creator has an influence on solving and that can be considered in the context of ethics, but still from a human perspective, not from the technology itself.

Another area of discussion on the ethicality of AI is the issue of pushing humans/employees out of structures of economic organizations. It is indicated that Artificial Intelligence, by replacing human resources, has a negative impact on employees' self-esteem, confidence in decisions and activity, and finally, the assessment of their own impact on the functioning of the company. This is a negative phenomenon for enterprises, which may negatively affect their flexibility in the dynamically chang ing economic reality, where decisions are rarely zeros and ones. It is important that the replacement of human work by AI technology contributes to the change in the labor market. Some professions are becoming unnecessary; it is said that medium competences are losing their value and the demand for high and low competences is increasing. According to the authors of this chapter, this may contribute to an increase in human dependence on technology, which in itself may be used by groups of people for unethical purposes. It should also be stated here that unethicality will be the result of human activity, and not the AI itself.

The issues discussed in this work are part of the current discussions. We realize that research on the issue of AI ethics must continue. For now, however, they should focus on human ethics in the reality of the increasing level of use and increasingly common creation of solutions based on Artificial Intelligence. This will have a positive impact on the future development of technology, where its application possibilities will undoubtedly expand significantly and, consequently, its impact on society will increase. The authors of this chapter consider differences in the approach to ethics depending on the geographical place of human existence as one of the most important research and practical problems in this respect. What is considered unethical in one community is not always recognized as such in another.

REFERENCES

Algorytmy – dyskryminacja i wykluczenie oraz ich społeczne konsekwencje, publication available at: https://swps.pl/centrum-prasowe/informacje-prasowe/24585-algorytmy-dyskryminacja-i-wykluczenie-oraz-ich-spoleczne-konsekwencje (access: 20.02.2024).

Bampton R., Cowton Ch. J. (2002). The teaching of ethics in management accounting: Progress and prospects, *Business Ethics: A European Review*, 11(1), 1–103.

Bécue A., Praça I., Gama J. (2021). Artifcial intelligence, cyber-threats and Industry 4.0: Challenges and opportunities, *Artificial Intelligence Review*, 54, 3849–3885.

Bellman R. (1978). *An Introduction to Artificial Intelligence*, Boyd & Fraser Publishing Company.

Bendiab G., Hameurlaine A., Germanos G., Kolokotronis N., Shiaeles S. (2023). Autonomous vehicles security: Challenges and solutions using blockchain and artificial intelligence, *IEEE Transactions on Intelligent Transportation Systems*, 24(4), 3614–3637.

Bhambri P., Gupta, S. (2005). A survey & comparison of permutation passibility of fault-tolerant multistage interconnection networks. In National Conference (Abstract Book) (p. 13). Presented at the "Application of Mathematics in Engineering & Technology" conference at MIMIT, Malout held during March 26–28, 2005.

Bhambri P., Gupta S., Bhandari A. (2005). Soft computing techniques. In UGC Sponsored National Conference (Abstract Book) (p. 100). Presented at the "Emerging Computing Technologies" conference at HRMMV, Jalandhar held during September 29–30, 2005.

Bhambri P., Singh I. (2005). Electrical actuation systems. In National Conference (Abstract Book) (p. 8). Presented at the "Application of Mathematics in Engineering & Technology" conference at MIMIT, Malout held during March 26–28, 2005.

Bo L., Peng Q., Bo L., Shuai D., Jingen L., Jiquan P., Jinfeng Y., Bowen Z. (2023). Trustworthy AI: From principles to practices, *ACM Computing Surveys*, 55(9), 2–27.

Buck B., Morrow J. (2018). AI, performance management and engagement: Keeping your best their best, *Strategic HR Review*, 17(5), 261–262.

Cave S., Dihal K. (2019). Hopes and fears for intelligent machines in fiction and reality, *Nature Machine Intelligence*, 1, 74–78.

Chanda S. S., Banerjee D. N. (2022). Omission and commission errors underlying AI failures, *AI & Society*, 1–10.

Chojnowski M. (2022). Etyka sztucznej inteligencji, wprowadzenie, Centrum Etyki Technologii, Instytut Humanites, Warszawa, 8–15.

Derave Ch., Genicot N., Hetmanska N. (2022). The risks of trustworthy Artificial Intelligence: The case of the European travel information and authorisation system, *European Journal of Risk Regulation*, 13(3), 389–418.

Etyka, Internetowa Encyklopedia PWN, available at: https://encyklopedia.pwn.pl/haslo/etyka;3898956.html (access: 11.2.2024).

Fjeld J., Achten N., Hilligoss H., Nagy A. Ch., Srikumar M. (2020). Principled Artificial Intelligence: Mapping consensus in ethical and rights-based approaches to principles for AI, Berkman Klein Center Research Publication, 1, 2–49.

Garri Kasparow: Ludzie są większym zagrożeniem niż sztuczna inteligencja, available at: https://www.rp.pl/nowe-technologie/art19079911-garri-kasparow-ludzie-sa-wiekszym-zagrozeniem-niz-sztuczna-inteligencja (access: 10.02.2024).

Hamilton R.H., Davison H.K. (2022). Legal and ethical challenges for HR in machine learning, *Employee Responsibilities and Rights Journal*, 34, 19–39.

Kaplan A., Haenlein M. (2020). Rulers of the world, unite! The challenges and opportunities of artificial intelligence, *Business Horizons*, 63(1), 37–50.

Kisiel A. (2014). Etyka a zarządzanie – wybrane problemy, *Studia Ekonomiczne Uniwersytetu Ekonomicznego w Katowicach*, 180(1), 86–91.

Kumięga P. (2016). Etyczne wątpliwości w zarządzaniu organizacjami zautomatyzowanymi, (in.:) *Etyka w organizacji: zarządzanie.* kultura, polityka, ed. by Ł. Burkiewicz, J. Kucharski, Wydawnictwo WAM, Kraków, 387.

Miernicki M., Ng I. (2021). Artificial intelligence and moral rights, *AI & Society*, 36, 319–329.

Mittelstadt, B. (2019). Principles alone cannot guarantee ethical AI, *Nature Machine Intelligence*, 1, 501–507.

Nath R., Sahu V. (2020). The problem of machine ethics in artificial intelligence, *AI & Society*, 35, 103–111.

Nowakowski M. (2022), Etyka w świecie sztucznej inteligencji. Status quo iprzyszłość, cz. 1, available at: https://pl.linkedin.com/pulse/etyka-w-%C5%9Bwiecie-sztucznej-inteligencji-status-quo-i-1-michal (access: 20.02.2024).

Piecuch I. (2024). Raport iTWIZ, W jaki sposób liderzy cyfryzacji w Polsce wykorzystują algorytmy AI, 4xAI: prawo, etyka, biznes i technologia, 42.

Raisch S., Krakowski S. (2021). Artificial Intelligence and management: The automation–augmentation paradox, *Academy of Management Review*, 46(1), 1–11.

Rana R., Bhambri P. (2024). Artificial Intelligence and Internet of Things (IoT) facilitated digital twin for Industry 4.0 application domains. In S. Rani, P. Bhambri, S. Kumar, P. K. Pareek, & A. A. Elngar (Eds.), *AI-Driven Digital Twin and Industry 4.0: A Conceptual Framework with Applications* (1st ed., pp. 168–181). CRC Press. https://doi.org/10.1201/9781003395416

Rattan M., Bhambri P., Shaifali. (2005). Information retrieval using soft computing techniques. In National Conference (pp. 58–60) on Bio-informatics Computing. Presented at TIET, Patiala held during March 18–19, 2005.

Russell S., Norvig P. (2003). *Artificial Intelligence: A Modern Approach*, 2nd edition, Prentice Hall.

Singh M.P., Bhambri P. (2024). AI-Driven digital twin conceptual framework and applications. In S. Rani, P. Bhambri, S. Kumar, P. K. Pareek, & A. A. Elngar (Eds.), *AI-Driven Digital Twin and Industry 4.0: A Conceptual Framework with Applications* (1st ed., pp. 127–141). CRC Press. https://doi.org/10.1201/9781003395416

Tunçer A. S. (2015). Importance of human resources training in an information age, *European Journal of Social Science Education and Research*, 2(1), 1–4.

Uczenie maszynowe co to jest?, https://www.sap.com/poland/products/artificial-intelligence/what-is-machine-learning.html (access: 10.01.2024).

Walczak-Duraj D. (2017). Współczesne tendencje zmian kultury zarządzania organizacją, ed. by: D. Walczak-Duraj, R. Knap, *Humanizacja Pracy*, 3(289), 10.

Wittenberg A. (2022). Sztuka kontra AI, czyli o nieetycznych skutkach rozwoju sztucznej inteligencji, available at: https://forsal.pl/lifestyle/technologie/artykuly/8606440,sztuka-kontra-ai-nieetyczne-skutkih-rozwoju-sztucznej-inteligencji.html (access: 22.02.2024).

Wytyczne w zakresie etyki dotyczące godnej zaufania sztucznej inteligencji, available at: https://www.europarl.europa.eu/meetdocs/2014_2019/plmrep/COMMITTEES/JURI/DV/2019/11-06/Ethics-guidelines-AI_PL.pdf (access: 21.02.2024).

Zalewski T. (2020). Definicja sztucznej inteligencji, (in.:) Prawo sztucznej inteligencji, ed. by L. Lai, M. Świerczyński, Monografie Prawnicze, Warszawa, 2.

5 Building Smart Cities through IoT Entrepreneurship

Tomasz Turek

5.1 INTRODUCTION

Today's cities are becoming increasingly intelligent which means that they are progressively saturated with Information and Communication Technologies. These solutions record an increasing number of parameters that can be identified in the complex and multifaceted organism that is the city. It is clear that the collected data has no value in itself, for their value emerges only during the action of processing and extracting the resulting information from them, which raises the level of awareness on a specific topic or in a specific area. Proper interpretation of the obtained results allows for more efficient management of the city at many levels. It applies both to strategic decisions (which have a long-time horizon) and to day-to-day decisions (which have a short-time horizon). The data obtained can be used by people in management processes, but equally important is the continuous exchange of information resources with other devices and elements of the urban infrastructure which contributes to the self-control of the city.

5.2 THE ESSENCE OF THE SMART CITY

The concept of the "smart city" emerged in the early 1990s. The book *The Technopolis Phenomenon: Smart Cities, Fast Systems, Global Networks* (Gibson et al., 1992) points to the development of new types of cities called technopolises that use advanced Information and Communication Technologies to attract investment, create new jobs, and foster innovation. Another milestone in the development of the Smart City concept was set by Robert E. Hall in his publication *The Vision of Smart City* (Hall et al., 2000). His idea was that the cities of the future would be safe, environmentally friendly and efficient because all elements of the built environment would be designed and maintained using advanced and integrated electronic sensors linked through networks to computer systems. These systems were said to have collected the data, and with the help of appropriate algorithms, supported decision-making processes. Since then, the idea has begun to become a reality as more and more cities around the world are incorporating the concept of "smart" into their development strategy. Examples of cities that have become "smart" and have been at

 DOI: 10.1201/9781003484226-5

TABLE 5.1
Examples of Cities that have become "smart" and have been at the Top of the Rankings

City	Rank 2023	Rank 2021	Rank 2020	Rank 2019
Zurich	1	1	1	1
Oslo	2	2	2	2
Canberra	3	–	-	-
Copenhagen	4	5	3	4
Lausanne	5	4	-	-
London	6	3	10	3
Singapore	7	7	7	10

the top of the rankings for years are Zurich, Oslo, Canberra, Copenhagen, Lausanne, London, and Singapore (imd.cld.bz 2023) (Table 5.1).

Later in this chapter will be shown the rules for calculating the above rankings, but first it is worth considering what a smart city is today. For the purposes of this chapter, it is acceptable to employ the definition that it is a place (city) which uses Information and Communication Technologies to improve public services and make better use of resources, while having the least possible (negative) impact on the environment. Thus, it should be noted that the concept of "smart city" should be considered in at least four dimensions: technological, public services, resources, and environmental (Bhambri and Khang, 2024).

A similar approach is proposed in the publication "Everything You Wanted to Know about Smart Cities" (Monathy, 2016) whose authors define "smart city" as components and characteristics (Figure 5.1).

Smart cities are considered from the perspective of four layers, which are components, attributes, themes, and infrastructure. Cities can have different constituents (infrastructure, buildings, transport, etc.) that are at different levels of "smartness." Another layer is attributes, which in turn relate to sustainability, quality of life, urbanisation, and smartness (Rana and Bhambri, 2024). In cities defined as "smart," there are key themes that influence development and integration. The fourth layer is infrastructure, which includes physical elements and Information and Communication Technology (ICT), which is the nervous system of the city. Services are delivered through both physical infrastructure and ICT.

The drive to be "smart" stems from the need to compete with other cities. A kind of benchmarking of cities, copying and adapting models that have worked elsewhere, makes living in a smart city more attractive and enjoyable. Some of the key benefits associated with the concept include:

- **Governance:** In smart cities, decision-making at strategic, tactical, and operational levels is more efficient due to the ability to gather more data and information. Information resources are made available and analyzed in

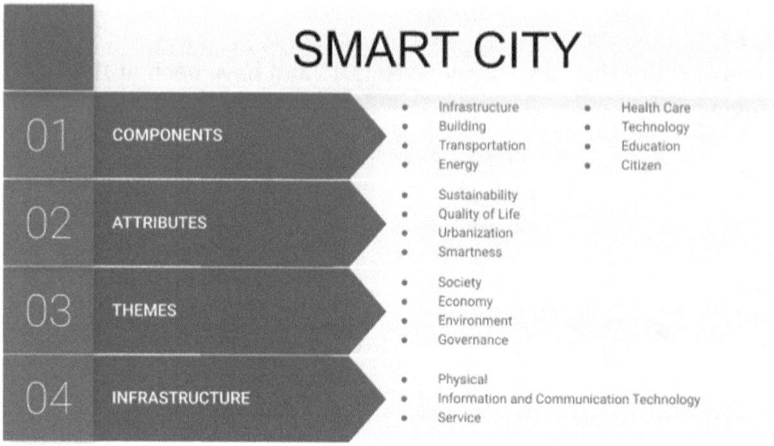

FIGURE 5.1 Smart city components and characteristics.

real time. The quality of data and information, as well as its reliability and accuracy, also promotes optimal management.

- **Attractiveness:** Smart cities attract people through their features. Young people in particular appreciate ubiquitous and useful technology. Smart cities also attract investors, which makes job opportunities more attractive, and the modern economy also allows people to work remotely. Attractive and safe cities make it possible to be online, move quickly, and be mobile.
- **Safety:** Smart cities are safe due to the widespread use of cameras, sensors, and access systems (Singh and Bhambri, 2024). This is not a desire for surveillance, but a concern for a shared sense of security. By analyzing data from traffic routes, traffic and communications can be controlled. This makes them smooth, friendly, safe, and it saves time.
- **Environment:** Smart cities take care of their surroundings as electronic systems collect and analyze environmental data and try to respond to worrying signals in real time. These signals concern air quality, water quality, forests, parks, etc. Environmental issues are also linked to the rational use of resources and the sustainability development.
- **Adaptation:** The interweaving of the different layers of the smart city (Figure 5.1) means that cities are constantly adapting to the needs of their inhabitants. Residents become prosumers of sorts, shaping their environment, services, and components through their behavior, signals, and responses.

In this context, reference can once again be made to the IMD Smart City Index (2023) which lists and offers statistics of cities that are the most "smart." Zurich has been at the top of the list for years. Figure 5.2 shows an excerpt from the report containing the answers of respondents – Zurich residents – in selected areas.

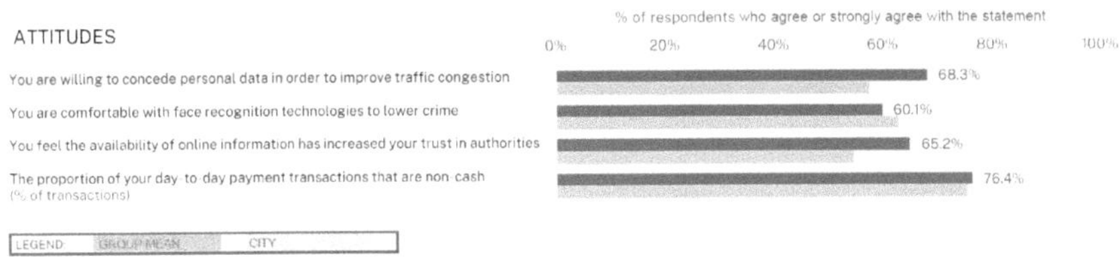

FIGURE 5.2 Smart city attitudes of the city of Zurich (source: IMD Smart City Index 2023).

Without technological solutions, the idea of a smart city would remain just a concept, impossible to implement (Sharma et al., 2020). At the core of such a system is the Internet of Things (IoT). It is thanks to the things scattered around the city: sensors, cameras, devices, readers, and detectors that events and many parameters can be recorded. This data then needs to be sent to the relevant data centers for processing and the result of this action is an added value that helps to turn the idea into a living organism of the smart city.

5.3 DEFINING IOT FROM A CITY MANAGEMENT PERSPECTIVE

As in many cases, it is impossible to give a single, coherent definition of the Internet of Things. This is because the technology has been developing for many years. In fact, devices connected to the network have been appearing almost since the beginning of the internet and were not typical computers or peripherals. Instead, they were equipped with network interfaces, sensors and controllers. It is believed that the first device to be considered IoT, which was the idea of John Romkey (Romkey, 2017), was a toaster that was switched on by a command from a computer. Since then, the development of the IoT has accelerated rapidly. The number of connected devices is estimated to be around 8 billion in 2010 and 20 billion in 2020. Predictions, including from Gartner Inc., suggest that there could be as many as 70 billion devices by 2025. The term IoT itself was introduced by Kevin Ashton in 1999 (Ashton, 2009). At the time, the term was associated with RFID technology, the supply chain and the internet.

Today, Internet of Things (IoT) can be defined as a **concept that takes into account the ubiquitous presence of various things and objects in the environment that are able to interact with each other and with other things and objects through wireless and wired connections to create new services, values, and purpose**.

There are a number of factors that have contributed to the proliferation of IoT that are mainly technological innovations that enable the production of low-power devices, miniaturization and network standards that support IoT (Baccelli, 2021). In general, the IoT concept can be presented as a model (Figure 5.3).

The above model means that a number of sensors examine certain physical characteristics of the environment (e.g. temperature, light intensity, noise, humidity) and convert them into digital form (Singh et al., 2005). Through networking and communication, the digital data can be sent to other systems and devices for processing,

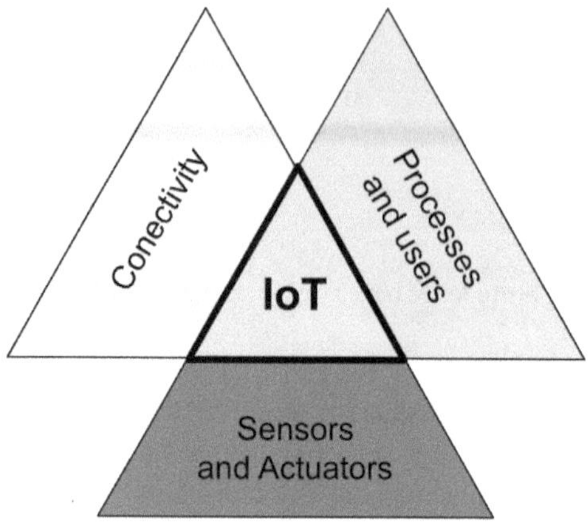

FIGURE 5.3 The IoT concept (source courtesy: Salazar, Silvestre, 2017).

whose results can add value to processes and users, and analysis at various levels. Other systems, users and processes can also send data to a set of actuators, which in turn influence the physical characteristics of their environment (Patel et al., 2016).

Modern IoT is not just about reading data from sensors, processing, analyzing, and controlling actuators. In fact, technological developments are making it possible to harness the power of cloud computing and machine learning. This expands the possibilities of the IoT. Cloud computing and data analytics include improved machine learning applications which operate at a new level of Artificial Intelligence. IoT also incorporates the notion of sensing and data analytics that drive remote control (OECD, 2016).

Understood in this way, the definition and model of the IoT enables a range of benefits to be exploited in many areas of human's lives, such as for instance health, manufacturing, education, smart home, and their relationship with the environment. However, in the perspective of this chapter of the work, the question "How does IoT affect the implementation of the smart city concept?" must be asked. The answer is simple and obvious because the smart city concept is virtually impossible without the Internet of Things. These are devices with sensors and actuators, connected to a network and communicating with each other, that makes it possible to collect quantitative data on the state of various physical and mathematical values, as well as data of a qualitative nature. These data, processed according to specific rules, in turn make it possible to generate a certain added value which allows us to add the adjective "smart" (Dawood, 2020).

The sensor systems in the city make it possible to analyze the state of air pollution. Other IoT devices, such as cameras, influence the safety of residents and traffic optimisation. Another example are devices that automatically switch on street and pavement lighting. The examples are endless. IoT is, therefore, one of the critical infrastructures without which the smart city concept cannot be fully realized (Sunyaev, 2020). In the following sections of this chapter, there is reference to several areas of urban functioning that are supported by modern internet-based technologies. These are monitoring the infrastructure of utilities provided to residents, energy management, public safety, sustainability, and smart services. These areas have been chosen because of their relevance to the urban experience.

5.4 AUTOMATIC MONITORING OF THE INFRASTRUCTURE OF UTILITIES SUPPLIED TO RESIDENTS

One of the essential feature of urban life is access to utilities: electricity, gas, water, sewerage, and the internet. The problems faced by residents of villages and smaller towns are virtually non-existent. Modern cities have a well-developed infrastructure, so the inhabitants of residential areas are not concerned about access to utilities (Bhambri et al., 2005). Those who live or plan to live in their own homes also have few problems. Widespread and continuous access to utilities is so natural and obvious that sometimes all it takes is a minor water main break or a damaged power line for residents to feel a great deal of anxiety and discomfort. That is why modern cities care about the quality of their infrastructure and work closely with their utility providers.

The most common use of IoT for infrastructure management and utilities are:

- Infrastructure condition monitoring.
- Utility infrastructure management.
- Consumption billing.

Utility infrastructure condition **monitoring** refers to the continuous examination and analysis of the condition of the transmission networks. In the case of electricity networks, parameters such as current, voltage and power factors are analyzed on an ongoing basis. Failures and unexpected or unplanned higher than normal consumption are also detected.

Energy network **management** allows future energy consumption to be planned and influences cost containment. Areas of undesirable energy consumption are also diagnosed. Management also enables strategic planning for the development of network infrastructure for new settlements, industrial plants, or commercial premises.

Metering automatically reads the amount of energy used by residents, businesses, or public institutions. In many cities, the days of consumers going from house to house to read meters are over. An example of the architecture of a power system monitoring system is shown in Figure 5.4.

The monitoring model presented takes into account the following elements: smart energy meters, RS 485 network, PLC, LAN/WAN network, and Supervisory Control and Data Acquisition (SCADA). The smart meter takes advantage of IoT capabilities. In addition to standard functions such as measuring voltage, current, frequency, etc., it has the ability to communicate via network protocols (e.g. RS 485). This allows real-time parameter readings and data transfer to other devices in the system. The data first goes to the PLC and then it travels via wide area and/or local area networks to the SCADA

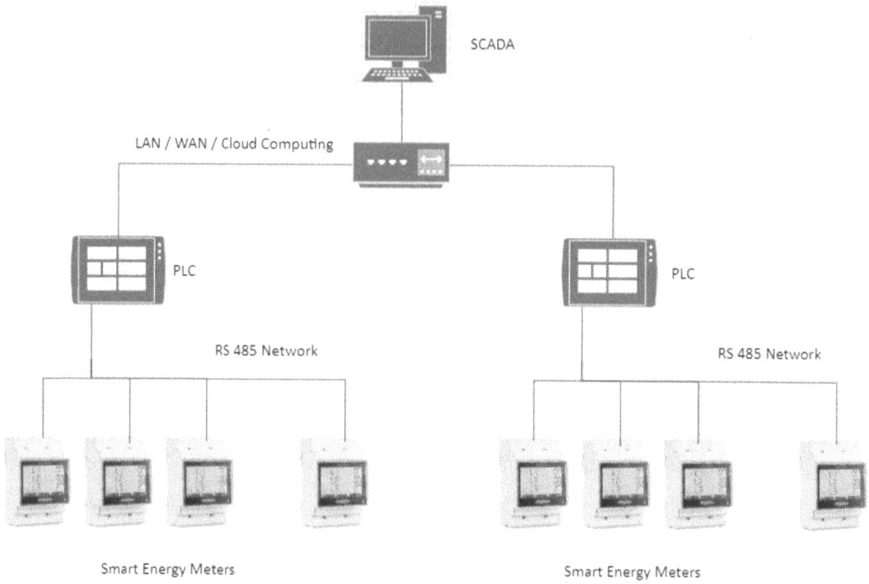

FIGURE 5.4 Energy monitoring system architecture (source courtesy: instrumentationtools .com).

controller. SCADA systems are used to monitor and control physical processes in industry and infrastructure on a large scale and over long distances. In this energy monitoring system, SCADA is used to visualise current energy meter readings and to collect historical data for future use. Similar solutions to the energy monitoring system can be found in any other utility. One example is the concept of using IoT to monitor and ensure the quality of water supplied to homes (Velayudhan et al., 2022) (Figure 5.5).

Another issue is the interaction between the utilities. They provide their services within the city and therefore share the same space. The routes of individual lines may have different relationships to each other. Sometimes there may be a collision along the route. In other cases, a certain "symbiosis" is allowed (Bhambri and Gupta, 2005). For example, power lines may provide a natural environment for fiber-optic internet or television. In still other situations, the routing of lines may not overlap with other utilities (e.g. gas and electricity).

Cities can use sophisticated mapping and geographic information systems (GIS) to support utility management. In this case, IoT devices that are part of the infrastructure of utility providers are other elements that improve the functioning of such solutions.

5.5 IOT IN PUBLIC SAFETY

One of the key factors influencing the well-being of residents is a sense of safety. It is not without reason that security has been identified as one of the basic human needs. This is illustrated by Maslow's famous hierarchy of needs.

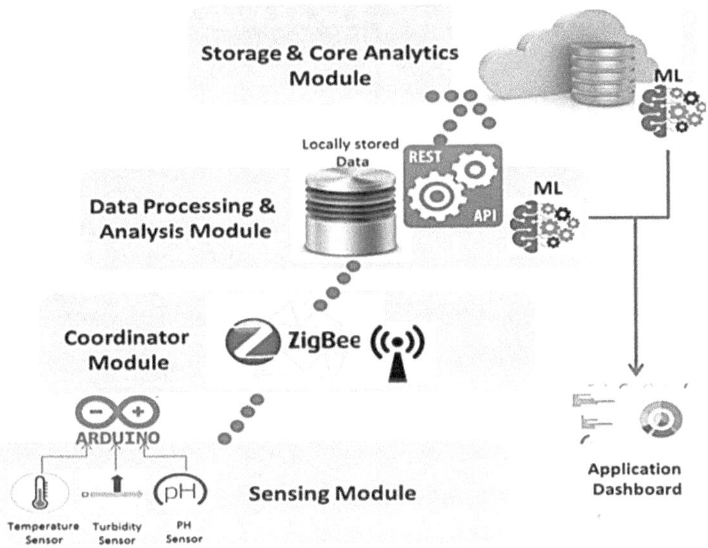

FIGURE 5.5 IoT–water quality management system (source courtesy: Jan et al., 2021).

Many people are able to live in a less attractive place (in terms of opportunities to get a good job, living conditions, etc.) as long as they feel safe. Parents do not want to worry about their children, what they will do and who they will meet at school, how they will spend the day after school or where they will play. Many property advertisements mention a safe and quiet location, which are seen as important attributes.

For this reason, city managers are making efforts in the area of safety. This aspect is formally included in many cities' strategies and a key technology in this area is IoT.

The most obvious application for IoT in public safety is video surveillance. Virtually every modern city has one. Where appropriate, public systems can be supported by CCTV installed in shops, banks, businesses, and private homes. It is important to recognise that the pace of technological progress in video surveillance is very fast. Today's systems use Artificial Intelligence to enable automatic facial recognition, identification of abnormal situations, and analysis and prediction of threats. Many activities that used to be performed by humans can now be carried out without human intervention. Traditional solutions often require tedious work over many hours. For example, surveillance systems can now identify object types: people, objects, motion, etc. (Figure 5.6).

However, monitoring systems are not the only solutions and areas affecting urban safety. It should be noted that IoT can also be used in:

- **Traffic management:** Many cities face problems of congestion and heavy traffic. By equipping major road junctions and intersections with equipment that automatically analyses traffic, it is possible to optimise routes,

FIGURE 5.6 Example of object detection (https://tryolabs.com/guides/video-analytics-guide).

synchronise traffic lights and dynamically manage lanes (Figure 5.7). Traffic optimisation contributes to shorter travel times, reduced emissions and less urban pollution.

- **Accident prevention:** Traffic optimisation is one of the factors contributing to accident reduction. However, vehicle-to-vehicle (V2V) and vehicle-to-infrastructure (V2I) solutions play a very important role in this area. V2V refers to all IoT systems and solutions that enable relationships to be established between vehicles. Modern cars have active cruise control to prevent accidents. Vehicles also use communication solutions (GPS, WiFi, LTE) to send data to (and from) other vehicles about road conditions, the type of cargo being carried, etc., V2I, on the other hand, are IoT solutions that allow vehicles to communicate with elements of the road infrastructure. For example, it is possible to send a message about an accident, an oil spill, etc.
- **Detecting accidents and mishaps:** Many vehicles, machines and equipment are now equipped with various types of sensors and devices, e.g. GPS, accelerometers, noise meters. Similar sensors can be found in devices worn by people: smartphones, smartwatches, clothing. Data from the sensors is transmitted, recorded and analyzed in real time. Any unusual behavior (e.g. sudden acceleration, sudden loss of speed, lack of communication) can be identified as an accident.
- **Accident reporting and assistance:** Identifying an accident by analyzing IoT-derived data can result in an automatic call for help. IoT systems can also help emergency services reach the scene of an accident efficiently and accurately. Emergency systems can be used to set up signals to inform other

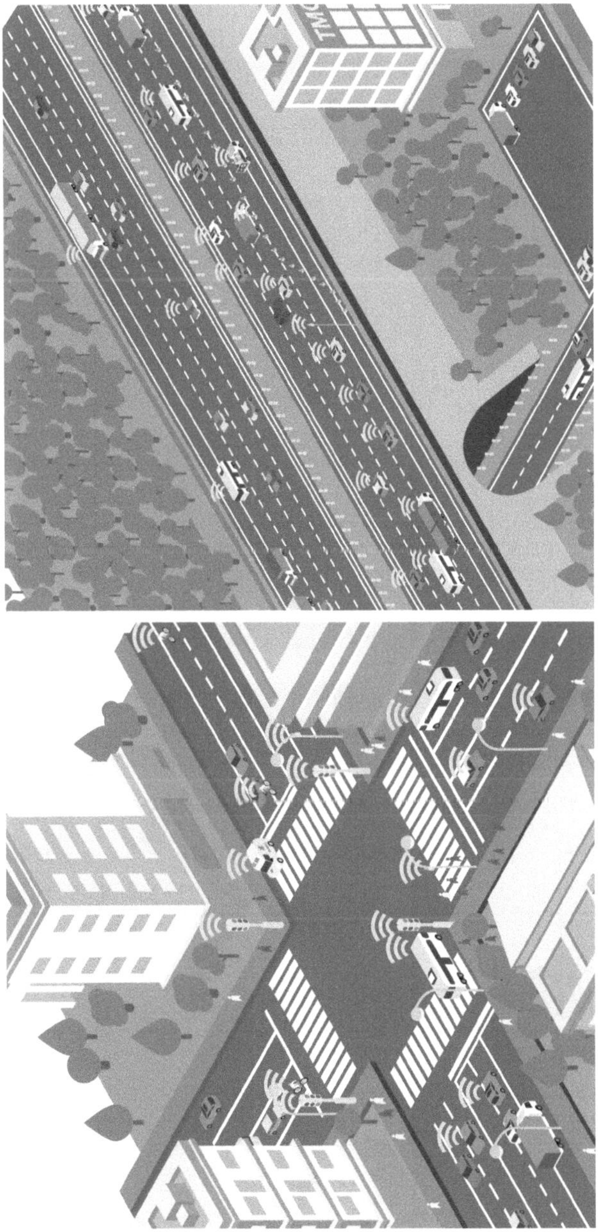

FIGURE 5.7 IoT in traffic management (https://www.kapsch.net/).

road users. GPS shows the location of the accident with a high degree of accuracy. Where possible, sensors monitor the victim's vital signs.

- **Crisis management.** Cities, by their very nature, have large populations and are therefore vulnerable to emergency situations. Examples of such situations include a riot in the city, a major traffic accident, a fire or a flood. The Covid-19 pandemic was also a crisis situation. IoT devices in the city not only make it possible to monitor crisis situations, but also support society with information. They can indicate patterns of behavior, provide information on evacuation routes, etc.
- **Forecasting natural disasters:** Many cities are located in areas prone to natural disasters. These can include the threat of floods, earthquakes, volcanic eruptions, tsunamis, storms, landslides or mining damage. Scientists have been able to predict the occurrence of such phenomena for many years. Depending on the threat, cities are setting up early warning systems (Samadi, 2022). At the same time, they are proposing models for the behavior of residents. One example is the flood warning system shown in Figure 5.8.

Ubiquitous technology, monitoring and IoT devices that continuously analyze millions of events and parameters make it possible to efficiently manage the security of the city. Of course, there will always be skeptics who see this as a loss of privacy.

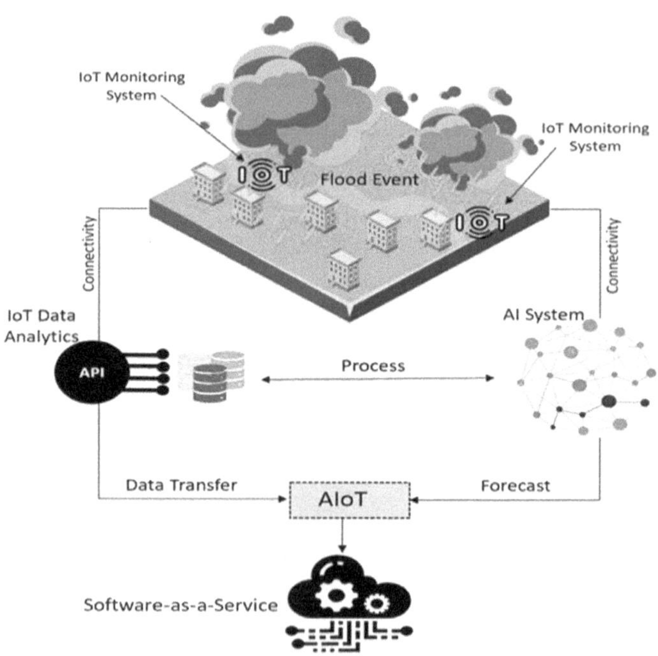

FIGURE 5.8 Components of an IoT flood warning system (source: Samadi, 2022).

This slight loss of privacy is a cost that must be born in order to achieve a higher level of security. Weighing the two, it must be concluded that people gain more than lose by living in safer and smarter cities.

5.6 SUSTAINABLE CITIES USING IOT

The idea of sustainable cities derives directly from the concept of sustainable development (Ojo-Fafore, 2019). The sustainable city is a concept that smart cities aim to achieve. The main characteristics of sustainable cities are the presence of green spaces, zero-emission transport, the use of renewable energy sources, the preservation of walkability, and the rational use of energy.

IoT makes it possible to support the concept of a sustainable city (Liu, 2018). As indicated in the previous sections of this chapter, cities are saturated with all kinds of devices and sensors that monitor and analyze certain physical values in the city in real time. By drawing conclusions and adding specific behavioral models, it is possible to contribute to the quest for sustainability (Sharma and Jangirala, 2022). Selected areas for supporting the concept of a sustainable city using IoT are presented in Figure 5.9.

Figure 5.9 shows four areas where IoT can support the concept of a smart, sustainable city. The first area is intelligent energy consumption. IoT solutions can automate the process of switching city lights on and off, adjusting them according to need and time of day. Then there is the smart home concept. IoT-managed devices in the home contribute directly to optimizing energy consumption in the home and indirectly to energy consumption in the city (Pandiyan et al., 2023). Cities are increasingly using renewable energy sources due to IoT solutions that switch from traditional to renewable energy sources.

Another area is intelligent transport and IoT can contribute to the rational use of it. This is because devices (e.g. GPS, Google Map, etc.) can determine optimal travel routes, taking into account the avoidance of green zones. The same devices

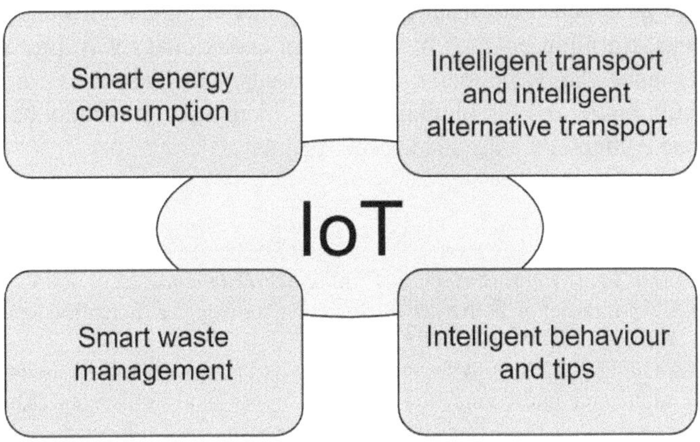

FIGURE 5.9 IoT in smart sustainable city.

can suggest cheaper and less carbon-intensive public transport journeys. In addition, most alternative transport solutions are based on IoT. For example, urban bike rental systems.

The third area identified is intelligent waste management. IoT-based solutions make it possible to display the correct bins and segregate waste. Most cities have also implemented apps that notify residents of collection deadlines for certain types of waste. These apps also have an educational effect. They make residents aware and encourage them to behave appropriately. The IoT also has a control function. Sensors installed at landfills and collection points verify that the right waste has been placed in the right container.

5.7 SELECTED TRENDS FOR IOT IN THE SMART CITY

Predicting the future is always difficult. However, it is important to realise that the prevailing trend around AI will also influence the area of IoT use in the smart city. The services and solutions described in this chapter will therefore become even more saturated with AI. In particular, it can be expected to see autonomous cars and autonomous transport. Today's vehicles are so saturated with IoT and AI solutions that they are, in principle, already capable of autonomous driving. The main obstacle seems to be citizens' beliefs about the inadequacy of the technology. Another predicted trend also relates to transport. With parking problems in big cities and the rising cost of vehicle maintenance, shared mobility is gaining popularity. IoT has a big role to play here. Smartphone apps make it easier to find a vehicle (car, bike), pay for the service, unlock the device and operate it.

A major trend in smart and sustainable cities will undoubtedly be zero-emission buildings. This is due to current trends and rising maintenance costs. Sensors and actuators will collect physical data from the building and transmit it to a control center. Building management systems will adjust emission parameters and, where possible, use renewable energy sources.

The development of AI and related machine learning will undoubtedly change the way cities are governed. Yet this does not mean that city presidents and councillors will disappear overnight but step by step, in an evolutionary way, preparations for digital governance should be made. The smart city will therefore be a network of different technologies and IT solutions that will intertwine. It will not be possible to say where one technology ends and another begins.

REFERENCES

Ashton, K. (2009). That "Internet of Things" thing. *RFiD Journal*, 22, 97–114.

Baccelli, E. (2021). Internet of Things (IoT). Societal Challenges & Scientific Research Fields for IoT, White Book, INRIA, nr 05.

Bhambri, P., & Gupta, S. (2005). A survey & comparison of permutation passibility of fault-tolerant multistage interconnection networks. In National Conference (Abstract Book) (p. 13). Presented at the "Application of Mathematics in Engineering & Technology" conference at MIMIT, Malout held during March 26–28, 2005.

Bhambri, P., Gupta, S., & Bhandari, A. (2005). Soft computing techniques. In UGC Sponsored National Conference (Abstract Book) (p. 100). Presented at the "Emerging Computing Technologies" conference at HRMMV, Jalandhar held during September 29–30, 2005.

Bhambri, P., & Khang, A. (2024). The human-machine nexus with art-making generative AIs. In S. Hai-Jew (Ed.), *Making Art With Generative AI Tools* (pp. 73–85). IGI Global. https://doi.org/10.4018/979-8-3693-1950-5.ch.

Dawood, A. (2020). Internet of Things (IoT) and its applications: A survey. *International Journal of Computer Applications,* 175, 975–8887. https://doi.org/10.5120/ijca2020919916.

Gibson, D.V., Kozmetsky, G., & Smilor, R.W. (1992). *The Technopolis Phenomenon: Smart Cities, Fast Systems, Global Networks.* Lanham, MD: Rowman & Littlefield Publishers.

Hall, R., Bowerman, B., Braverman J., Taylor, J., Todosow, H., & Wimmersperg, U. (2000). The vision of a smart city. Presented at the 2nd International Life Extension Technology Workshop.

IMD Smart City Index 2023. https://imd.cld.bz/Smart-City-Index-20233/6/.

Jan, F., Min-Allah, N., & Düştegör, D. (2021). IoT based smart water quality monitoring: Recent techniques, trends and challenges for domestic applications. *Water,* 13(13), 1729. https://doi.org/10.3390/w13131729.

Liu, L. (2018). IoT and a sustainable city. *Energy Procedia,* 153, 342–346. https://doi.org/10.1016/j.egypro.2018.10.080.

Mohanty, S. (2016). Everything you wanted to know about smart cities. *IEEE Consumer Electronics Magazine,* 5, 60–70. https://doi.org/10.1109/MCE.2016.2556879.

OECD. (2016). The Internet of Things. Seizing the benefits and addressing the challenges, OECD Digital Economy Papers, No. 252.

Pandiyan, P., Saravanan, S., Usha, K., Kannadasan, R., Alsharif, M.H., & Mun-Kyeom, K. (2023). Technological advancements toward smart energy management in smart cities. *Energy Reports*, 10, 648–677. https://doi.org/10.1016/j.egyr.2023.07.021.

Patel, K., Patel S., Scholar, P., & Salazar, C. (2016). Internet of Things-IOT: Definition, characteristics, architecture, enabling technologies, application & future challenges. *International Journal of Engineering Science and Computing*, 6(5). https://doi.org/10.4010/2016.1482.

Rana, R., & Bhambri, P. (2024). Artificial Intelligence and Internet of Things (IoT) facilitated digital twin for Industry 4.0 application domains. In S. Rani, P. Bhambri, S. Kumar, P.K. Pareek, & A.A. Elngar (Eds.), *AI-Driven Digital Twin and Industry 4.0: A Conceptual Framework with Applications* (1st ed., pp. 168–181). CRC Press. https://doi.org/10.1201/9781003395416.

Romkey, J. (2017). Toast of the IoT: The 1990 Interop Internet Toaster. *IEEE Consumer Electronics Magazine*, 6(1), 116–119. https://doi.org/10.1109/MCE.2016.2614740.

Salazar, J., & Silvestre, S. (2017). Internet of things, Czech Technical University of Prague, Faculty of Electrical Engineering.

Samadi, S. (2022). The convergence of AI, IoT, and big data for advancing flood analytics research. *Frontiers in Water,* 4. https://doi.org/10.3389/frwa.2022.786040.

Sharma, P., & Jangirala, S. (2022). Internet of Things for sustainable urbanism. *Journal of Physics: Conference Series,* 2236, 012008. https://doi.org/10.1088/1742-6596/2236/1/012008.

Sharma, R., Bhambri, P., & Sohal, A.K. (2020). Energy bio-inspired for MANET. *International Journal of Recent Technology and Engineering*, 8(6), 5580–5585. https://www.ijrte.org/wp-content/uploads/papers/v8i6/F8522038620.pdf. https://doi.org/10.35940/IJRTE.F8522.038620.

Singh, M., Singh, P., Kaur, K., & Bhambri, P. (2005). Database security. In Proceedings of National Conference (pp. 35–41). Presented at the "Future Trends in Information Technology" conference at SJPMLIET, Radaur, Yamuna Nagar held during March 6–7, 2005.

Singh, M.P., & Bhambri, P. (2024). AI-Driven digital twin conceptual framework and applications. In S. Rani, P. Bhambri, S. Kumar, P.K. Pareek, & A.A. Elngar (Eds.), *AI-Driven Digital Twin and Industry 4.0: A Conceptual Framework with Applications* (1st ed., pp. 127–141). CRC Press. https://doi.org/10.1201/9781003395416.

Sunyaev, A. (2020). The Internet of Things. In *Internet Computing*. Springer, Cham. https://doi.org/10.1007/978-3-030-34957-8_10.

Velayudhan, N.K., Pradeep, P., Rao, S.N., Devidas, A.R., & Ramesh, M.V. (2022). IoT-Enabled water distribution systems – A comparative technological review, *IEEE Access*, 10, 101042–101070. https://doi.org/10.1109/ACCESS.2022.3208142.

6 Cybersecurity for Sustainable Entrepreneurship

Aleksandra Ptak and Tomasz Lis

6.1 INTRODUCTION

With dynamic technological development and increasing global ecological awareness, companies around the world face the challenge of adapting to the changing business landscape. In the face of growing threats related to both cybercrime and the increasing visible effects of climate change, companies are forced to revise their strategies and economic practices.

At its core, sustainable entrepreneurship transcends the boundaries of traditional business paradigms, focusing not only on maximizing profits, but also on balancing these goals with concern for the environment, community, and ethical values. The concept of sustainable entrepreneurship sets a new standard for companies, requiring them to make business decisions that take into account a wide range of stakeholders, including not only shareholders, but also employees, customers, local communities, and the natural environment. To be truly sustainable, modern enterprises must take actions that not only minimize their negative impact on the planet, but also contribute to social justice, the creation of sustainable jobs, and the protection of personal and corporate data (Zgrzywa-Ziemak, 2019).

Cybersecurity is a set of activities aimed at reducing the risk associated with information technology and functioning in cyberspace. As part of these activities, it focuses on ensuring process security, conducting preventive activities such as preventing cyber incidents and building awareness of appropriate online practices, and using appropriate technologies and solutions that protect data (Górny, 2016). In the face of more and more advanced cyberattacks, protecting data and IT infrastructure has become an indispensable priority for every organization, regardless of its size or industry. In today's world, where the flow of information plays a key role in the functioning of enterprises, threats such as hacker attacks, data leaks or ransomware pose a real challenge to the security and stability of business activities.

Hacker attacks are becoming more complex and sophisticated, using advanced penetration techniques that can bypass traditional security systems (Rattan et al., 2005). Data leaks, especially regarding confidential information of customers or employees, can lead not only to direct financial losses related to compensation and

DOI: 10.1201/9781003484226-6

penalties for violating data protection regulations, but can also destroy the company's reputation and customer trust for many years. Ransomware, i.e. malicious software that blocks access to data or systems, can paralyze a company's operations and result in huge financial losses and operational downtime (Money, 2024).

The introduction of effective cybersecurity strategies is, therefore, becoming a key element of sustainable entrepreneurship, enabling companies not only to protect their interests and assets but also to demonstrate social responsibility and care for the business environment in which they operate. Companies that take measures to ensure the security of their data and IT infrastructure not only protect their profits, but also look after the well-being of their stakeholders and support the creation of a safer and more sustainable business environment for all market participants. Further, implementing effective cybersecurity strategies can provide additional benefits to businesses beyond providing protection against cyberattacks. Improving data security and IT infrastructure can also increase operational efficiency by reducing the risk of downtime due to security incidents and facilitating compliance with data protection regulations. Additionally, taking actions for cybersecurity can increase the trust of customers, business partners and investors, which may contribute to an increase in the company's value on the market (Morganstanley, 2023). In this way, cybersecurity becomes a key element of business strategy that not only supports sustainable entrepreneurship, but also lays the foundation for stability and operational success in the digital era.

In this chapter, the authors focus on analyzing the role of cybersecurity in the context of sustainable entrepreneurship and present the main challenges and strategies that companies can use to effectively integrate cybersecurity into their sustainable business practices (Sharma et al., 2020). By understanding this synergy, companies can not only minimize the risk of financial and reputational loss, but also support the creation of a more secure, resilient and sustainable business environment that fosters innovation and long-term socio-economic development.

6.2 ENTREPRENEURSHIP IN THE CONCEPT OF SUSTAINABLE DEVELOPMENT

P.F. Drucker defines entrepreneurship as a way of behaving by an entrepreneur and an enterprise, based on the readiness and ability to undertake and creatively solve new problems. In this perspective, entrepreneurship includes the ability to use emerging opportunities and challenges and flexible adaptation to changing conditions. Drucker also emphasizes that the key tool in entrepreneurial activity is innovation, which creates new opportunities to generate wealth from existing resources (Drucker, 2006). T. Listwan defines entrepreneurship as an activity focused mainly on starting new projects, creatively solving problems, taking risks, identifying and using opportunities and flexible adaptation to the changing environment (Listwan, 2006). Within the framework of economic theory, entrepreneurship can be perceived as a type of work of a specific nature or as the fourth factor of production, equivalent to land, capital, and labor. Key characteristics of entrepreneurship include the ability to identify needs, refine concepts, and the willingness to take risks (Bryant, 2015).

Entrepreneurship is a complex research area; it covers not only the activities of a single entity, but also the functioning of the entire organization or system. Entrepreneurship and economic growth are an integral part of the modern business landscape. However, as technology progresses and society evolves, more and more importance is attached to sustainable development and ethical business conduct (Wach & Głodowska, 2022).

The essence of sustainable development is to combine economic and social growth by ensuring a balance in the use of both renewable and non-renewable resources, along with improving the quality of life in a healthy environment. Economic development should strive for more efficient use of raw materials and other natural resources, rationalization of energy and labor consumption, as well as the promotion of ecological technologies and the protection of natural and cultural heritage (Singh et al., 2005). Sustainable development can be described as a humanitarian concept of global progress that prioritizes improving the quality of life and well-being of humanity while also taking into account the long-term effects of economic industrial development (Mensah & Casadevall, 2019).

Sustainable companies focus their goals on implementing positive social, economic and environmental changes. They strive to strike a balance between making profits and making lasting changes that are consistent with the United Nations Sustainable Development Guidelines. These companies are being perceived as caring about their surroundings and community, which translates into greater interest from customers and potential investors. Their actions contribute to strengthening the brand and increasing its profitability. There are many benefits to adopting sustainable business practices. First of all, it builds a positive image of the company, attracting customers who prefer products and services based on sustainable values. Secondly, sustainable companies are often more economically efficient and perform better in changing mar ket conditions. Such enterprises can support the development of local communities by creating jobs and engaging in community activities (Zgrzywa-Ziemak, 2019). It is important to emphasize that striving for sustainable development is not an end in itself, but a means to achieve real and lasting benefits for both business and society.

6.3 CYBERSECURITY IN THE ERA OF DIGITALIZATION

Over two decades ago, the world entered an era in which the ability to acquire, process information and create knowledge became key factors of success in the economic sphere (Verhoef et al., 2021). The implementation of new solutions in company management is now an essential element of their further development. The process of digitization of information and the increasing penetration of modern ICT into all spheres of life and work result in flexibility in structures and workplaces. The use of Information and Communication Technologies ICT leads to improved customer comfort, flexibility and efficiency of the company's internal processes, which ultimately contributes to the overall coherent development of industrial and business informatization. That's why more and more companies recognize digital technologies as a key element of their business strategy (Basl & Sąsiadek, 2012). However, as technology advances, new online security challenges also emerge. Cybersecurity is, therefore, becoming a key issue that requires constant attention and appropriate actions from both individuals and organizations.

In the digital world where personal, financial and business data is stored and processed online, cybercrime threats are becoming more complex and dangerous (Bhambri et al., 2005). Hacking attacks, identity theft, computer viruses and other forms of cyberattacks can lead to serious consequences such as data loss, financial damage and even damage to reputation and trust.

Reports conducted by institutions monitoring cybersecurity show that the greatest threat to companies is not a single attack, but a series of unrelated incidents, which may include attempts to extort personal data, false bank invoices, infected software, money theft, unauthorized financial transactions, extortion ransom, and disruption of services (Gruszecki, 2020).

The "ENISA Threat Landscape 2021" report, prepared by the European Cybersecurity Agency, provides an overview of key cybersecurity threats. Among them were mentioned (ENISA, 2021):

- Ransomware: Malware that blocks access to data and requires a ransom payment.
- Malware: Software that can damage or steal information.
- Cryptojacking: Mining cryptocurrencies using devices belonging to others.
- Threats related to e-mail.
- Personal data breaches and other data threats.
- Distributed Denial of Service (DDoS) attacks that can disrupt data availability and integrity (Blog Google, 2022).
- Propagating disinformation and false information.
- Other threats such as configuration errors, physical infrastructure incidents, and natural disasters.
- Supply chain attacks that can lead to business interruption.

To effectively defend against cyberattacks, companies must develop appropriate organizational criteria, policies and procedures. Having a solid background in digital security is an absolute necessity. In the face of growing threats and criminal activities in cyberspace, the lack of adequate data protection calls into question the reliability and stability of operation of modern enterprises (pb, 2024).

According to the Polish incident response team – CERT, in 2022, 322,479 reports of cybersecurity incidents were recorded. CERT Poland made a careful classification, based on which 115,164 reports were selected, of which 39,683 cybersecurity incidents were registered. For comparison with the previous year, a total of 116,071 reports were registered in 2021. Of all the reports, 65,586 were selected, based on which a total of 29,483 unique cybersecurity incidents were registered. This represents an increase of as much as 178% comparing 2022 with 2021. Among the incident categories listed in the 2022 report, the most common are (Raport, 2022):

- Computer fraud – 88.2%.
- Malware – 8.6%.

- Burglary – 0.9%.
- Offensive and illegal content – 0.8%.
- Vulnerable services – 0.5%.

Bitdefender's 2023 report shows a noticeable increase in the use of ransomware by hackers, which not only encrypts data but also steals it. Social engineering methods are becoming more and more sophisticated, making phishing attacks increasingly difficult to recognize. Almost all cybersecurity experts (99%) are concerned about these changes. According to a study by the Bitdefender team, the main problems for IT teams are: (Bitdefender, 2023):

- Security vulnerabilities.
- Zero-day exploits.
- Attacks on supplychains.
- Ransomware and social engineering attacks.

Over 72% of respondents noticed an increase in the sophistication of phishing attacks in their companies. Professionals' concerns are justified, as more than half of them (52%) have experienced data breaches related to cybersecurity incidents in the last year. The most of data breaches occurred in the United States (75%), followed by the United Kingdom (51%), Germany (49%), and France (42%). Many security professionals (over 40%) have been instructed to maintain confidentiality in the event of a data leak, even though they are legally required to report it. In the USA, this rate reaches up to 71%. However, only 15% of respondents in Germany and 27% in France followed the instructions and kept information about the data breach secret.

In light of the above data, there is a clear increase in the number of reported cybersecurity incidents compared to previous years. This dynamic development of cyber threats highlights the importance of an effective cyber security strategy for companies (Bhambri & Khang, 2024). It covers a comprehensive enterprise approach to preventing, detecting, and responding to and protecting against cyberspace threats. To develop an enterprise cybersecurity strategy, it is necessary to identify several key issues. To do this, it is necessary to answer the question of what constitutes an object of protection. In the first place will certainly be databases, workstations, and other devices connected to the network. It is necessary to define all elements precisely and understandably. Another important question is what types of threats should be defended against. The strategy should be comprehensive and provide protection against all types of threats. This especially applies to phishing, smishing, document loss, DoS and DDoS attacks, ransomware, bots, trojans, and other malicious software. However, it is necessary to carefully analyze which types of attacks are particularly dangerous for a given company (Rana & Bhambri, 2024). The last important question to ask is what protective measures will be taken. The strategy should identify all available methods that will be used to ensure security. At the same time, it should be checked whether they are sufficient to effectively counteract threats. In the cybersecurity strategy, it is important not only to take into account current threats, but also potential ones that may occur in the future and threaten the company's operations. Therefore, it is recommended to set strategies and goals for different time horizons. The good strategy includes planned changes for the coming months and general directions for the coming years. There are many ways to increase security, such as regular software updates, using strong passwords to authorize access to devices and networks, backing up data and storing it in various locations, and using biometric methods for identification. However, it is important to strike a balance when developing cybersecurity best practices. Online security should not limit business's growth in any way (bp, 2024).

As part of fundamental security principles in a company, the following steps can be adopted (Bank, 2018):

- Creating strong passwords.

Passwords should be long enough and contain both upper and lower case letters, numbers and special characters. It is important to avoid using the same passwords for different databases and to change existing passwords regularly. Employees should also be aware of the need to avoid using private passwords for company systems and saving them as text files on computers or visible notes.

- Use a VPN when using public networks.

With more and more employees working remotely or hybridly and traveling on business trips, using a company VPN when connecting to public Wi-Fi networks in coffee shops, airports, and other public places is crucial to maintain company data security.

- Participation in cybersecurity training and courses.

Employees must be aware of various threats, such as phishing, pharming, or social engineering techniques used by hackers. It is also necessary to familiarize them with methods of recognizing fake emails that are not spam.

6.4 CYBERSECURITY IN THE CONTEXT OF SUSTAINABLE DEVELOPMENT OF ENTERPRISES

For businesses of all sizes and industries, cybersecurity has evolved to become an essential element of business strategy. Their approach to cybersecurity is based on balancing two main goals: ensuring the protection of data and infrastructure against cyberattacks and maintaining compliance with the principles of sustainable development. Effectively managing cybersecurity risk in the context of ESG (environmental, social, and governance) provides companies with the opportunity to secure their operations, customers and reputation, while meeting their social and environmental goals. Investing in modern technological solutions not only reduces the negative impact on the environment, but also minimizes the risk of cyberattacks, which translates into stability and business success (Erp-view, 2023).

First, sustainable businesses understand that cybersecurity is a key element in ensuring business continuity and protecting the sensitive data of their customers and employees. That's why they invest in advanced protective technologies, implement data security policies and regularly train staff in cybersecurity awareness. Secondly, sustainable companies strive to minimize the negative impact of their activities on the environment and society. In the context of cybersecurity, this means, among others, ensuring the energy efficiency of IT systems and minimizing emissions of pollutants related to their online activities. Additionally, sustainable companies take action to promote ethical standards in cybersecurity, e.g. by working with non-governmental organizations dealing with digital rights and personal data protection (kpmg, 2024).

In the face of constant changes, it is important to maintain regulations that support companies in ensuring safety and minimizing potential risks. Examples of such initiatives include the General Data Protection Regulation (GDPR), the revised Network and Information Systems Directive (NIS2), the Digital Operational Resilience Act (DORA), the Sustainable Finance Disclosure Regulation (SFDR) and the Corporate Sustainability Reporting Directive (CSRD). These regulations are crucial to ensuring stability and fairness in the business environment (ey, 2023).

Sustainable enterprises' approach to cybersecurity is based on a balance between data and infrastructure protection and compliance with the principles of sustainable development, which helps build customer trust, increase resistance to cyberattacks and promote the values of socially responsible business.

6.5 EMPLOYEE EDUCATION IN CYBERSECURITY

Today, employee knowledge and awareness are key components of IT security. Simply having the right equipment and security measures is not enough the people who work with them must be aware of the basic principles of network security.

Building the safety culture among employees at all levels of the organization is essential in today's reality. Regular training, phishing testing, and promoting cybersecurity awareness by leaders are key to ensuring that every team member is prepared for possible threats. Well-prepared training allows everyone to understand the topic of cybersecurity and makes it easier to remember (kicb, 2024). An educated and aware of potential threats employee is no longer the weakest link in the system, but becomes the first line of defense.

According to a 2021 study commissioned by Sophos, conducted on groups of 800 employees in three countries: Poland, Hungary, and the Czech Republic, only half of the companies provided training in the field of cyber threats, and large companies were more likely to care about training employees in the field of cybersecurity. In Poland, among employees in companies with over 250 employees, 44% received support from the IT department and training related to cyber threats. In medium-sized companies employing 50 to 250 people, 26% of respondents had access to such support, while in small companies (up to 50 employees), 32% of employees could count on support. The largest companies were much less likely to skip training and technical support altogether – only 15% of employees were left without help, compared to 27% in companies with up to 50 people (Sophos, 2021).

Awareness of cyber threats among staff in Poland (89%) is higher than in the Czech Republic or Hungary, where 80% and 73% know what the types of risk for the company may be. The threat most frequently indicated by employees is the leak of company information or credentials (38%). Slightly fewer respondents (37%) identified fake e-mails and malicious links in messages sent to employees. Every third person believes that the risk may be the carelessness of colleagues online and the unintentional downloading of malware from suspicious websites. Only the fourth place (27%) is taken by criminals taking advantage of vulnerabilities in the system. One-fifth of staff believe that threats may result from the actions of dissatisfied employees or inadequately secured and outdated equipment. The fewest indications concerned the carelessness of business partners and subcontractors (18%).

Among Polish employees, one in ten feel unsure what to do if they encounter a suspicious email, situation, or application while at work. This phenomenon is more common in small- and medium-sized companies (10% and 13%, respectively) than in large enterprises (5%). It is worth emphasizing that this percentage is higher than in the case of employees in Hungary or the Czech Republic, where the lack of knowledge of appropriate procedures was noticed in 8% of respondents. Despite this, the majority of employees (58%) declare that they would report such an event immediately. However, one in three (32%) would simply delete the suspicious email. It is worth noting that people employed in larger companies would be much more likely to report such an event to the appropriate person (72%), as opposed to employees of small- and medium-sized companies (50% and 57%, respectively). Employees of small companies would decide to delete a suspicious e-mail themselves more often than those in medium-sized and large companies (40%, 30% and 23%, respectively).

In Polish companies, the most frequently used security measure is data access control. As many as 71% of employees have access only to the files they need for work. It is primarily the largest enterprises employing over 250 people which apply such rules to a greater extent – 20 percentage points more than in the case of small companies (82% and 62%, respectively). Polish companies also use multi-factor authentication, which includes authorization via SMS, token or application, more often than Czech and Hungarian ones. However, only half of the companies use this verification method (in the Czech Republic 39%, in Hungary 38%). A similar percentage of companies use encrypted connections such as VPN (53%). An interesting phenomenon is that the attitude towards multi-factor authentication is almost identical regardless of the size of the company.

6.6 PROVIDERS OF SOLUTIONS IN THE CYBERSECURITY SECTOR

The largest companies in the cybersecurity industry are based mainly in the United States. According to data for 2020, suppliers such as Cisco, Palo Alto Networks, Fortinet, Check Point, Symantec, IBM, and McAfee were at the top of this list (Nomios, 2023).

Cisco

The solutions offered by Cisco are dedicated mainly to small- and medium-sized companies, and are characterized by quick malware detection and the ability to detect almost 100% of attacks in just a few minutes (dnarynkov, 2022). Their solutions include a set of security products and services including firewalls, intrusion prevention systems (IPS), secure internet gateways, cloud solutions, and the latest security automation technologies. In the dynamic world of cybersecurity, the emergence of SASE (secure access service edge) plays a crucial role. Cisco is positioned as a Visionary in the Gartner Magic Quadrant for Single Vendor SASE. However, it continues to develop in this aspect; in 2023 Cisco started significant investments in the SASE area, and in 2024 it announced even more innovations. One of the more interesting improvements Cisco has made to SASE solutions is integration with SecureX, their XDR (extended detection and response) platform. This integration provides robust zero trust security features, taking security posture management to the next level (Cisco, 2024).

Palo Alto Networks

The company has an established reputation as a leader in the volatile SASE market. Their SASE solutions are carefully crafted to protect an organization by providing defense against a variety of cyber threats, including malware, ransomware and phishing attacks. In June 2023, Palo Alto Networks launched Prisma Cloud 3.0, a stable cloud security platform that provides comprehensive protection for cloud-native applications and workloads. In September of the same year, Cortex XDR Platform was introduced – an innovative security platform that efficiently integrates data from the entire security environment, providing organizations with extensive detection and response (XDR) capabilities. One month later, in October 2023, the Prisma Cloud platform was enhanced with code to cloud intelligence. This innovative functionality provides a single, trusted source of application risk information, benefiting developers, operations and security teams. The latest version, called the Darwin release, introduces a number of new capabilities that help organizations better address security threats early in the production lifecycle. Additionally, they integrate changes across the entire cloud workflow (Murdoch, 2024).

Fortinet

The company stands out for its advanced, consistent and automated defense systems against the most advanced cyber threats. The foundation of Fortinet's security is the Fortinet Security Fabric, a consistent and integrated platform that provides comprehensive protection for the entire digital infrastructure. This framework integrates network, security and analytics to create a comprehensive shield that offers unparalleled protection against cyber threats. Fortinet offers the FortiSASE standard. Although their SASE platform continues to evolve, it ensures the security of the hybrid workforce through a modern Zero Trust architecture, ensuring network visibility and security throughout the organization. Beyond SASE, Fortinet's Zero Trust portfolio is growing rapidly with new tools supporting micro-segmentation, identity and access management (IAM), and security information and event management (SIEM). Their progress in Zero Trust earned them a leadership position in the Forrester Wave for Zero Trust Edge Solutions for Q3 2023 (Fortinet, 2024).

Check Point

Check Point Software Technologies is a leading provider of an AI-powered, cloud-delivered cybersecurity platform. The Infinity platform revolutionizes the efficiency and precision of cybersecurity by offering the most advanced detection rates available on the market. Its functions not only enable the prediction of potential threats, but also significantly shorten the response time by acting proactively. Infinity's integrated platform offers companies the flexibility to secure any type of IT environment, ensuring consistent visibility, excellent threat protection, and significant savings. It consists of four key components: CloudGuard for cloud protection, Quantum for network security, Harmony for securing the work environment, and Infinity Core Services for unified security operations. Advanced real-time threat protection powered by machine intelligence covers all possible attack vectors – from computer code to the cloud, networks, users, email communications and Internet of Things (IoT) devices. It uses over 50 Artificial Intelligence engines to deliver the highest-class protection against the most sophisticated attacks, including zero-day attacks, phishing, and domain name system (DNS) attacks (checkpoint, 2024).

Symantec

Symantec's solutions, now part of Broadcom, have a long history in the market and continue to evolve to meet changing challenges. It provides protection for all end devices, both traditional and mobile, using technology such as prevention against attacks and breaches, limiting the scale of attacks, detection, and response. All these mechanisms are based on the Global Intelligence Network. Symantec's Single-Agent offers flexible implementation and management options, including full cloud, on-premises and hybrid solutions. Symantec Identity Security supports the implementation of new applications and increases efficiency in any environment. This allows to quickly respond to the organization's current business needs. The solution guarantees constant management, authentication and registration of privileged account activity. It also ensures enforcement of the principle of minimum privileges and provides an audit and reporting trail, enabling compliance with legal regulations. A comprehensive zero trust system allows for positive authentication and authorization of users before granting them access to company resources and data. The system constantly monitors their activity and can take appropriate security actions if suspicious behavior is detected (Symantec, 2020).

6.7 CONCLUSIONS

Specialists report unfavorable forecasts – it is estimated that by 2025, global losses related to cybercrime activities will reach US $10.5 trillion annually. By comparison, cyberattacks cost the world $9.5 trillion in 2024 and $8 trillion in 2023 (Expressvpn, 2024).

The increase in costs associated with cybercrime has occurred mainly because, in addition to the actual damage or destruction of data, entrepreneurs now have to take into account increasingly higher sums for file decryption, loss of productivity, business disruption due to attacks, and the expenditure of time and money needed to minimize the damage. There are also costs associated with law enforcement investigations, lawsuits, and image damage for companies (Simplilearn, 2024).

Cybersecurity in the company is a key issue for its proper functioning, so it should be taken care of in all possible aspects, from network security to employee education. Cybersecurity is the process that requires constant monitoring, adaptation to new threats and regular strategy updates. In the digital age, strategic thinking about cybersecurity is essential to ensuring the long-term success and security of a business. By taking these steps, organizations can not only increase their resilience to cyberattacks, but also ensure that their defense efforts are fully integrated with business goals and strategy. A strategic approach to cybersecurity allows to optimize expenses in this area, transforming them from perceived operational costs into a strategic investment. By focusing on business priorities and identifying key assets, companies can allocate financial resources more effectively, securing the most valuable elements of their digital infrastructure (Pb, 2024). This approach not only reduces the financial risk associated with cyberattacks, but also supports the sustainable development of the organization, ensuring that every money spent on cybersecurity is an investment in the company's future. In this way, cybersecurity is no longer perceived as a financial burden and becomes a key element of growth and innovation strategies.

REFERENCES

Bank.pl. (2018). Cyberbezpieczeństwo w firmie: pięć zasad, które powinien znać każdy pracownik, available at: https://bank.pl/cyberbezpieczenstwo-w-firmie-piec-zasad-ktore-powinien-znac-kazdy-pracownik/ (access: 10.02.2024).

Basl, J., & Sąsiadek, M. (2012). Innowacyjna rola ICT w społeczeństwie i w gospodarce. W: R. Knosala (red.), *Innowacje w zarządzaniu i inżynierii produkcji* (s. 759–768). Opole: Oficyna Wydawnicza Polskiego Towarzystwa Zarządzania Produkcją.

Bhambri, P., Gupta, S., & Bhandari, A. (2005). Soft computing techniques. In UGC Sponsored National Conference (Abstract Book) (p. 100). Presented at the "Emerging Computing Technologies" conference at HRMMV, Jalandhar held during September 29–30, 2005.

Bhambri, P., & Khang, A. (2024). The human-machine nexus with art-making generative AIs. In S. Hai-Jew (Ed.), *Making Art with Generative AI Tools* (pp. 73–85). IGI Global. https://doi.org/10.4018/979-8-3693-1950-5.ch004.

Bitdefender.pl. (2023). Wyzwania związane z cyberbezpieczeństwem, przed którymi stoją firmy w 2023 roku, available at: https://bitdefender.pl/wp-content/uploads/2023/04/Wyzwania-zwiazane-z-cyberbezpieczenstwem.pdf (access: 13.03.2024).

Blog.google. (2022). New insights for defending the software supply chain, available at: https://blog.google/technology/safety-security/new-insights-for-defending-the-software-supply-chain/ (access: 10.03.2024).

Bryant, P. T. (2015). Entrepreneurship and organizations. In J. D. Wright (Ed.), *International Encyclopedia of the Social & Behavioral Sciences* (2nd ed., pp. 681–685). Elsevier. https://doi.org/10.1016/B978-0-08-097086-8.73005-7

Checkpoint.com. (2024). The leading cyber security platform, available at: https://www.checkpoint.com/infinity/ (access: 20.02.2024).

Cisco.vom. (2024). Cisco SecureX. A simplified security experience, available at: https://www.cisco.com/c/en_be/products/security/securex/index.html (access: 06.03.2024).

Dnarynkow.pl. (2022). Spółki, które dadzą zarobić tysiące procent? Cyberbezpieczeństwo sektorem dekady! available at: https://dnarynkow.pl/sektor-cyberbezpieczenstwa-najciekawszy-temat-inwestycyjny-dekady/ (access: 17.23.2024).

Drucker, P. F. (2006). *Innovation and Entrepreneurship.* New York: Harper Business.

ENISA.europa.eu. (2021). ENISA threat landscape, available at: https://www.enisa.europa.eu/publications/enisa-threat-landscape-2021/@@download/fullReport (access: 17.03.2024).

Erp-view.pl. (2023). Cyberbezpieczeństwo w kontekście ESG: zrównoważona i bezpieczna przyszłość w cyfrowym świecie, available at: https://www.erp-view.pl/bezpieczenstwo-it/30656-cyberbezpieczenstwo-w-kontekscie-esg-zrownowazona-i-bezpieczna-przyszlosc-w-cyfrowym-swiecie.html (access: 22.02.2024).

Expressvpn.com. (2024). The true cost of cyber attacks in 2024 and beyond, available at: https://www.expressvpn.com/pl/blog/the-true-cost-of-cyber-attacks-in-2024-and-beyond/ (access: 11.02.2024).

Ey.com. (2023). Wyzwania Cyberbezpieczeństwa 2022 Perspektywa EY, available at: https://www.ey.com/pl_pl/cybersecurity/wyzwania-cyberbezpieczenstwa-2022 (access: 13.03.2024).

Fortinet. Fortinet accelerate 2024. https://www.fortinetaccelerate.com/home

Górny, P., & Krawiec, J. (2016). Cyberbezpieczeństwo – podejście systemowe, OBRONNOŚĆ. Zeszyty Naukowe 2(18).

Gruszecki, P. (2020). Cyberbezpieczeństwo, czyli niezbędna odpowiedź przedsiębiorców na wiele zagrożeń z różnych źródeł, available at: https://www.dzp.pl/blog/ip/cyberbezpieczenstwo-czyli-niezbedna-odpowiedz-przedsiebiorcow-zagrozen-roznych-zrodel/ (access: 03.03.2024).

Kicb.pl. (2024). Szkolenia z zakresu cyberbezpieczeństwa, available at: https://kicb.pl/nasza-oferta/szkolenia-z-cyberbezpieczenstwa/#Szkolenia%20Z%20Cyberbezpiecze%C5%84stwa (access: 07.03.2024).

Kpmg.com. (2024). Cyberbezpieczeństwo w kontekście ESG: Zrównoważona i bezpieczna przyszłość w cyfrowym świecie, available at: https://kpmg.com/pl/pl/home/media/press-releases/2023/10/cyberbezpieczenstwo-w-kontekscie-esg-zrownowazona-i-bezpieczna-przyszlosc-w-cyfrowym-swiecie.html (access: 13.02.2024).

Listwan, S. J., Cullen, F. T., & Latessa, E. J. (2006). How to prevent prisoner re-entry programs from failing: Insights from evidence-based corrections. Corrections Today, 70(3), 19-26. Retrieved from https://www.uc.edu/content/dam/uc/ics/docs/ListwanCullenLatessaHowToPrevent.pdf"

Mensah, J., & Casadevall, S. R. (2019). Sustainable development: Meaning, history, principles, pillars, and implications for human action: Literature review. *Cogent Social Sciences*, 5(1), 1–21.

Money.pl. (2024). Firmowe dane na celowniku. Jak przetrwać w erze cyberzagrożeń? available at: https://www.money.pl/gospodarka/firmowe-dane-na-celowniku-jak-przetrwac-w-erze-cyberzagrozen-6968029873793920a.html (access: 09.03.2024).

Morganstanley. (2023). AI and cybersecurity: A new era, available at: https://www.morganstanley.com/articles/ai-cybersecurity-new-era (access: 10.03.2024).

Murdoch Webster, Paloalto Networks, available at: https://mwtg.com.au/palo-alto-networks (access: 10.03.2024).

Nomios.pl. (2023). Najważniejsze firmy zajmujące się cyberbezpieczeństwem, które warto obserwować w 2024 roku, available at: https://www.nomios.pl/aktualnosci/firmy-zajmujace-sie-cybersecurity-2024/ (access: 07.03.2024).

Pb.pl. (2024). Cyberbezpieczeństwo firmy: jak tworzyć i skutecznie egzekwować strategię cybersecurity? available at: https://www.pb.pl/konferencje/it/cyberbezpieczenstwo-firmy-jak-tworzyc-i-skutecznie-egzekwowac-strategie-cybersecurity-118358 (access: 15.02.2024).

Rana, R., & Bhambri, P. (2024). Artificial Intelligence and Internet of Things (IoT) facilitated digital twin for Industry 4.0 application domains. In S. Rani, P. Bhambri, S. Kumar, P. K. Pareek, & A. A. Elngar (Eds.), *AI-Driven Digital Twin and Industry 4.0: A Conceptual Framework with Applications* (1st ed., pp. 168–181). CRC Press. https://doi.org/10.1201/9781003395416.

Raport roczny z działalności CERT POLSKA z 2022 r., Krajobraz bezpieczeństwa polskiego internetu, available at: https://cert.pl/uploads/docs/Raport_CP_2021.pdf (access: 13.03.2024).

Rattan, M., Bhambri, P., & Shaifali. (2005). Institution for a sustainable civilization: Negotiating change in a technological culture. In National Conference (Abstract Book) (p. 7). Presented at the "Technical Education in Globalized Environment- Knowledge, Technology & The Teacher" conference at SBBSIET, Padhiana, Jalandhar held during February 25–26, 2005.

Sharma, R., Bhambri, P., & Sohal, A. K. (2020). Energy bio-inspired for MANET. *International Journal of Recent Technology and Engineering*, 8(6), 5580–5585. https://www.ijrte.org/wp-content/uploads/papers/v8i6/F8522038620.pdf. https://doi.org/10.35940/IJRTE.F8522.038620.

Simplilearn. (2024). 20 Emerging cybersecurity trends to watch out in 2024, available at: https://www.simplilearn.com/top-cybersecurity-trends-article (access: 23.02.2024).

Singh, M., Singh, P., Kaur, K., & Bhambri, P. (2005). Database security. In Proceedings of National Conference (pp. 35–41). Presented at the "Future Trends in Information Technology" conference at SJPMLIET, Radaur, Yamuna Nagar held during March 6–7, 2005.

Sophos. (2021). The state of ransomware 2021. Sophos News. https://news.sophos.com/en-us/2021/04/27/the-state-of-ransomware-2021/

Symantec. (2020). Symantec named a leader, available at: https://symantec-enterprise-blogs.security.com/blogs/product-insights/symantec-named-leader-2019-gartner-magic-quadrant-secure-web-gateways-and-2019-magic (access: 12.03.2024).

Verhoef, P. C., Broekhuizen, T., Bart, Y., Bhattacharya, A., Qi Dong, J., Fabian, N., & Haenlein, M. (2021). Digital transformation: A multidisciplinary reflection and research agenda. *Journal of Business Research*, 122, 889–901.

Wach, K., & Głodowska, A. (2022). Entrepreneurship research in economics and management: Understanding the term and research trends. *International Entrepreneurship Review*, 8(4), 7–24.

Zgrzywa-Ziemak, A. (2019). Model zrównoważenia przedsiębiorstwa, Oficyna Wydawnicza Politechniki Wrocławskiej Wrocław.

7 Future Prospects and Trends in Digital Entrepreneurship Based on Sustainable Development

Klaudia Smoląg and Iwona Chomiak-Orsa

7.1 INTRODUCTION

Promoting access to information and communication technologies (ICT), its continuous development and increasingly broader functionality have significantly influenced economic and social life. Moreover, ICT has significantly contributed to the digitization of many areas of activity of every organization operating in the market. The ongoing digital transformation in all areas of life affects not only the change of business models or the development of remote learning and work but also forces the creation of new operating strategies. Additionally, the need to understand digital transformation and evaluate organizational performance is becoming a key success factor in the digital economy (Ahmad et al., 2021).

The good side of computerization also has its dark side – urbanization and industrialization are factors that have a negative impact on our natural environment. Both of these phenomena – digital transformation and sustainable entrepreneurship – are perceived as important factors determining the functioning of all organizations and entire societies. Hence, more and more attention is being paid to the relationship between digitalization and sustainable entrepreneurship (Xu et al., 2022). Digital transformation brings many benefits but it is also a source of certain threats. A challenge for digital transformation are areas related to sustainable development (Diaz & Montalvo, 2022). Therefore, an important task is not only a focused approach on how to use ICT, but also how these solutions can affect the functioning of economic, social, and environmental life.

Taking into account the assumptions and requirements of sustainable development in enterprise, development strategies force managers to seek a balance between the expenditure that should be incurred on pro-ecological activities together with planned investments in new technologies that do not necessarily fit into the

DOI: 10.1201/9781003484226-7

guidelines for sustainable development activities. The problem of finding a balance between these often contradictory directions is a challenge for modern management of organizations. The above observations, as well as the analysis of the literature, contributed to the formulation of the research question to which this chapter is intended to answer.

How does scientific literature present trends and prospects for the development of digital entrepreneurship?
How important are sustainability aspects in digital entrepreneurship strategies?

The chapter presents literature research based on bibliometric analysis and content analysis of scientific materials that constitute seminar studies. In addition, the main assumptions regarding sustainable development and digital transformation are characterized.

7.2 LITERATURE BACKGROUND

7.2.1 Trends Identification

The development of technology is closely correlated with the increase in the number of enterprises operating in the cyber space and creating their value proposition for customers based on the use of ICT, as well as increasingly advanced digital technologies (Haseeb et al., 2019; Šimelytė et al., 2021; Liszczyk & Chomiak-Orsa, 2024). This trend has been observed in business practice since the mid-1990s and has been accelerating over time. Creating new technological solutions involved the creation of new groups of enterprises that produced this technology, implemented technological solutions on the market, and provided services directly related to this technology such as consulting and training complementary to the developed technology; or indirectly related such as implementing new digital solutions into business processes not directly related to the new technology. The ongoing changes in the functioning of organizations as well as the creation of new, previously unknown business solutions determine the increased interest of scientists in ongoing business transformations. Especially publications in the area of management and quality indicate ongoing changes and evolutions in business models, entrepreneurship trends, and innovations created and implemented in enterprises (Abdurazzakov et al., 2020; Ślusarczyk et al., 2020; Jelonek & Chomiak-Orsa, 2018).

Therefore, in order to identify the level of interest in the topics covered in this chapter, the authors have conducted a bibliometric analysis of the scope of the literature (Bhambri et al., 2005). Given the proliferation of databases hosting scientific papers and the subsequent rise in publication duplication across these databases, the authors specifically concentrated their attention on Scopus, a single database.

In the first step, the authors searched the database in two separately defined thematic areas, namely "digital entrepreneurship" and "sustainable innovations" in the context of the sustainable development goals defined in the 2030 agenda. As the

FIGURE 7.1 First step in Scopus searching. Source: Author's rendition.

result of this search, they obtained information on the number of publications that were included in the Scopus database on January 24, 2024. In order to narrow down the results and verify publications that were dedicated to both the issue of digital entrepreneurship in the context of sustainable development and innovative technological solutions, a database query was constructed containing defined concepts that correlate with this topic. As a result of the above action, 10,453 publications were identified. The combination of the two areas as well as the quantitative results obtained in the process of processing records in the Scopus database are presented in Figure 7.1.

In order to verify the results obtained from the first stage of the search, the authors narrowed the analysis to reviewing the given concepts in terms of "title-abstract-keywords." This stage of verifying the content of the Scopus database corrected the number of publications to 8,777. The initial verification of the publications returned as a result of the search showed that a significant part of the publications indicated by the given criteria are in disciplines other than management and quality sciences, which are covered by the analysis of the considerations presented in the article. Therefore, another limitation that was defined in the query to the Scopus database was the definition of the field of management and quality science. As a result of the above narrowing, 1,054 publications were obtained. Problems indicating the importance of the development of digital entrepreneurship and information and communication technologies in determining this development are extremely topical and are becoming more and more important with the emergence of newer technological solutions used in this area (Sharma et al., 2020). This information aided in deciding that only publications that have been included in the database since 2020 will be taken into account in further, detailed biblometric analysis. The process of searching the database, the definition of individual queries, as well as the obtained search results, are presented in Table 7.1.

The number of publications indicated as a result of the third step of the research procedure was assessed in the context of the number of publications in particular years. Figure 7.2 presents the trend in the number of publications in the thematic scope. As a result of the analysis, the authors found that since 2019, a spectacular increase in interest in the research topic (as seen in Table 7.1) among authors has been observed.

TABLE 7.1
The Bibliometric Analysis Process of the Scopus Database and the Number of Publications (results 01/24/2023)

Step	Query Construction	Numer of Publications
Step 1	"sustainable development" OR "sustainable development goals" OR "sustainable entrepreneurship" OR "sustainable digitisation" OR "sustainable industry 4.0" OR "sustainable digital innovation" OR "green deal" OR "agenda 2030"	**357,048**
	digitisation OR "digital transformation" OR "digitisation process" OR "digitisation of services" OR "sustainable digital infrastructure" OR "digital economy" OR "digital competence" OR "digital exclusion" OR "cloud computing" OR "Internet of Things" OR "artificial intelligence" OR "5G" OR "augmented reality" OR "digital campus"	**974,702**
	(("sustainable development" OR "sustainable development goals" OR "sustainable entrepreneurship" OR "sustainable digitisation" OR "sustainable industry 4.0" OR "sustainable digital innovation" OR "green deal" OR "agenda 2030") AND (digitisation OR "digital transformation" OR "digitisation process" OR "digitisation of services" OR "sustainable digital infrastructure" OR "digital economy" OR "digital competence" OR "digital exclusion" OR "cloud computing" OR "Internet of Things" OR "artificial intelligence" OR "5G" OR "augmented reality" OR "digital campus"))	**10,453**
Step 2	TITLE-ABS-KEY ((("sustainable development" OR "sustainable development goals" OR "sustainable entrepreneurship" OR "sustainable digitisation" OR "sustainable industry 4.0" OR "sustainable digital innovation" OR "green deal" OR "agenda 2030") AND (digitisation OR "digital transformation" OR "digitisation process" OR "digitisation of services" OR "sustainable digital infrastructure" OR "digital economy" OR "digital competence" OR "digital exclusion" OR "cloud computing" OR "Internet of Things" OR "artificial intelligence" OR "5G" OR "augmented reality" OR "digital campus"))) AND (LIMIT-TO (DOCTYPE, "ar") OR LIMIT-TO (DOCTYPE, "cp"))	**8,777**
Step 3	TITLE-ABS-KEY ((("sustainable development" OR "sustainable development goals" OR "sustainable entrepreneurship" OR "sustainable digitisation" OR "sustainable industry 4.0" OR "sustainable digital innovation" OR "green deal" OR "agenda 2030") AND (digitisation OR "digital transformation" OR "digitisation process" OR "digitisation of services" OR "sustainable digital infrastructure" OR "digital economy" OR "digital competence" OR "digital exclusion" OR "cloud computing" OR "Internet of Things" OR "artificial intelligence" OR "5G" OR "augmented reality" OR "digital campus"))) AND (LIMIT-TO (DOCTYPE , "ar") OR LIMIT-TO (DOCTYPE, "cp")) AND (LIMIT-TO (SUBJAREA, "BUSI"))	**1,054**

(*Continued*)

TABLE 7.1 (CONTINUED)
The Bibliometric Analysis Process of the Scopus Database and the Number of Publications (results 01/24/2023)

Step	Query Construction	Numer of Publications
Step 4	TITLE-ABS-KEY ((("sustainable development" OR "sustainable development goals" OR "sustainable entrepreneurship" OR "sustainable digitisation" OR "sustainable industry 4.0" OR "sustainable digital innovation" OR "green deal" OR "agenda 2030") AND (digitisation OR "digital transformation" OR "digitisation process" OR "digitisation of services" OR "sustainable digital infrastructure" OR "digital economy" OR "digital competence" OR "digital exclusion" OR "cloud computing" OR "Internet of Things" OR "artificial intelligence" OR "5G" OR "augmented reality" OR "digital campus"))) AND (LIMIT-TO (SUBJAREA , "BUSI")) AND (LIMIT-TO (DOCTYPE , "ar") OR LIMIT-TO (DOCTYPE , "cp")) AND (LIMIT-TO (PUBYEAR , 2020) OR LIMIT-TO (PUBYEAR , 2021) OR LIMIT-TO (PUBYEAR, 2022) OR LIMIT-TO (PUBYEAR, 2023) OR LIMIT-TO (PUBYEAR, 2024))	**692**

Source: Author's compilation

The graphical presentation of the stage of research presented in Figure 7.2 indicates that in the years 1995–2018, there was little interest in the thematic area analyzed in the article. In the years 2015–2019, small upward trends were observed, but after 2019, a significant increase in interest in this topic is clearly visible, which could be caused by the publication of the UN resolution. Therefore, the authors

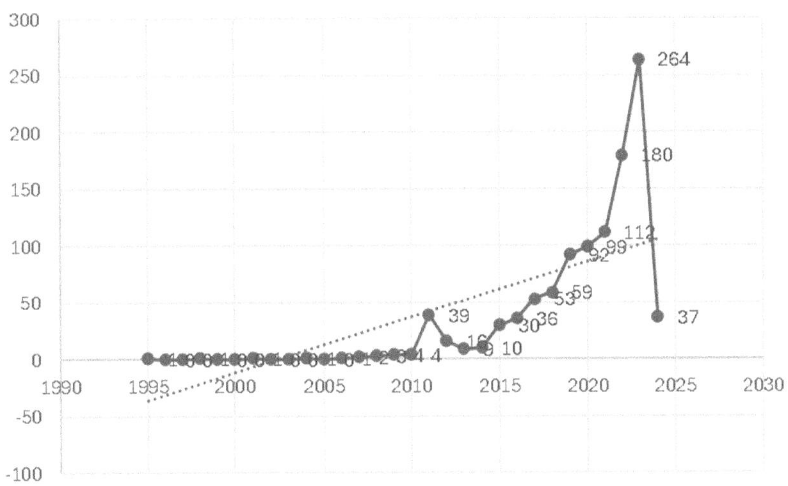

FIGURE 7.2 Trend in the number of publications in the analyzed area. Source: Author's rendition.

decided to further analyze articles that were published after this year. The adopted research procedure resulted in the identification of 692 articles (Step 4 in research procedure), which were qualified for further bibliometric processing.

7.2.2 Cluster Analysis

The next step was to cluster keywords that were defined by the authors from the articles found in the Scopus database. The first keyword clustering analysis yielded 2,273 keywords. The VOSviewer program used for visualization automatically created a network of clusters for 2,100 words. The visualization is presented in Figure 7.3.

As a result of the analysis procedure, it was found that the distribution of publications and the wide spectrum of defined keywords required the adoption of another limitation, which is the assessment of the distribution of keywords using a filter indicating the creation of clusters of articles in which keywords were repeated in at least five publications. The applied limitation reduced the number of listed keywords from 2,273 to 74. The visualization of the above stage is presented in Figure 7.4.

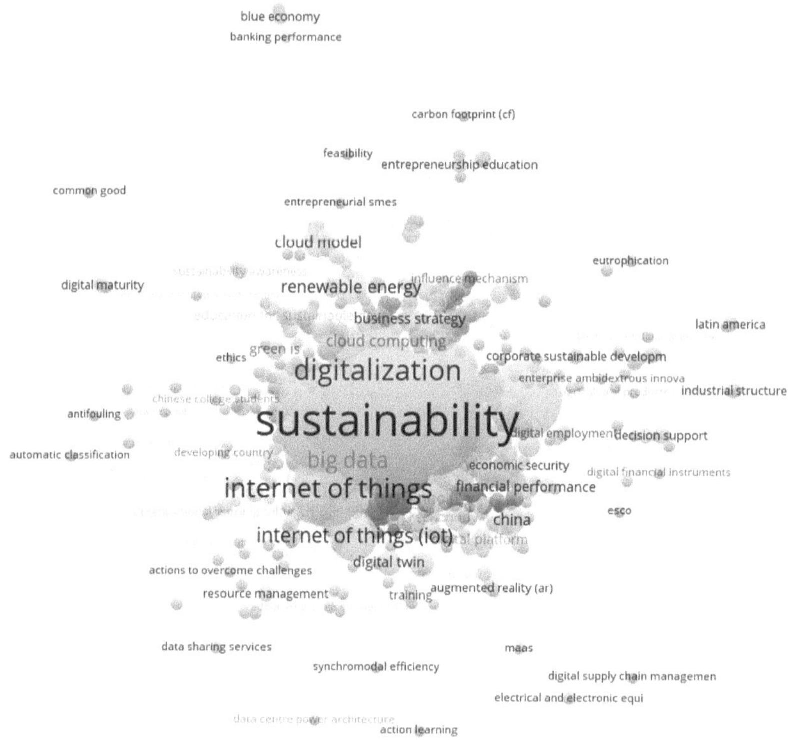

FIGURE 7.3 The result of the first keyword clustering. Source: Author's rendition using VOSviewer software.

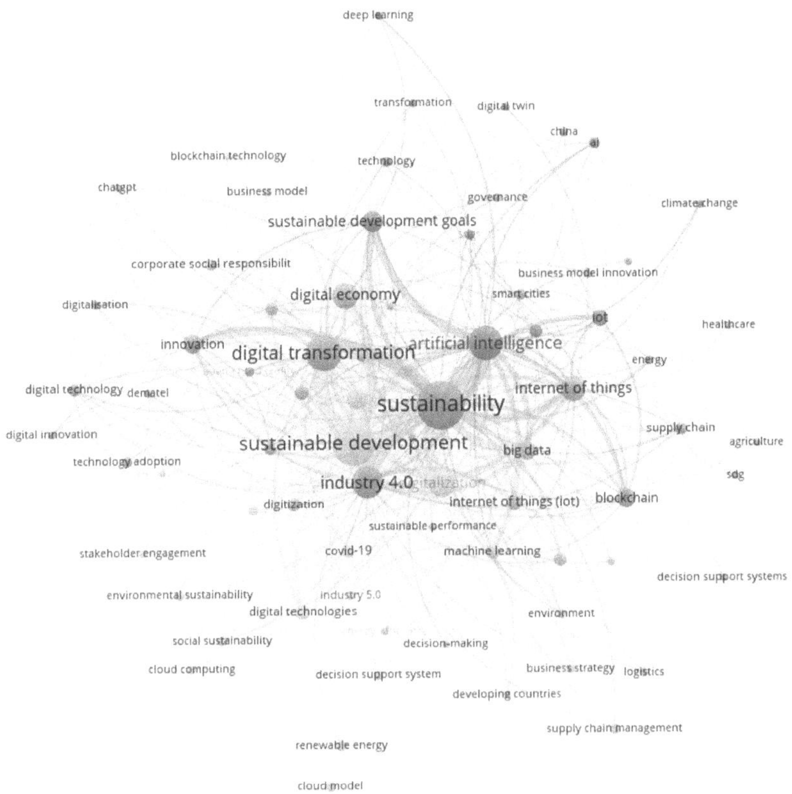

FIGURE 7.4 The result of the second keyword clustering with limits. Source: Author's rendition using VOSviewer software.

As a result of the keyword analysis of the identified range of articles, it can be concluded that most of the identified articles refer to demonstrating the relationships between various information and communication technologies such as artificial intelligence, Internet of Things (IoT), cloud computing (CC), big data (BD), machine learning, or blockchain, in relation to digital transformation that meets the requirements of sustainable development or sustainable innovation (Hudaszek in., 2024; Małowiecki & Chomiak-Orsa, 2024; Rutkowska & Sulich, 2020).

The next stage of the bibliometric analysis was the identification of the countries from which the authors who published their publications in the Scopus database were affiliated at least twice. For the entire pool of 93 identified countries, using the two-repeat filter, 50 countries were identified from which authors who wrote at least two articles in the period 2019–2024 came from. Clustering of publications from the perspective of the countries from which most publications come from in the analyzed area is presented in Figure 7.5.

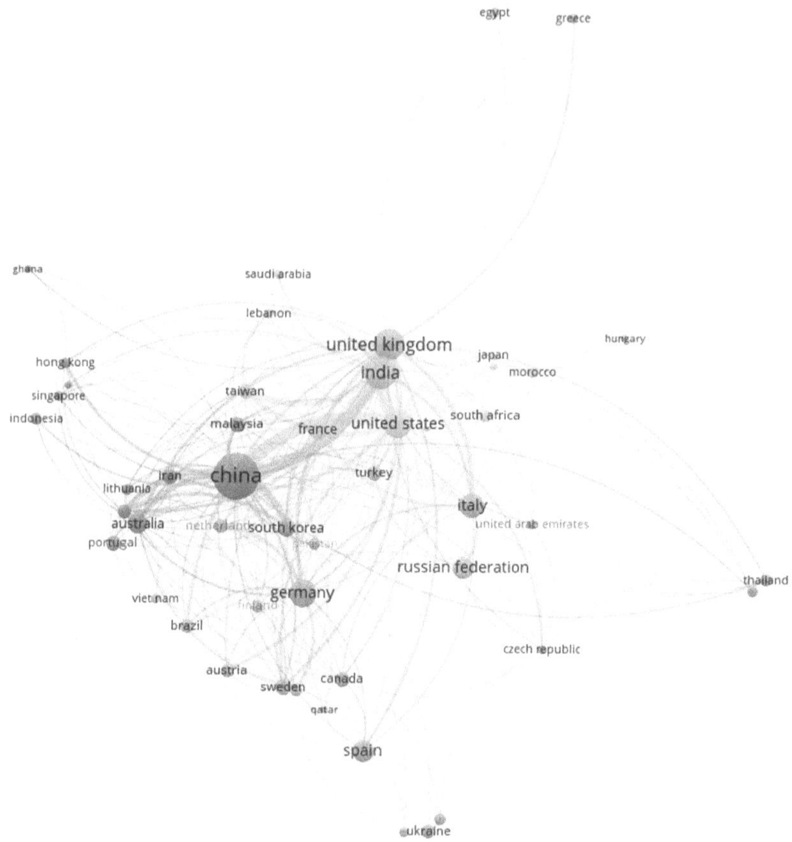

FIGURE 7.5 The result of the clustering of publications from the perspective of the "author's country" criterion. Source: Author's rendition using VOSviewer software.

As can be seen from the cluster network presented in Figure 7.5, most publications were developed in China, followed by India and Great Britain. A significant number of publications were also created in the United States, Germany, and France.

7.2.3 Most Frequently Cited Publications – Content Analysis

The last stage of the biblometric analysis was to search for publications with the status of seminal publications, i.e., those that are most often cited and indicated by other authors. From the statistically analyzed database of articles, five publications were selected that could be given the status of "seminal publications" in the discussed thematic area. Table 7.2 presents an analysis of the five most frequently cited works from the analyzed set of publications.

A detailed analysis of the text showed that despite the efforts of the authors discussing the development of digital entrepreneurship in the context of sustainable development, the research gap regarding the identification of the determinants of

TABLE 7.2
Analysis of the Five most Frequently Cited Works from the Analyzed set of Publications

Authors	Number of Citation	Conclusions from Detailed Text Analysis
Nižetić et al., 2020	**359**	The authors of the article point out the possibilities offered by the use of IoT technology in achieving sustainable development effects by enterprises using this technology. As a result of ongoing changes in the natural environment, the devastation of biological balance and excessive consumption of our planet's natural resources, technologies such as IoT may be a technology that will significantly improve resource management processes. However, technological development, despite the unquestionable benefits it may bring, also carries threats. Therefore, the implementation of IoT should take place in processes that are subject to continuous monitoring, while ensuring environmental protection and protecting the harmful effects of technology. The authors drew attention to the need to undertake research efforts assessing the effects of technology implementation in the context of sustainable development goals. In the article, the authors discussed the results in such applications as: • The use and implementation directions of IoT technology in the use of energy and the environment. • The creation of alternative solutions supporting a smart city equipped with IoT. • Building e-health systems through the use of IoT in supporting and protecting life. • Implementing IoT technology in transport and low-emission products. The main results of introducing the environment to the output that can be obtained through the use of IoT tools, as well as through the use of IoT solutions (Xu et al., 2022).
Li et al., 2020	**345**	The indicated seminar publication presents a wide spectrum of scientists' interests regarding the use in entrepreneurship and the implementation of digital technologies that are the basis of the Industry 4.0 concept. Moreover, the publication presents the results of an empirical study whose aim was to answer the question: how do digital technologies affect economic and environmental performance in the new era of Industry 4.0? Chinese manufacturing companies took part in the study. The results show that digital technologies influence both economic and environmental outcomes. The summary concluded that mediating effects are enhanced when there is a high degree of environmental dynamism. In the article, the authors summarized the results of their research by assessing factors such as: • The impact of digital technologies with particular emphasis on IoT, CC, BD on changes in China's economic results, • Assessed the impact of digital technologies on the efficiency of business processes. • Recommended technological solutions that can promote sustainable development. • Assessed selected technology solutions from the perspective of their impact – positive and negative on environmental aspects.

(Continued)

TABLE 7.2 (CONTINUED)
Analysis of the Five most Frequently Cited Works from the Analyzed set of Publications

Authors	Number of Citation	Conclusions from Detailed Text Analysis
Di Vaio et al., 2020	**314**	The article contains a systematic review of the literature in the area of the use of artificial intelligence in the creation of sustainable business models (SBM). The aim of the authors' literature research was to identify whether the use of artificial intelligence tools has an impact on the management patterns of organizations striving for sustainable use of resources, consistent with the goals of the UN agenda. In the article, the authors emphasized the role of knowledge management as the next stage of development and implementation of artificial intelligence in modern organizations. The bibliometric analysis of scientific articles from 1990–2019 conducted by the authors shows that the challenge related to IT innovations concerns not only technological aspects but primarily covers ethical, social, economic, and legal areas. Summarizing the research, the authors emphasized the importance of artificial intelligence in creating sustainable business models in such contexts as: • The relationship between the implementation and development of artificial intelligence and the creation of sustainable business models. • Identifying the research gap in scientific publications devoted to AI and sustainable business models, respectively. • The impact of AI development on the implementation of goal 12 of the UN resolution and other goals. The main result of the publication was to draw attention to the need to correlate scientific research on the use of AI in creating sustainable business models with the expectations of business practice, which should strive for sustainable operations.

(Continued)

TABLE 7.2 (CONTINUED)
Analysis of the Five most Frequently Cited Works from the Analyzed set of Publications

Authors	Number of Citation	Conclusions from Detailed Text Analysis
Kristoffersen et al., 2020	**293**	In the article, the authors analyzed the importance of digital technologies in the development of the circular economy. The UN resolution and guidelines on sustainable development have contributed to increasing the emphasis on finding solutions that will allow for the optimal and sustainable use of resources while ensuring the growth of the economic potential of enterprises. Despite the importance of these topics, the authors identified, as a result of literature research, a significant research gap in this area. They diagnosed the lack of systematicity and uniformity in approaches, regulations, and implementation of digital technologies, which are also supposed to bring sustainable effects. The authors of the article, wishing to fill the above cognitive gap, conducted a systematic review of the literature. As a result of the research, they defined the concept and guidelines for creating a smart circular economy. The obtained research results include: • Defining the concept of a smart circular economy. • Creating a set of solutions for a unified "understanding" of the concept of ideas. • Identifying the cognitive gap between the current UN guidelines and the availability of its updates, primarily through the use of digital technologies. • Defined strategic activities that are necessary for the actual implementation of executive activities. The paper presents the research conducted by the authors, who aim to address the challenges associated with the deliberate use of digital technology in the creation and implementation of the Smart Circular Economy concept.

(Continued)

TABLE 7.2 (CONTINUED)
Analysis of the Five most Frequently Cited Works from the Analyzed set of Publications

Authors	Number of Citation	Conclusions from Detailed Text Analysis
Fatimah et al., 2020	275	The authors of the article identified and described the challenges faced by countries' economies as a result of implementing the principles of sustainable development, especially in the context of respect for natural resources and carrying out recycling activities. The authors referred the considerations presented in the article to the problems that Indonesia has to deal with. Describing Indonesia as a research example in the article, they defined a number of problems that must be dealt with by the economies of developing countries that have so far applied ineffective policies for the use of natural resources with a simultaneous lack of attention to recycling processes. For the purposes of the publication, the authors conducted a comprehensive systematic review of the literature, conducted intensive focus group discussions and direct observations in Indonesian cities. As a result of the identified problems and the research conducted, the authors defined the main goal to fill this empirical gap – to create a sustainable and intelligent nationwide waste management system using Industry 4.0 technologies. The solution proposed in the article includes: • Identification of key business processes related to waste management. • Logistics processes. • Processing and disposal processes. • Identification of technological solutions. • Possibilities of implementing digital technologies. • Modeling IoT solutions that collect, report, and monitor all processes included. The model solution proposed by the authors is directly in line with the implementation of the UN goals: good health and well-being (goal 3); clean water and sanitation (goal 6); economic growth and decent work (goal 8); responsible consumption and production (goal 12); and climate action (goal 13).

Source: Author's compilation

digital entrepreneurship, as well as its sustainable context, has still not been filled (Bhambri, 2024). The issue of the impact of the development of digital technologies on entrepreneurship is so broad that most publications are separate analyses of individual phenomena or problems that arise in this context (Rattan et al., 2005). This perspective is further broadened by the context of sustainable development, which, being a separate and important problem, determines many approaches to the issue defined.

As shown in the previous sections of the study, digitalization influences the creation of new business models that are the embodiment of digital entrepreneurship. Moreover, the creation of new business solutions often creates the groundwork for the creation of further business solutions that are a response to the needs created as a result of digital evolution.

7.3 SUSTAINABLE DEVELOPMENT AND DIGITAL TRANSFORMATION

7.3.1 Digital Transformation

Digital transformation is perceived as a digital revolution that has significantly changed not only the way organizations operate, but also significantly changes how we function in society today, work, learn, and spend our free time. The rapid development of various types of ICT solutions and their growing availability, as well as increasingly extensive functionality, is associated with the generation of new solutions, structures, and models that could provide a competitive advantage, and efficient and effective use of resources or support activities aimed at sustainable development (Shanmuga and Bhambri, 2024). The widespread use of digital technologies creates a new digital order in all spheres of social and economic life. The integration of digital technologies with business processes determines not only changes within the organization, including operating strategy and business processes, but also influences external systemic conditions regarding new market trends, competitive changes, and social attitudes, which include, among others, digital trust or technology adoption (Satalkina & Steiner, 2020). Digital transformation brings many benefits, including improved business models, communication channels, decision-making, and the growing possibility of remote work and learning (Diaz & Montalvo, 2022; Chomiak-Orsa & Martusewicz, 2023; Chomiak-Orsa & Smolag, 2024; Petri & Chomiak, 2024). On the other hand, digital transformation poses specific challenges to organizations and society, which primarily include digital competences (Bhambri & Bhandari, 2006). The need to develop new digital skills allows for the effective and efficient use of digital solutions that are constantly evolving. Digital competences are an important factor shaping the quality of life in the digital world of both individuals and entire organizations (Brunetti et al., 2020). It is worth emphasizing that digital transformation requires IT specialists and adapting their competences to market needs (Gudanowska et al., 2020; Szajt & Smoląg, 2020).

Digital transformation is multidisciplinary because, as Verhoef et al. emphasize, it involves changes in strategies, organization, information technology, supply chains, and marketing (Verhoef et al., 2021). The need for digital transformation is reinforced by several factors, including (Al Mashalah et al., 2022; Feroz et al., 2021; Verhoef et al., 2021):

- Development of ICT, which include primarily the internet, mobile technologies, cloud computing, Internet of Things, augmented reality (AR), Big Data, 3D printing, artificial intelligence, blockchain and process robotization.
- Development of e-commerce and the related intensity and global scope of competition.
- Changing consumer behavior.

Digital transformation brings fundamental changes not only in the functioning of organizations themselves, but also in society (Trzaska et al., 2021). Moreover, it is perceived as a process of introducing new business models and digital platforms (Verhoef et al., 2021; Feroz et al., 2021). This is what distinguishes digital transformation from digitalization, which concerns automated procedures and tasks such as the conversion of analog information into digital and adding digital components to product or service offerings (Feroz et al., 2021).

Digital transformation is included in the European program The Road to a Digital Decade, which sets digital goals for EU countries until 2030. These goals were divided into four areas:

- Strengthening digital skills and education.
- Secure and sustainable digital infrastructure.
- Digital transformation in business area.
- Digital processes in public services.

All identified areas are aimed at achieving digital transformation to empower citizens and businesses in line with EU values. The implementation of the program takes into account annual cooperation involving both the Commission and the Member States (https://commission.europa.eu/strategy-and-policy/priorities-2019-2024/europe-fit-digital-age/europes-digital-decade-digital-targets-2030_pl).

The next important step in Europe's digital transformation is the Declaration of Digital Rights and Principles, which was signed on December 5, 2022. Important elements of this declaration related to sustainable development include principles relating to the operation of digital entities in a responsible and safe manner and drawing attention to the fact that digital devices support sustainability and ecological transformation.

7.3.2 Digital Entrepreneurship versus Sustainable Development

The idea of sustainable development is that "the needs of the present generation can be met without compromising the opportunities of future generations to meet them" (https://www.gov.pl/web/rozwoj-technologia/zrownowatyny-rozwoj). The goal of sustainable development is to ensure economic development while protecting social and environmental balance (https://eur-lex.europa.eu/PL/legal-content/glossary/sustainable-development.html; Silvestre & Ţîrcă, 2019; Sulich & Zema, 2018). This requires implementation of economic goals related to making profits, which is necessary for the operation and success of the enterprise. Sustainable development assumes that these goals will be achieved in an ethical manner and with respect for the community in which the organization operates and the natural environment. In turn, social goals concern caring for other people, promoting diversity, equality, and fairness in the workplace or education. Environmental goals refer to minimizing the negative impact of human and organizational activities on the natural environment, which includes, among others, reducing greenhouse gas emissions, waste, and pollution, and efficient use of resources, e.g. water (https://www.kozminski.edu.pl/pl/review/zrownowazony-rozwoj-firm-na-czym-tak-naprawde-polega).

In 2015, the 2030 Agenda for Sustainable Development was developed by the leaders of 193 UN member countries. This document contains 17 sustainable development goals, which indicate the most important challenges divided into five areas: people, planet, prosperity, peace, and partnership (https://kampania17celow.pl/agenda-2030/; https://sdg.gov.pl/o_sdg/).

Sustainable development and digital transformation are an important determinant related to the functioning of organizations and entire societies in the modern world. Progressing digitalization, on the one hand, can significantly support the implementation of sustainable development goals, but on the other hand, it contributes to the degradation and intensification of the use of natural resources. An example of this would be the use of ICT to improve the quality and availability of educational services. ICT has influenced the development of remote services and changed the approach to learning processes and educational needs. It has contributed to the creation of new business models – offering remote educational services, the provision of which does not require material resources such as lecture rooms, and may significantly increase the availability of this type of service (Chomiak-Orsa & Smoląg, 2024). On the other hand, however, the growth of social problems resulting from the intensification of the feeling of isolation is accelerating as remote educational services devalue the creation of interpersonal bonds. This also applies to enterprises and the increasingly popular remote work. Moreover, as mentioned earlier, digitalization forces the development of digital competences and skills, which is crucial for achieving the sustainable development goals (Diaz & Montalvo, 2022).

In turn, digital infrastructure is related to the production of electronic devices, which are manufactured using expensive, often rare raw materials, some of which cannot be recycled and pose a huge challenge to the idea of sustainable development. The issue of disposal of electronic equipment is a huge problem for the environment, because electrical and electronic equipment waste is one of the fastest growing waste streams, of which less than 40% is recycled. (https://www.europarl.europa.eu/news/pl/headlines/society/20201208STO93325/zuzyty-sprzet-elektryczny-i-elektroniczny-w-ue-fakty-i-liczby-infografika).

REFERENCES

Abdurazzakov, O., Illés, B. C., Jafarov, N., & Aliyev, K. (2020). The impact of technology transfer on innovation. *Polish Journal of Management Studies*, 21(2), 9–23. https://doi.org/10.17512/pjms.2020.21.2.01.

Ahmad, A., Alshurideh, M., Al Kurdi, B., Aburayya, A., & Hamadneh, S. (2021). Digital transformation metrics: A conceptual view. *Journal of management Information and Decision Sciences*, 24(7), 1–18.

Al Mashalah, H., Hassini, E., Gunasekaran, A., & Bhatt, D. (2022). The impact of digital transformation on supply chains through e-commerce: Literature review and a conceptual framework. *Transportation Research Part E: Logistics and Transportation Review*, 165, 102837. https://doi.org/10.1016/j.tre.2022.102837.

Bhambri, P. (2024). Wallets and transactions. In D. Darwish (Ed.), *Decentralizing the Online Experience with Web3 Technologies* (pp. 90–106). IGI Global. https://doi.org/10.4018/979-8-3693-1532-3.ch004.

Bhambri, P., & Bhandari, A. (2006). Different protocols for wireless security. In National Conference on Advancements in Modeling and Simulation at LLRIET, Moga held during January 20–21, 2006.

Bhambri, P., Gupta, S., & Bhandari, A. (2005). Soft computing techniques. In UGC Sponsored National Conference (Abstract Book) (p. 100). Presented at the "Emerging Computing Technologies" conference at HRMMV, Jalandhar held during September 29–30, 2005.

Brunetti, F., Matt, D. T., Bonfanti, A., De Longhi, A., Pedrini, G., & Orzes, G. (2020). Digital transformation challenges: Strategies emerging from a multi-stakeholder approach. *The TQM Journal*, 32(4), 697–724.

Chomiak-Orsa, I., & Martusewicz, J. (2023). Creating good practice in effective sustainability management by implementing the EFQM model. Procedia *Computer Science*, 225, 3517–3526. https://doi.org/10.1016/j.procs.2023.10.347.

Chomiak-Orsa, I., & Smolag, K. (2024). Remote learning technologies in achieving the fourth sustainable development goal. In S. Nowaczyk, et al. Artificial Intelligence. ECAI 2023 International Workshops. ECAI 2023. Communications in Computer and Information Science, vol 1948. Cham: Springer. https://doi.org/10.1007/978-3-031-50485-3_12.

Diaz, R., & Montalvo, R. (2022). Digital transformation as an enabler to become more efficient in sustainability: Evidence from five leading companies in the Mexican market. *Sustainability*, 14(22), 15436. https://doi.org/10.3390/su142215436.

Di Vaio, A., Palladino, R., Hassan, R., & Escobar, O. (2020). Artificial intelligence and business models in the sustainable development goals perspective: A systematic literature review. *Journal of Business Research*, 121, 283–314.

Fatimah, Y. A., Govindan, K., Murniningsih, R., & Setiawan, A. (2020). Industry 4.0 based sustainable circular economy approach for smart waste management system to achieve sustainable development goals: A case study of Indonesia. *Journal of Cleaner Production*, 269, 122263.

Feroz, A. K., Zo, H., & Chiravuri, A. (2021). Digital transformation and environmental sustainability: A review and research agenda. *Sustainability*, 13(3), 1530. https://doi.org/10.3390/su13031530.

Gudanowska, A., Kononiuk, A., Siderska, J., & Dębkowska, K. (2020). Uwarunkowania ucyfrowienia procesów produkcji i wzrostu kompetencji cyfrowych społeczeństwa. Oficyna Wydawnicza Politechniki Białostockiej. https://doi.org/10.24427/978-83-66391-23-9.

Haseeb, M., Hussain, H. I., Ślusarczyk, B., & Jermsittiparsert, K. (2019). Industry 4.0: A solution towards technology challenges of sustainable business performance. *Social Sciences*, 8(5), 154. https://doi.org/10.3390/socsci8050154.

Hudaszek, K., Chomiak-Orsa, I., & AL-Dobai, S. A. M. (2024). Green hardware infrastructure for algorithmic trading. W S. Nowaczyk, P. Biecek, N. C. Chung, M. Vallati, P. Skruch, J. Jaworek-Korjakowska, S. Parkinson, A. Nikitas, M. Atzmüller, T. Kliegr, U. Schmid, S. Bobek, N. Lavrac, M. Peeters, R. van Dierendonck, S. Robben, E. Mercier-Laurent, G. Kayakutlu, M. L. Owoc, ... V. Dimitrova (Red.), *Artificial Intelligence. ECAI 2023 International Workshops* (s. 194–200). Springer Nature Switzerland. https://doi.org/10.1007/978-3-031-50485-3_19.

Jelonek, D., & Chomiak-Orsa, I. (2018). The application of ICT in the area of value co-creation mechanisms support as a determinant of innovation activities. *International Journal of Ambient Computing and Intelligence*, 9(2), 32–42. https://doi.org/10.4018/IJACI.2018040103.

Kristoffersen, E., Blomsma, F., Mikalef, P., & Li, J. (2020). The smart circular economy: A digital-enabled circular strategies framework for manufacturing companies. *Journal of Business Research*, 120, 241–261.

Li, Y., Dai, J., & Cui, L. (2020). The impact of digital technologies on economic and environmental performance in the context of industry 4.0: A moderated mediation model. *International Journal of Production Economics*, 229, 107777.

Liszczyk, K., & Chomiak-Orsa, I. (2024). The importance of eco-commerce in the context of sustainable development: A case study analysis. W S. Nowaczyk, P. Biecek, N. C. Chung, M. Vallati, P. Skruch, J. Jaworek-Korjakowska, S. Parkinson, A. Nikitas, M. Atzmüller, T. Kliegr, U. Schmid, S. Bobek, N. Lavrac, M. Peeters, R. van Dierendonck, S. Robben, E. Mercier-Laurent, G. Kayakutlu, M. L. Owoc, ... V. Dimitrova (Red.), *Artificial Intelligence. ECAI 2023 International Workshops* (s. 223–229). Springer Nature Switzerland. https://doi.org/10.1007/978-3-031-50485-3_23.

Małowiecki, A., & Chomiak-Orsa, I. (2024). Fairlearn parity constraints for mitigating gender bias in binary classification models – Comparative analysis. W S. Nowaczyk, P. Biecek, N. C. Chung, M. Vallati, P. Skruch, J. Jaworek-Korjakowska, S. Parkinson, A. Nikitas, M. Atzmüller, T. Kliegr, U. Schmid, S. Bobek, N. Lavrac, M. Peeters, R. van Dierendonck, S. Robben, E. Mercier-Laurent, G. Kayakutlu, M. L. Owoc, ... V. Dimitrova (Red.), *Artificial Intelligence. ECAI 2023 International Workshops* (s. 148–154). Springer Nature Switzerland. https://doi.org/10.1007/978-3-031-50485-3_13

Nižetić, S., Šolić, P., Gonzalez-De, D. L. D. I., & Patrono, L. (2020). Internet of Things (IoT): Opportunities, issues and challenges towards a smart and sustainable future. *Journal of Cleaner Production*, 274, 122877.

Petri, M., & Chomiak, I. (2024). Towards a sustainable future: The use of electrical energy in smart cities. W S. Nowaczyk, P. Biecek, N. C. Chung, M. Vallati, P. Skruch, J. Jaworek-Korjakowska, S. Parkinson, A. Nikitas, M. Atzmüller, T. Kliegr, U. Schmid, S. Bobek, N. Lavrac, M. Peeters, R. van Dierendonck, S. Robben, E. Mercier-Laurent, G. Kayakutlu, M. L. Owoc, ... V. Dimitrova (Red.), *Artificial Intelligence. ECAI 2023 International Workshops* (s. 216–222). Springer Nature Switzerland. https://doi.org/10.1007/978-3-031-50485-3_22

Rattan, M., Bhambri, P., & Shaifali. (2005). Institution for a sustainable civilization: Negotiating change in a technological culture. In National Conference (Abstract Book) (p. 7). Presented at the "Technical Education in Globalized Environment- Knowledge, Technology & The Teacher" conference at SBBSIET, Padhiana, Jalandhar held during February 25–26, 2005.

Rutkowska, M., & Sulich, A. (2020). Green Jobs on the background of Industry 4.0. *Procedia Computer Science*, 176, 1231–1240. https://doi.org/10.1016/j.procs.2020.09.132.

Satalkina, L., & Steiner, G. (2020). Digital entrepreneurship and its role in innovation systems: A systematic literature review as a basis for future research avenues for sustainable transitions. *Sustainability,* 12(7), 2764. https://doi.org/10.3390/su12072764.

Shanmuga, S. M., & Bhambri, P. (2024). Bone marrow cancer detection from leukocytes using neural networks. In P. Bhambri, S. Rani, & M. Fahim (Eds.), *Computational Intelligence and Blockchain in Biomedical and Health Informatics* (pp. 307–319). CRC Press, Taylor & Francis Group. https://doi.org/10.1201/9781003459347.

Sharma, R., Bhambri, P., & Sohal, A. K. (2020). Energy bio-inspired for MANET. *International Journal of Recent Technology and Engineering*, 8(6), 5580–5585. https://www.ijrte.org/wp-content/uploads/papers/v8i6/F8522038620.pdf. https://doi.org/10.35940/IJRTE.F8522.038620.

Silvestre, B. S., & Ţîrcă, D. M. (2019). Innovations for sustainable development: Moving toward a sustainable future. *Journal of Cleaner Production*, 208, 325–332.

Šimelytė, A., Tvaronavičienė, M., Holmen, R. B., Burinskas, A., & Razminienė, K. (2021). Knowledge and technology transfer as driving force for social innovations. *Polish Journal of Management Studies*, 23(2), 512–536. https://doi.org/10.17512/pjms.2021.23.2.31.

Ślusarczyk, B., Tvaronavičienė, M., Haque, A. U., & Oláh, J. (2020). Predictors of Industry 4.0 technologies affecting logistic enterprises' performance: International perspective from economic lens. *Technological and Economic Development of Economy*, 26(6), 1263–1283. https://doi.org/10.3846/tede.2020.13376.

Sulich, A., & Zema, T. (2018). Green jobs, a new measure of public management and sustainable development. *European Journal of Environmental Sciences*, 8(1), 69–75. https://doi.org/10.14712/23361964.2018.10

Szajt, M., & Smoląg, K. (2020). Competences of engineers 4.0 and their knowledge on the Industry 4.0 conception. In *Education Excellence and Innovation Management: A 2025 Vision to Sustain Economic Development during Global Challenges* (red.) Soliman Khalid S.

Trzaska, R., Sulich, A., Organa, M., Niemczyk, J., & Jasiński, B. (2021). Digitalization business strategies in energy sector: Solving problems with uncertainty under industry 4.0 conditions. *Energies*, 14(23), 7997.

Verhoef, P. C., Broekhuizen, T., Bart, Y., Bhattacharya, A., Dong, J. Q., Fabian, N., & Haenlein, M. (2021). Digital transformation: A multidisciplinary reflection and research agenda. *Journal of Business Research*, 122, 889–901. https://doi.org/10.1016/j.jbusres.2019.09.022.

Vial, G. (2019). Understanding digital transformation: A review and a research agenda. *The Journal of Strategic Information Systems,* 28(2), 118–144.

Xu, G., Hou, G., & Zhang, J. (2022). Zrównoważona przedsiębiorczość cyfrowa: perspektywa możliwości cyfrowych poprzez orientację na innowacje cyfrowe w celu tworzenia wartości społecznej i środowiskowej. *Zrównoważony rozwój,* 14(18), 11222. https://doi.org/10.3390/su141811222.

8 Blockchain and Sustainable Supply Chains

Dariusz Dudek

8.1 INTRODUCTION

In today's global world, where environmental concerns and ethical business practices are becoming increasingly important, supply chains play a vital role in facilitating the movement of both products and services from manufacturers to final clients. They constitute a complex and extensive system consisting of many actors, processes, and stages that must be managed effectively to ensure operational and economic fluidity. However, traditional supply chains often face a number of challenges, such as lack of transparency and exposure to counterfeiting and fraudulent practices (Zheng et al. 2019).

Current economies place increasing emphasis on environmental consciousness and ethical accountability, which require innovative solutions to increase the supply chain effectiveness (Seuring, Müller 2008). In this context, blockchain technology is emerging as a promising tool that can help break through these constraints and provide efficiency improvements in sustainable supply chains (Ivanov 2020).

This chapter explores the blockchain technology integration in supply chains in the context of sustainability. Analyzing present trends, benefits, and challenges related to the use of blockchain technology in supply chains, the chapter will enable a better understanding of the technology's contribution to enhancing the effectiveness and openness of global supply chains.

8.2 THE MAIN CHALLENGES IN SUPPLY CHAIN MANAGEMENT

To function effectively and maintain competitiveness in the market, companies are forced to face new challenges that require constant monitoring, innovation, and cooperation between various supply chain participants. Among the main issues facing sustainable supply chains are various types of disruptions, lack of transparency, and the need to reduce over-consumption of natural resources.

DOI: 10.1201/9781003484226-8

8.2.1 Disruptions to the Sustainable Supply Chain

Over 56% of companies around the world experience supply chain disruptions annually (BCI-Business Continuity Institute 2019), which puts them in a very difficult situation (Katsaliaki et al. 2022) and sometimes even prevents them from functioning completely. Global supply chain networks are particularly vulnerable to high levels of turbulence and uncertainty. The 2020 Covid-19 outbreak closed international borders and closed many key facilities, revealing the weakness of global supply chains (Esteve, Alvarez 2021). Many organizations have realized that the resilience of the supply chain to disruptions is very important, and many enterprises have begun reconfiguring their supply chains to be ready for new types of disruptions (Carr, Ribeiro-Soriano 2020).

One of the primary obstacles in the context of sustainable supply chain management is identifying potential threats to sustainable development and developing a strategy for managing them. Research on this topic is paying increasing attention to identifying and managing disruptions and understanding the influence of sustainable growth on the flexibility and resilience of the supply chain. It is also crucial to take into account the role of technology and advancement in the process of improving supply chain management. Technological development, including the growing importance of the Internet of Things (IoT), Artificial Intelligence (AI) and blockchain technology, creates both new opportunities and new challenges for management of supply chains (Ivanov, Dolgui 2020). In the current conditions, companies must skillfully use new technologies in supply chain management (Kim, Lee 2022) to adapt them to increase the supply chain's resistance to various types of disruptions (Choi, Wong 2020).

8.2.2 Visibility in Supply Chains

With the turbulent and changing market environment and the growing complexity of supply chains, a new concept focusing on supply chain transparency has emerged. The aim is to improve decision-making, respond effectively to challenges (Machado et al. 2021), and be able to provide more information about their own supply chains (Schäfer 2023). Organizations are adopting supply chain transparency practices to reap the benefits of increasing product and company trust, and strengthening customer loyalty (Pinnington et al. 2023). This is because a supply chain based on timely and accurate data improves its performance and minimizes the risk of disruptions and quality problems.

Supply chain transparency comprises two elements: visibility and disclosure. Visibility entails the precise identification and gathering of data from every link within the supply chain (Singh et al. 2004). Disclosure, conversely, involves conveying this information, both internally and externally, at the necessary or desired level of detail. In light of reports concerning forced labor, food contamination, and human rights violations, consumers are concerned that their purchasing decisions have a major impact on their communities, the world, and above all, their health (Khan et al. 2022).

Governments have instituted new regulations mandating companies to conduct more thorough assessments of their suppliers (Shanmuga, Bhambri 2024). In addition to consumer worries and regulatory mandates, a deeper comprehension of the supply chain leads to improved management and adherence to corporate principles and values, supporting the process of identifying carbon footprints throughout the supply chain.

To ensure full transparency, companies must consider every point in the supply chain, from sourcing inputs and transporting goods to finished products and services as they are delivered to customers (Gardner et al. 2019). Transparent supply chains provide managers and decision-makers with real-time information about inventory levels, production schedules, and customer demand. With accurate data, companies can easily identify bottlenecks and make informed decisions, which ultimately increases operational efficiency (Jahani et al. 2023). By sharing information about suppliers and their operating practices, companies can also identify potential risks related to quality, sustainability, and compliance (Weber et al. 2020). This allows them to reduce disruptions in supply chains, ensuring on-time deliveries and possibly mitigating potential risks.

8.2.3 The Excessive Consumption of Natural Resources

Excessive consumption of natural resources poses a significant challenge to creating and maintaining sustainable supply chains (United Nations 2015). Its consequences are multidimensional and include environmental, social, and economic aspects (Bell 2023), so it is important to identify the factors contributing to overconsumption and propose potential solutions.

Although organizations try to decrease their environmental footprint and meet the expectations of stakeholders (Borja et al. 2019), we are still dealing with excessive consumption of natural resources, which leads to serious environmental degradation (United Nations Environment Program 2020). Cutting down trees for the needs of the paper industry and agriculture (Bell 2023) not only destroys the habitats of many animal and plant species but also causes climate change, limiting the ability to sequester carbon dioxide (World Wildlife Fund 2021). Moreover, the extraction of minerals and fossil fuels pollutes the air, water, and soil, causing irreversible damage to entire ecosystems (Bell 2023).

The depletion of natural resources also has profound social consequences (Singh et al. 2005). Excessive consumption not only deepens social inequalities but also threatens the cultural heritage of indigenous peoples, thus leading to the displacement of indigenous peoples (Prasad, Grint 2019). Moreover, the exploitation of natural resources can fuel conflict and disrupt social harmony as limited resources become objects of desire and contention (Mol, Carter 2019).

Excessive consumption of natural resources also poses serious economic challenges for supply chains. As resources become scarcer, their prices increase, which negatively affects production costs. This, in turn, affects the affordability of goods and services, thereby reducing consumers' purchasing power (Attinasi et al. 2021).

To ensure the right conditions, supply chains must adopt more sustainable practices that reduce resource dependence and promote circular economy principles.

Consumer demand also contributes to the overconsumption of natural resources in supply chains, driving production and leading to overexploitation. Marketing efforts often emphasize rapid product obsolescence and unrelenting consumerism, which further fuels this cycle (Bauer et al. 2019). In turn, poor governance allows companies to exploit legal loopholes and conduct environmentally harmful activities with impunity. Finally, short-term profit maximization often takes precedence over long-term sustainability, leading to short-sighted decision-making in supply chains (Hall et al. 2018).

Solving the problem of excessive consumption of natural resources requires a multifaceted approach (Rana, Bhambri 2024). First, increased awareness and education about sustainable consumption and production is essential. Governments, non-governmental organizations, and businesses must work together to promote environmental awareness among consumers. Second, stringent regulations and enforcement mechanisms need to be implemented to curb unsustainable practices. Governments should encourage companies to adopt sustainable activities, while imposing restrictions on companies that do not take environmental issues into account (Jain, Bhambri 2006). There is also a need for widespread adoption of circular economy rules that are focused on maximizing the use of resources while minimizing waste and promoting sustainable supply chains.. Achieving a balance between economic growth and the well-being of our planet is a key element for achieving the long-term viability of supply chains and the well-being of future generations.

Modern enterprises face a number of challenges in supply chain management that affect their efficiency, competitiveness, and ability to adapt to changing market conditions. To meet these requirements, enterprises should skillfully implement effective management methods and concepts. It is also important to identify and solve various challenges related to supply chain oversight, as well as to adapt the management methods used to the specificity of a given enterprise.

8.3 CRITICAL ATTRIBUTES OF BLOCKCHAIN TECHNOLOGY

Supply chains are the backbone of modern economies, connecting manufacturers, suppliers, distributors, and retailers to ensure service efficiency and the smooth flow of goods. However, this dependency comes with a number of challenges such as lack of transparency, low efficiency, and potential fraud. Many authors also emphasize the urgent need to solve these problems due to the growing demand for sustainable and ethical practices (Duan et al. 2022).

Due to the involvement of various stakeholders in the supply chain, it becomes virtually unattainable to maintain complete and accurate documentation of the route of goods from production to consumption. A lack of transparency not only makes it difficult to ensure accountability, but it also makes it harder to identify and fix any problems. To solve these problems, blockchain technology offers a novel approach. Based on the principles of decentralization and transparency, blockchain provides solutions where all stakeholders have real-time equal access to information (Tapscott 2016), while limiting counterfeiting and unethical practices.

8.3.1 Characteristics of Blockchain Technology

Blockchain technology is a way of structuring data invented by the creator of Bitcoin, Satoshi Nakamoto (Nakamoto 2008). Blockchain is referred to as distributed ledgers or distributed ledger. It is a kind of electronic list in which all transactions between users are recorded chronologically. Blockchain is an innovative technology that is based on blocks of data containing encrypted information about transactions made on the internet (Nakamoto 2008). Each block comprises a distinct number of transactions and is linked chronologically to previous blocks, creating an irrefutable chain (Tapscott 2016). The process of creating subsequent blocks involves finding the appropriate padding value so that the hash of the entire block meets certain criteria (Antonopoulos 2014). Once a block is completed, secure code is created to connect it to subsequent blocks.

Blockchain is a decentralized and distributed system, which means that there is no central entity controlling all data (Swan 2015). Each user has an identical copy of the database, and new blocks are added according to the consensus algorithm adopted by the network (Figure 8.1). This makes the system resistant to cyberattacks because it is more difficult for hackers to access multiple user accounts than to compromise a single database (Narayanan et al. 2016).

Blockchain can store a variety of information such as commercial transactions, purchases, and currency sales (Prusty 2018). It is also the basis for cryptocurrency technologies such as Bitcoin and Ethereum. Thanks to its structure and properties, blockchain opens up new possibilities in the field of secure exchange of information and values on the internet.

Blockchain technology offers an irreversible record of data in a manner that renders falsification impossible. This particular type of decentralized database requires time-stamped blocks and transactions and is, therefore, ideal for collecting important information about economic events, asset, or cash transfers. The data is cryptographically secured and the transactions are impossible to forge (Rattan et al. 2005). Once sent, changes cannot be made. Data that has already been stored in a block cannot be modified without authorization, because instead of changing one block, a change would have to be made to the entire chain.

8.3.2 Smart Contracts

Smart contracts are independent agreements that automatically execute and enforce transactions according to predefined conditions. By eliminating the need for intermediaries, smart contracts improve efficiency, reduce costs, and minimize the potential for errors or disputes. By increasing transparency, efficiency, and environmental impact, smart contracts can help create more resilient and responsible supply chains that meet the needs of current and future generations. By implementing these solutions into the supply chain, we can verify the authenticity of products or ensure compliance with sustainable development standards (Iansiti, Lakhani 2017).

Smart contracts can also help improve the efficiency of supply chain operations, while offering significant cost reductions (Kim, Laskowski 2019). For example, smart contracts can streamline the payment process between suppliers and manufacturers by automatically releasing funds after goods are delivered and verified (Groschopf et al. 2021). Using real-time data from IoT devices and external sources, smart contracts can minimize a carbon footprint of supply chains by optimizing the route of goods or reducing unused transport capacity (Saini et al. 2023). A smart contract can also be programmed to reward consumers for returning used products for recycling, thereby promoting closed-loop supply chain models and reducing the consumption of primary resources (Upadhyay et al. 2021).

8.4 THE APPLICATION OF BLOCKCHAIN TECHNOLOGY IN ECO-FRIENDLY SUPPLY CHAINS

Modern supply chains often involve multiple suppliers, manufacturers, warehouses, transportation, and sales outlets around the world. They become difficult to manage due to their complexity and geographical dispersion. A big threat is the variability of demand for products, which can lead to inefficiencies in supply chains such as overstocks or out-of-stocks. In dynamic supply chains, particular emphasis should be placed on assessing supply risks, caused by, among others, climate disasters, political disruptions, international trade issues, or other factors causing supply disruptions or delays. One of the most important challenges is the increasing requirements regarding the environment and sustainable development. Increasing pressure from customers and their expectations for a company to operate in alignment with the sustainable development principles will significantly influence the design of supply chains and the selection of suppliers, which will require effective management and innovation, concepts, and strategies.

Consumers are increasingly reaching for environmentally friendly products from ecological sources (Sun, Shi 2022), forcing companies to comply with ethical practices and regulatory compliance. Therefore, tracing the route and origin of products is of great importance. Visibility and transparency across the supply chain also fosters greater stakeholder trust, which is key to building long-term relationships with suppliers, customers, and regulators (The Morningstar Sustainalytics 2022).

Strategies to achieve clarity and transparency in the supply chain require close cooperation and information exchange through access to open communication channels. Seamless data exchange can be achieved through modern technology and collaborative platforms, thereby promoting transparency across organizational boundaries (Tse, Chiu 2018). Advanced technologies such as blockchain, combined with the Internet of Things (IoT) and Big Data, enable the recording and analysis of huge amounts of supply chain data. These technologies increase visibility and traceability, ensuring transparency of the entire supply chain through real-time monitoring (Xu et al. 2017).

The integration of management methods combined with digitalization and the ongoing development of technology is revolutionizing supply chain management by offering solutions that enable immediate identification and elimination of emerging disruptions (Obora 2020).

Thanks to the utilization of blockchain technology, it is feasible to offer a transparent and verifiable record of transactions through real-time insight into the supply chain (Tapscott 2016). Since information entered into the blockchain cannot be modified or manipulated, it ensures data integrity and accuracy. This makes it possible to completely eliminate intermediaries and decrease the threat of deception and unethical practices. Moreover, blockchain can significantly increase trust and cooperation between supply chain participants, leading to more efficient and reliable operation (Iansiti, Lakhani 2017). Blockchain integration in supply chains can also contribute to achieving sustainable development goals by promoting environmental responsibility. By using blockchain, companies can not only support the sustainable origin of raw materials, but also influence greater compliance with regulations, responsible mining, and fair trade. Additionally, by introducing smart contracts into the supply chain, various processes can be completely automated, eliminating errors and speeding up operations for all parties involved.

Another aspect that contributes to the sustainable development of supply chains using blockchain technology is process automation, thanks to smart contracts. Smart contracts as autonomous programs can perform specific actions when certain conditions are met (Ethereum Whitepaper 2023), which leads to a reduction in bureaucracy and human errors, as well as an increase in efficiency. Smart contracts can fully and automatically monitor product storage and transportation conditions, issuing an alert if certain parameters are violated. This avoids product waste and ensures that appropriate quality standards and the sustainability of the supply chain are maintained.

Managing a sustainable supply chain is becoming an increasingly complex task, requiring consideration of a broad spectrum of factors and challenges. A significant issue is to understand how sustainable development can affect various aspects of companies' operations and the interconnections among partners in the supply chain (Smith 2019). The transition to more sustainable business practices may lead to changes in supply models, the selection of new suppliers or transport methods, which in turn may affect the flexibility and resilience of the entire supply chain (Johnson, Wang 2020).

Implementing blockchain technology into physical business processes in an enterprise can be an extremely difficult and labor-intensive activity, but the benefits in the form of streamlined operations, increased trust, and customer loyalty can make enterprises actively implement these solutions.

8.5 EXAMPLES OF BLOCKCHAIN-BASED SUSTAINABLE SUPPLY CHAIN PROJECTS

In response to the challenges of ensuring sustainable development, more and more companies have begun to explore the potential applications of blockchain technology to improve management of supply chains. Sustainable supply chains are crucial in the fight against climate change, as global trade currently accounts for over 70% of global carbon emissions (International Energy Agency 2023). Thus, the products we consume every day are responsible for the greatest damage to our environment, as well as for the ongoing economic marginalization of farmers and producers.

The case studies presented below show how blockchain can be used to enhance transparency, traceability, and ethical practices in supply chains, ultimately contributing to a more sustainable and responsible global economy.

8.5.1 An Innovative XCEED Platform in Renault Car Production

Currently produced cars should meet not only rigorous safety standards, but also environmental protection standards. Meeting all required standards and regulations requires tracking huge amounts of information from many sources, because the average supply chain in the car manufacturing industry includes approximately 30,000 parts, which additionally come from hundreds of suppliers.

Over the past few years, Renault has made a significant technological leap by implementing blockchain technology into its extensive automotive supply chain ecosystem. Car manufacturers must keep up with constantly evolving standards to comply not only with current regulations but also to prepare for future requirements and standards such as cybersecurity or vehicle and parts recycling.

The use of blockchain technology for supply chain management has enabled Renault to offer real-time compliance certification to partners, regulators, and customers. Decentralized ledger technology enables the sharing and tracking of information among multiple users, providing access control and complete visibility, while maintaining the confidentiality of their data for each party to the transaction. Users and transactions undergo verification and are stored by blockchain, which not only accelerates the exchange of information and increases efficiency, but also creates a network of trust between all transaction participants (IBM 2024).

8.5.2 Blockchain in Pharmaceutical Supply Chains

New regulations aimed at enhancing the security of the pharmaceutical supply chains in the United States forced the search for new solutions that would increase transparency and trust through the capacity to verify, track, and distribute prescribed medications among all its participants.

Actions taken by four organizations (IBM, KPMG, Merck, and Walmart) resulted in a solution based on blockchain technology (FDA DSCSA 2020). Blockchain technology, with its shared ledger, immutable data, and inherent traceability of drugs, has the potential to tackle the myriad challenges of the intricate pharmaceutical supply chain (Wilson 2015). An important issue was the implementation of a system that would enable quick warning of partners in the supply chain in the event of a drug withdrawal, along with detailed identification of the batch affected by the problem. Until now, the notification process was largely manual and fragmented, using separate systems, which increased response times to drug recalls. Moreover, the process itself was extremely expensive and time-consuming due to the lack of uniform warning standards among business partners. In the case of a product recall, enterprises had to depend on cooperation to swiftly identify the recalled product and notify partners downstream in the supply chain. The use of blockchain technology and the specificity provided by serial product marking allows for the eliminatation of unnecessary communication and prevents the quarantine of an important product, which ultimately leads to a reduction in the amount of pharmaceutical waste.

Tests have demonstrated that blockchain is the optimal technology to fulfill the requirements of the pharmaceutical supply chain. Blockchain furnishes a unified record of product movement, linking diverse organizations and systems, and enhances patient safety by triggering product alerts and enhancing visibility to pertinent supply chain partners during product investigations or recalls. The implementation of blockchain technology provides improved recall functionality and accelerates the communication process associated with notifying downstream partners about problematic inventories from three days to just ten seconds (FDA DSCSA 2020).

8.5.3 Bext360

Founded in 2016, Bext360 uses blockchain, Artificial Intelligence, and IoT to digitize global supply chains to improve sustainable practices for communities, consumers, and the environment. The mission of the company is to ensure transparency in global supply chains, enabling clients to know the exact origin and production standard of a given product. Bext360 focusses on supply chains for commodities such as fisheries, coffee, minerals, lumber, palm oil, and cotton to ensure that items are systematically tracked and monitored across the whole supply chain, from the producers to the final customers. The Bext360 blockchain software was launched to protect product manufacturers and suppliers from injustice by enabling product identification via an electronic token. This unique identifier allows the product to be tracked throughout the supply chain. After harvest, growers scan their products using image recognition technology and the Internet of Things and then assess their quality, assigning them a fair price (Bext360 2024). This information is then passed on to potential buyers in the supply chain so they can take appropriate action. Thanks to this solution, growers can receive payment in advance.

8.5.4 IBM Food Trust

IBM Food Trust is a blockchain-powered platform designed to enhance transparency and traceability in food supply chains. By using blockchain, the platform enables suppliers, retailers, and consumers to trace the path of food items from producer to consumer. Each transaction recorded on the blockchain is immutable, meaning it cannot be altered or removed, guaranteeing the reliability of the data and the assurance that goods are produced sustainably and meet certain quality and safety standards.

The main task for the Food Trust system is to improve companies' ability to identify food recall issues, detect outbreaks more quickly, and reduce customer risk (Nash 2018). Application of the solution at Walmart Inc. allowed for reduction of the time needed to track products from 6 days and 18 hours down to 2.2 seconds. Additionally, implementing a system based on blockchain technology guarantees high product quality, saves money, and promotes responsibility by quickly detecting and disposing of contaminated food before it reaches the consumer.

8.5.5 Devoleum Platform for Tracking the Food Chain

The Devoleum platform is used to track and manage the product supply chain, especially in the food sector. It uses blockchain technology to ensure transparency and security, enabling the registration of every transaction and every stage of production. Thanks to this, clients can trace the origin of products and companies can quickly respond to possible quality problems or threats. Devoleum also helps in the fight against counterfeits and supports work on minimizing waste in the supply chain. Devoleum utilizes interconnected open data from physical or digital supply chains, crafting narratives about products and illustrating the steps that contributed to their uniqueness and value in a clear and detailed manner (Mercuri et al. 2021). Devoleum is an open-source project enabling anyone to monitor the supply chain of companies within the agri-food sector. This sector was specifically selected due to its high sensitivity among customers regarding the disclosure of sustainability information (Saurabh, Dey 2021).

8.5.6 Provenance Application for Tracking the Supply Chain in the Cosmetics Sector

The Provenance platform is an innovative application based on blockchain technology that is changing the way consumers and producers perceive the supply chain (Provenance 2024). Operating mainly in the pharmaceutical and food sectors, it enables tracking of products from the manufacturer to the end customer. The tool offers consumers the ability to scan QR codes or other identifiers to obtain detailed information on the origin, quality, and compliance of goods with safety standards. Thanks to blockchain technology, Provenance ensures data immutability, which eliminates the risk of forgery and enables the sharing of reliable information about products. This tool not only ensures trust between customers and producers through transparency in the supply chain but also allows for faster response to possible quality problems or threats. As a result, Provenance contributes to improving safety standards, product quality, and consumer trust in brands.

SUMMARY

Given the highly dynamic and globalized business environment, supply chain management poses many challenges to companies that require innovative solutions and a high degree of flexibility. For this reason, companies need to constantly monitor changing market conditions, invest in cutting-edge technologies, and develop management strategies to effectively manage their supply chains and remain competitive in the market.

Integrating blockchain technology into supply chains can revolutionize the way companies operate, especially in terms of sustainability. By increasing transparency, reducing inefficiencies and promoting environmental responsibility, blockchain can contribute to achieving global sustainability goals.

The integration of blockchain technology can have a beneficial effect on the efficiency of enterprises and their supply chains. In this context, blockchain can be a

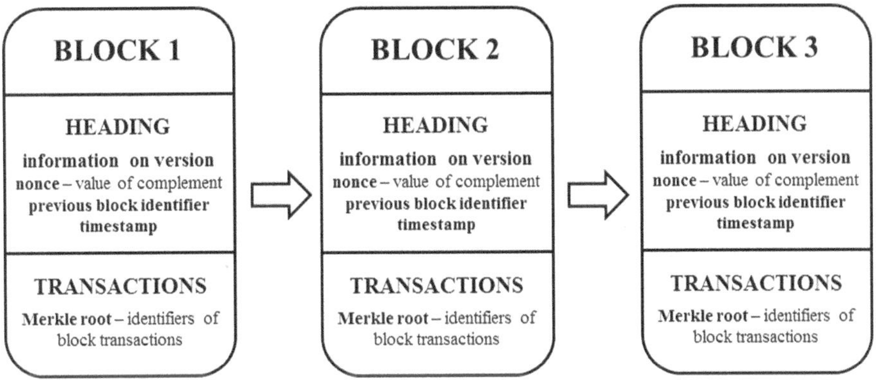

FIGURE 8.1 Simplified structure of blocks in blockchain technology. Source: Author analysis compiled using Nakamoto 2008.

catalyst for the advancement of innovative, sustainable supply chains, becoming, in combination with AI, the Internet of Things, Big Data and Cloud Computing, a key factor influencing the implementation of sustainable development goals. Blockchain can be highly advantageous as a coordination tool, facilitating the connection of multiple distributed databases. This technology offers a new dimension by providing transparency, which, when combined with the functionality of smart contracts and tokenization, can aid in achieving the sustainability objectives mentioned earlier. Moreover, the lack of intermediaries in blockchain technology helps to minimize transaction costs. Promoting a new business model can help reduce waste and offer sustainable products. This increases social responsibility, which may be a crucial component in attaining sustainable development goals.

However, to fully exploit the advantages of blockchain technology in sustainable supply chains, key challenges related to scalability, interoperability and regulation must be overcome. With continued progress and concerted efforts from academia, industry, and policymakers, blockchain has the potential to transform supply chains into more transparent, efficient and sustainable ecosystems (Figure 8.1).

REFERENCES

Antonopoulos, A. M. (2014). *Mastering Bitcoin: Unlocking Digital Cryptocurrencies.* O'Reilly Media.

Attinasi, M. G., Balatti, M., Mancini, M., Metelli, L. (2021) Supply chain disruptions and the effects on the global economy. *ECB Economic Bulletin*, Issue 8/2021. https://www.ecb.europa.eu/pub/economic-bulletin/focus/2022/html/ecb.ebbox202108_01~e8ceebe51f.en.html

Bauer, T., Wilke, R., Naderer, G. (2019). Planned obsolescence: A systematic literature review. *Sustainability*, 11(24), 7126. https://doi.org/10.3390/su11247126

BCI-Business Continuity Institute. (2019). Supply chain resilience 10 year trend analysis. 2009–2018. Zurich Insurance Group. https://www.b-c-training.com/img/uploads/resources/Supply-Chain-Resilience-10-year-trend-analysis.pdf

Bell, J. (2023). Understanding our apples: Renewable natural resource supply chains. *Supply Chain Management* 101, June 8, 2023, The University of Tennessee, Knoxville.

Bext360. (2024). https://www.bext360.com

Borja, A., Briz, J., Masdeu, M. (2019). The role of organizational culture in sustainable supply chain management: A systematic review. *Supply Chain Management: An International Journal*, 24(4), 621–639. https://doi.org/10.1108/SCM-06-2018-0222

Carr, A. S., Ribeiro-Soriano, D. (2020). Unpacking the impact of COVID-19 on supply chains: A conceptual framework for understanding firm vulnerability and resilience. *International Journal of Production Economics*, 233, 107916. https://doi.org/10.1016/j.ijpe.2021.107916

Choi, Y., Wong, C. (2020). The impact of emerging technologies on sustainable supply chain management: A systematic literature review. *International Journal of Production Economics*, 226, 107610.

Duan, Y., Aloysius, J. A., Mollenkopf, D. A. (2022). Communicating supply chain sustainability: Transparency and framing effects. *International Journal of Physical Distribution & Logistics Management*, 52(1), 68–87. https://doi.org/10.1108/IJPDLM-04-2020-0107

Esteve, M., Alvarez, E. M. (2021). COVID-19 and global governance of supply chains: The case of the European Union. *Globalizations*, 18(4), 615–632. https://doi.org/10.1080/14747731.2020.1873761

Ethereum Whitepaper 2023. https://ethereum.org/en/whitepaper

FDA (Food and Drug Administration) DSCSA (The Drug Supply Chain Security Act). Blockchain interoperability pilot project report. February, 2020.

Gardner, T. A., Benzie, M., Börner, J., et al. (2019). Transparency and sustainability in global commodity supply chains. *World Development*, 121, 163–177. https://doi.org/10.1016/j.worlddev.2018.05.025

Groschopf, W., Dobrovnik, M., Herneth, C. (2021). Smart contracts for sustainable supply chain management: Conceptual frameworks for supply chain maturity evaluation and smart contract sustainability assessment. Frontiers in Blockchain, 4, 506436. https://doi.org/10.3389/fbloc.2021.506436

Hall, J. J., Hutchinson, L. E., Langford, R. A. (2018). Sustainability of supply chains: A systematic literature review and future research agenda. *Journal of Cleaner Production*, 214, 155–166. https://doi.org/10.1016/j.jclepro.2018.12.153

Iansiti, M., Lakhani, K. R. (2017). The truth about blockchain. *Harvard Business Review*, 95(1), 118–127.

IBM. (2024). Driving auto supply chains forward with blockchain. https://www.ibm.com/case-studies/renault

International Energy Agency Report 2023 World Energy Outlook 2023. Executive summary. IEA Publications. www.iea.org

Ivanov, D. (2020). Blockchain in the supply chain: Benefits and challenges. *International Journal of Production Research*, 58(7), 2063–2065.

Ivanov, D., Dolgui, A. (2020). A digital supply chain twin for managing the disruption risks and resilience in the era of Industry 4.0. *Production Planning & Control*, 31(9), 700–714.

Jahani, H., Jain, R., Ivanov, D. (2023). Data science and big data analytics: A systematic review of methodologies used in the supply chain and logistics research. *Annals of Operations Research*. https://doi.org/10.1007/s10479-023-05390-7

Jain, V. K., Bhambri, P. (2006). *Fundamentals of Information Technology & Computer Programming* (4th reprint ed.). Katson Publishers. ISBN 81-88458-59-7.

Johnson, R., Wang, S. (2020). Sustainable supply chain management: Review and research opportunities. *International Journal of Production Economics*, 220, 107482.

Katsaliaki, K., Galetsi, P., Kumar, S. (2022). Supply chain disruptions and resilience: A major review and future research agenda. *Annals of Operations Research,* 319, 965–1002. https://doi.org/10.1007/s10479-020-03912-1

Khan, M., Parvaiz, G. S., Dedahanov, A. T., Abdurazzakov, O. S., Rakhmonov, D. A. (2022). The impact of technologies of traceability and transparency in supply chains. *Sustainability*, 14(24), 16336. https://doi.org/10.3390/su142416336

Kim, H. M., Laskowski, M. (2019). Toward an ontology-driven blockchain design for supply-chain provenance. *Sustainability*, 11(18), 4864.

Kim, H., Lee, S. (2022). The role of technology in sustainable supply chain management: A systematic literature review. *Sustainability*, 14(1), 68.

Machado, E., Scavarda, L. F., Caiado, R. G. G., Thomé, A. M. T. (2021). Barriers and enablers for the integration of Industry 4.0 and sustainability in supply chains of MSMEs. *Sustainability*, 13, 11664. https://doi.org/10.3390/su132111664

Mercuri, F., della Corte, G., Ricci, F. (2021). Blockchain technology and sustainable business models: A case study of devoleum. *Sustainability,* 13(10), 5619. https://doi.org/10.3390/su13105619

Mol, A. P. J., Carter, N. T. (2019). Natural resource governance in fragile states: The case of post-conflict Sierra Leone. *Development and Change*, 50(2), 409–431. https://doi.org/10.1111/dech.12442

The Morningstar Sustainalytics Corporate ESG Survey Report 2022. CSR and sustainability in transition. https://www.sustainalytics.com/corporate-esg-survey-report

Nakamoto, S. (2008). Bitcoin: A peer-to-peer electronic cash system. https://bitcoin.org/bitcoin.pdf

Narayanan, A., Bonneau, J., Felten, E., Miller, A., Goldfeder, S. (2016). *Bitcoin and Cryptocurrency Technologies: A Comprehensive Introduction.* Princeton University Press.

Nash, K. S. (2018). Walmart-led blockchain effort seeks farm-to-grocery-aisle view of food supply chain. https://www.wsj.com/articles/walmart-led-blockchain-effort-seeks-farm-to-grocery-aisle-view-of-food-supply-chain-1529946347

Obora, H. (2020). Integration of lean and agile approaches in supply chain management. In M. Kołodziejczak, I. Bednarskia-Wnuk, & I. Świątek-Barylska (Eds.), *Management Methods and Techniques. Inspirations for Theory and Practice* (pp. 71–87). Łódź: University of Łódź Publishing House. https://doi.org/10.18778/8220-435-3.04

Pinnington, B., Benstead, A., Meehan, J. (2023). Transparency in Supply Chains (TISC): Assessing and improving the quality of modern slavery statements. *Journal of Business Ethics*, 182, 619–636. https://doi.org/10.1007/s10551-022-05037-w

Prasad, A., Grint, K. (2019). The gender pay gap and the failure of corporate social responsibility: A path to excessive consumption, environmental degradation. *Organization*, 26(6), 929–950. https://doi.org/10.1177/1350508419844570

Provenance. (2024). https://www.provenance.org/case-studies

Prusty, N. (2018). *Blockchain Revolution for the Enterprise: Build Scalable Blockchain Applications with Privacy, Interoperability, and Permissioned Features.* Packt Publishing.

Rana, R., Bhambri, P. (2024). Healthcare computational intelligence and blockchain: Real-life applications. In P. Bhambri, S. Rani, & M. Fahim (Eds.), *Computational Intelligence and Blockchain in Biomedical and Health Informatics* (pp. 155–168). CRC Press, Taylor & Francis Group. https://doi.org/10.1201/9781003459347

Rattan, M., Bhambri, P., Shaifali. (2005). Information retrieval using soft computing techniques. In National Conference (pp. 58–60) on Bio-informatics Computing. Presented at TIET, Patiala held during March 18–19, 2005.

Saini, N., Malik, K., Sharma, S. (2023). Transformation of supply chain management to green supply chain management: Certain investigations for research and applications. *Cleaner Materials*, 7, 100172. https://doi.org/10.1016/j.clema.2023.100172

Saurabh, S., Dey, K. (2021). Blockchain technology adoption, architecture, and sustainable agri-food supply chains. *Journal of Cleaner Production*, 284, 124731.

Schäfer, N. (2023). Making transparency transparent: A systematic literature review to define and frame supply chain transparency in the context of sustainability. *Management Review Quarterly,* 73, 579–604. https://doi.org/10.1007/s11301-021-00252-7

Seuring, S., Müller, M. (2008). From a literature review to a conceptual framework for sustainable supply chain management. *Journal of Cleaner Production*, 16(15), 1699–1710.

Shanmuga, S. M., Bhambri, P. (2024). Bone marrow cancer detection from leukocytes using neural networks. In P. Bhambri, S. Rani, & M. Fahim (Eds.), *Computational Intelligence and Blockchain in Biomedical and Health Informatics* (pp. 307–319). CRC Press, Taylor & Francis Group. https://doi.org/10.1201/9781003459347

Singh, P., Singh, M., Bhambri, P. (2004). Interoperability: A problem of component reusability. In International Conference (Abstract Book) (p. 60). Presented at the "Emerging Technologies in IT Industry" at PCTE, Ludhiana on November 27, 2004.

Singh, P., Singh, M., Bhambri, P. (2005). Embedded systems. In U.G.C. Sponsored National Seminar (pp. 10–15). Presented at the "Embedded Systems" seminar at KMV, Jalandhar held during January 24–25, 2005.

Smith, J. (2019). Sustainable supply chain management: A literature review. *Journal of Cleaner Production*, 239, 118072.

Sun, Y., Shi, B. (2022). Impact of greenwashing perception on consumers' green purchasing intentions: A moderated mediation model. *Sustainability*, 14, 12119. https://doi.org/10.3390/su141912119

Swan, M. (2015). *Blockchain: Blueprint for a New Economy.* O'Reilly Media, Inc.

Tapscott, D., Tapscott, A. (2016). *Blockchain Revolution: How the Technology Behind Bitcoin is Changing Money, Business, and the World.* Penguin.

Tse, C. K., Chiu, A. P. (2018). Supply chain performance-driven blockchain architecture. *International Journal of Information Management*, 39, 80–92.

United Nations. (2015). Transforming our world: The 2030 agenda for sustainable development. https://sustainabledevelopment.un.org/post2015/transformingourworld

United Nations Environment Programme (UNEP). (2020). Emissions gap report 2020. https://www.unep.org/emissions-gap-report-2020

Upadhyay, A., Mukhuty, S., Kumar, V., Kazancoglu, Y. (2021). Blockchain technology and the circular economy: Implications for sustainability and social responsibility. *Journal of Cleaner Production*, 293. https://doi.org/10.1016/j.jclepro.2021.126130

Weber, S., Aulbach, C., Fingerle, C. (2020). Managing and controlling sustainability risks upstream: The role of buyer-supplier information sharing. *Journal of Purchasing and Supply Management*, 26(2), 100627. https://doi.org/10.1016/j.pursup.2019.100627

Wilson, M. D. (2015). American Health Policy, AHPI study shows clear connection between increased transparency and cost reductions. http://www.americanhealthpolicy.org

World Wildlife Fund (WWF). (2021). Living Planet Report 2020 – Bending the curve of biodiversity loss. https://www.worldwildlife.org/pages/living-planet-report-2020-bending-the-curve-of-biodiversity-loss

Xu, X., Weber, I., Staples, M., Zhu, L., Bosch, J., Bass, L., ... Pautasso, C. (2017). A taxonomy of blockchain-based systems for architecture design. In 2017 IEEE International Conference on Software Architecture (ICSA) (pp. 243–252). IEEE.

Zheng, Z., Xie, S., Dai, H., Chen, X., Wang, H. (2019). An overview of blockchain technology: Architecture, consensus, and future trends. In *Big Data* (pp. 229–257). Cham: Springer.

9 Protecting Sustainable Supply Chains from Cyber Threats

Pankaj Bhambri and Suresh Kumar

9.1 INTRODUCTION

Ensuring the resilience and integrity of global trade requires prioritizing the protection of sustainable supply chains against cyberattacks. As organizations become more dependent on interconnected digital networks for procurement, production, and distribution, they encounter an expanding range of cyber dangers that can interrupt operations, compromise sensitive data, and damage sustainability initiatives. To strengthen supply chains against cyber threats and ensure the continuity of sustainable practices and global economic stability, it is crucial to implement strong cybersecurity measures, promote collaboration among supply chain stakeholders, incorporate risk management frameworks, and utilize emerging technologies like blockchain and artificial intelligence.

9.1.1 Overview of the Intersection between Sustainability and Cybersecurity

The intersection between sustainability and cybersecurity underscores the critical need to address environmental, social, and governance (ESG) concerns alongside digital security imperatives. As organizations strive to achieve sustainability goals, they must recognize the interconnected nature of environmental stewardship, social responsibility, and cybersecurity resilience (Srinivasan and Oke, 2022). Cyberattacks not only pose immediate threats to data integrity and operational continuity but also have far-reaching consequences for sustainability efforts, ranging from environmental impact through data breaches to social repercussions due to supply chain disruptions (Pappas and Mikalef, 2021). Conversely, sustainable practices can enhance cybersecurity resilience by fostering a culture of transparency, accountability, and stakeholder engagement, which are foundational principles in both domains.

Moreover, the integration of sustainability and cybersecurity principles is increasingly recognized as a strategic imperative for businesses operating in an interconnected and digitally reliant global landscape. Organizations can reduce risks related to emerging threats, regulatory compliance, and reputational damage

DOI: 10.1201/9781003484226-9

while promoting environmental and social goals by integrating sustainability activities with cybersecurity strategy (Hsiao, 2024). This convergence enables businesses to leverage technology and innovation not only to enhance operational efficiency and mitigate environmental impact but also to fortify cyber defenses and promote responsible digital practices, thus creating synergies that drive long-term value creation and resilience in an ever-evolving business environment (Bhambri et al., 2005).

9.1.2 Importance of Protecting Supply Chains from Cyber Threats

Securing supply chains from cyberattacks is crucial for upholding economic stability, maintaining consumer trust, assuring uninterrupted operations, protecting intellectual property, adhering to regulations, and securing national security interests. Effective implementation of strong cybersecurity measures and mitigation of cyber threats necessitate collaboration and cooperation among businesses, governments, and other stakeholders. Protecting supply chains from cyber threats is crucial for several reasons (McNeill and Purvis, 2024):

- Economic impact: Supply chains are the backbone of economies. Any disruption due to cyber threats can lead to significant financial losses. A cyberattack can halt production, delay deliveries, and result in revenue loss for companies involved in the supply chain. This can have cascading effects on other businesses, causing a ripple effect throughout the economy.
- Customer trust and reputation: A breach in the supply chain has the potential to jeopardize confidential data such as customer information, intellectual property, and trade secrets (Singh et al., 2005). These breaches have the potential to undermine customer confidence and harm the reputation of the companies involved. Customers' trust in organizations' data protection capabilities may diminish, resulting in a decline in market share and revenue.
- Operational disruption: Cyberattacks on supply chains can disrupt operations, leading to delays in delivering goods and services. This can result in increased costs, decreased productivity, and missed deadlines (Finkelstein, 2020). In some cases, it may even lead to the shutdown of critical infrastructure, causing widespread disruption and chaos.
- Intellectual property theft: Supply chains often involve the exchange of sensitive intellectual property between different entities. Cyberattacks targeting supply chains can result in the theft of valuable intellectual property, including product designs, manufacturing processes, and proprietary algorithms. This can give competitors an unfair advantage and undermine the innovation and competitiveness of businesses.
- Regulatory compliance: Several sectors must adhere to strict regulatory mandates on data protection and cybersecurity (Bhambri and Khang, 2024). Failure to safeguard supply chains from cyber threats can lead to non-adherence to these requirements, resulting in legal sanctions, monetary penalties, and harm to one's reputation. Adhering to rules like as general data protection regulation (GDPR), health insurance portability

and accountability (HIPAA), and payment card industry data security standard (PCI-DSS) is crucial for upholding the confidence of consumers and stakeholders.

- National security: Supply chains are increasingly interconnected and globalized, making them vulnerable to cyberattacks from nation-states and other malicious actors. Disruption to critical supply chains such as those related to energy, healthcare, and defense can have severe implications for national security (Wu and Ivanov, 2021). Protecting supply chains from cyber threats is, therefore, essential for safeguarding national interests and ensuring resilience in the face of evolving cyber threats.

9.2 UNDERSTANDING DIGITAL RESILIENCE

Digital resilience refers to an organization's ability to anticipate, withstand, adapt to, and rapidly recover from digital disruptions, including cyberattacks, technological failures, and other digital threats. It involves proactive measures such as robust cybersecurity protocols, effective risk management strategies, redundant systems, and continuous monitoring to detect and mitigate threats early on (Choo, 2022). Additionally, digital resilience encompasses the agility to adapt and innovate in response to evolving digital challenges, as well as the capacity to quickly recover and resume normal operations following a disruption. By cultivating digital resilience, organizations can mitigate the impact of digital disruptions, maintain business continuity, safeguard critical assets and data, and sustain trust and confidence among stakeholders in an increasingly interconnected and digitalized world.

9.2.1 Definition of Digital Resilience and its Relevance to Sustainable Supply Chains

Digital resilience refers to an organization's ability to successfully predict, endure, adjust to, and bounce back from digital disturbances such as cyber assaults, technology breakdowns, data breaches, and other digital risks. It involves a proactive strategy for cybersecurity and risk management, together with the capability to swiftly recover and restart regular operations in challenging circumstances. Ensuring the sustainability and resilience of supply chains in the face of emerging digital threats requires the presence of digital resilience. By proactively addressing cybersecurity risks, protecting sensitive data, ensuring regulatory compliance, and maintaining business continuity, organizations can build trust, enhance collaboration, and sustainably manage their supply chain operations in an increasingly digitalized world.

In the context of sustainable supply chains, digital resilience is particularly relevant due to several factors:

- Supply chain complexity: Contemporary supply chains are intricately linked and transnational in nature, encompassing a multitude of participants, collaborators, and external suppliers. The aforementioned intricacy heightens the susceptibility to digital disruptions, given that the entire network could be impacted by even a vulnerable link in the chain (Qian and Wang, 2022). By instituting robust cybersecurity measures and ensuring the integrity and security of digital communications and transactions throughout the supply chain, digital resilience helps mitigate these risks.

- Data security and privacy: Sustainable supply chains are highly dependent on the transfer of sensitive data, such as product specifications, customer information, and environmental performance measurements. Safeguarding this data from cyber threats is crucial for upholding confidence and ensuring transparency across the supply chain. Digital resilience encompasses the utilization of encryption, access controls, and data protection techniques to prevent sensitive information from unlawful access or disclosure.
- Compliance and regulatory requirements: Sustainable supply chains must comply with several regulatory mandates pertaining to data protection, privacy, environmental sustainability, and social responsibility (Sharma et al., 2020). Failure to comply with these requirements may lead to legal sanctions, monetary penalties, and harm to one's reputation. Digital resilience is important for guaranteeing compliance by developing strong cybersecurity protocols and showing a dedication to safeguarding sensitive data and digital assets.
- Business continuity and resilience: Digital disruptions, including technological malfunctions and cyberattacks, have the potential to cause significant disruptions to supply chain operations, resulting in financial losses, delays, and other related complications. Digital resilience encompasses the implementation of backup procedures, redundant systems, and contingency plans to ensure business continuity and mitigate the effects of supply chain disruptions. Through the expeditious identification and mitigation of digital threats, organizations can effectively sustain the flow of products and services along the supply chain while minimizing downtime.
- Stakeholder trust and reputation: The establishment of sustainable supply chains is contingent upon the cooperation and confidence of a multitude of stakeholders such as investors, customers, suppliers, and regulators (Trivedi and Patel, 2022). Neglecting to safeguard against digital threats or experiencing a violation in data security can undermine confidence and harm the standing of the organization. Digital resilience contributes to the establishment of confidence and trust by showcasing a steadfast dedication to safeguarding confidential data, preserving the authenticity of digital transactions, and expeditiously addressing cyber incidents or threats.

9.2.2 Strategies for Building Digital Resilience in Supply Chain Operations

Building digital resilience in supply chain operations requires a multifaceted approach that integrates cybersecurity best practices, risk management strategies, and technology solutions. By adopting the strategies and integrating cybersecurity into their supply chain operations, organizations can build digital resilience and mitigate the risks posed by digital disruptions, ensuring the continuity, security, and sustainability of their supply chain operations. Here are some key strategies for organizations to consider:

- Risk assessment and management: It is essential to conduct exhaustive risk assessments in order to identify potential digital threats and supply chain vulnerabilities. Assessing the potential impact of digital disruptions on supply chain operations and evaluating the security posture of suppliers, third-party vendors, and other stakeholders consist of this. After the identification of risks, organizations have the ability to formulate risk management strategies that effectively prioritize and mitigate these hazards.
- Supplier cybersecurity due diligence: It is critical to evaluate the cybersecurity capabilities of third-party vendors and suppliers in order to guarantee the integrity of the entire supply chain. Organizations should establish clear cybersecurity requirements for suppliers, including minimum security standards, data protection protocols, and incident response procedures (Xu and Chen, 2022). Regular audits and assessments can help verify compliance with these requirements and identify areas for improvement.
- Cybersecurity training and awareness: Human error continues to be a paramount concern in the realm of cybersecurity vulnerabilities (Rattan et al., 2005). Implementing comprehensive training and awareness initiatives for all stakeholders, including employees and suppliers, can effectively cultivate a cybersecurity-oriented culture within the institution. Phishing awareness, password hygiene, data protection best practices, and incident response procedures should be included in the training curriculum.
- Implementing robust cybersecurity controls: Deploying robust cybersecurity controls and technologies is essential for protecting supply chain operations from digital threats. This includes implementing firewalls, intrusion detection systems, antivirus software, and encryption protocols to safeguard sensitive data and information assets. Additionally, organizations should establish access controls, authentication mechanisms, and monitoring systems to detect and prevent unauthorized access or malicious activities.
- Data encryption and protection: Encrypting sensitive data and information assets is essential for protecting an organization from unauthorized access or disclosure. They should implement encryption technologies to secure data in transit and at rest, both within their systems and across the supply chain. This helps mitigate the risk of data breaches and ensures the confidentiality and integrity of sensitive information.
- Incident response and business continuity planning: Developing comprehensive incident response and business continuity plans is critical for minimizing the impact of digital disruptions on supply chain operations. Organizations should establish clear procedures for detecting, reporting, and responding to cybersecurity incidents, as well as for restoring operations and recovering from disruptions (Lee and Marciniak, 2021). Regular testing and simulation exercises can help validate the effectiveness of these plans and ensure readiness to respond to cyber threats effectively.

- Continuous monitoring and threat intelligence: Implementing continuous monitoring and threat intelligence capabilities allows organizations to proactively detect and respond to emerging cyber threats in real time. This involves monitoring network traffic, system logs, and user activities for signs of suspicious behavior or unauthorized access. Leveraging threat intelligence feeds and cybersecurity information-sharing initiatives can also provide valuable insights into evolving cyber threats and helping organizations adapt their defenses accordingly.
- Collaboration and information sharing: Establishing alliances and promoting cooperation with other companies, industry groups, and government agencies can strengthen the ability of the supply chain to withstand and overcome digital challenges. By sharing cybersecurity best practices, threat intelligence, and incident response resources, companies can collaboratively tackle shared challenges and enhance their collective defenses against cyber threats.

9.3 CYBER RESILIENCE IN SUSTAINABLE SUPPLY CHAINS

Cyber resilience in sustainable supply chains refers to the ability of organizations to effectively anticipate, withstand, adapt to, and recover from cyber threats and disruptions while maintaining their commitment to environmental, social, and economic sustainability (de Haes and van Grembergen, 2022). This entails integrating robust cybersecurity measures, risk management strategies, and technology solutions into supply chain operations to safeguard sensitive data, protect critical assets, and ensure the continuity of sustainable practices. By prioritizing cyber resilience, organizations can mitigate the risks posed by digital threats, enhance trust and transparency throughout the supply chain, and sustainably manage their operations in an increasingly interconnected and digitalized world.

9.3.1 Exploring the Concept of Cyber Resilience within the Context of Sustainability

The concept of cyber resilience within the context of sustainability underscores the intersection between digital security and the broader goals of environmental, social, and economic sustainability. In an increasingly interconnected and digitalized world, organizations face a growing array of cyber threats that can jeopardize their ability to achieve sustainable outcomes. Cyber resilience in this context involves not only protecting sensitive data and critical infrastructure from cyberattacks but also ensuring that sustainability initiatives are resilient to digital disruptions. For example, a sustainable supply chain may rely on digital technologies to monitor and optimize energy usage, track the environmental impact of products, and promote fair labor practices (Acquisti and Friedman, 2021). Cyber resilience ensures that these digital systems are secure, reliable, and resilient to cyber threats, thus safeguarding the organization's ability to maintain sustainable practices and meet its sustainability goals.

Cyber resilience contributes to the overall resilience and effectiveness of sustainability efforts by mitigating the risks of disruption and ensuring continuity in the face of cyber threats. For instance, a cyberattack targeting a renewable energy infrastructure could disrupt the supply of clean energy, undermining efforts to reduce carbon emissions and combat climate change. By implementing robust cybersecurity measures and building resilience into digital systems, organizations can minimize the impact of cyber threats on sustainability initiatives and maintain the integrity and effectiveness of their sustainability programs. Ultimately, integrating cyber resilience into sustainability strategies enables organizations to navigate the complex challenges of an interconnected world while advancing their goals of environmental stewardship, social responsibility, and economic prosperity securely and sustainably.

9.3.2 Case Studies Illustrating Effective Cyber Resilience Measures in Supply Chains

Case studies demonstrate that effective cyber resilience in supply chains requires a proactive approach to cybersecurity, robust incident response capabilities, transparent communication, and collaboration among stakeholders (Raghavan, 2023). By learning from past incidents and implementing best practices, organizations can strengthen their resilience to cyber threats and safeguard the integrity, security, and continuity of supply chain operations. Several case studies provide insights into effective cyber resilience measures in supply chains:

- Maersk Cyber Attack Recovery (2017): In June 2017, Maersk, the world's largest container shipping company, fell victim to the NotPetya ransomware attack, which encrypted its IT systems and disrupted operations globally. Maersk's response to the attack showcased effective cyber resilience measures. The company swiftly activated its incident response plan, isolated infected systems, and communicated transparently with customers and stakeholders. Maersk also relied on backup systems and manual processes to continue essential operations while restoring its IT infrastructure. The incident highlighted the importance of regular backups, robust incident response plans, and transparent communication in mitigating the impact of cyberattacks on supply chain operations.
- Target Data Breach (2013): Target, a prominent retail company in the United States, experienced a significant data breach in late 2013, which resulted in the unauthorized access and potential compromise of the personal and financial data belonging to millions of consumers. The attack occurred due to the compromised credentials of a third-party vendor, which emphasizes the interconnectedness of supply chain cybersecurity. Target's response to the incident involved deploying advanced cybersecurity measures, including the adoption of multifactor authentication, the segregation of network access, and the enhancement of monitoring capabilities. The occurrence highlighted the significance of managing risks from external parties and working together to enhance cybersecurity throughout the supply chain network.
- Honda Cyber Attack (2020): In June 2020, Honda experienced a cyberattack that disrupted its global operations, including manufacturing plants

and customer service operations (Ghosh, 2023). The attack was attributed to the Snake ransomware group, which targeted Honda's internal systems. Honda's response to the attack emphasized transparency, with the company promptly disclosing the incident and providing updates to customers and stakeholders. The company also implemented measures to contain the attack, restore affected systems, and enhance cybersecurity defenses. The incident highlighted the need for proactive threat detection, rapid incident response, and collaboration with cybersecurity experts to mitigate the impact of cyberattacks on supply chain operations.

9.4 THREAT DETECTION MECHANISMS

Threat detection mechanisms encompass a range of proactive strategies, including network monitoring, anomaly detection, and behavioral analysis, to detect suspicious activities, unauthorized access attempts, and emerging cyber threats. By utilizing cutting-edge technologies like artificial intelligence and machine learning, organizations can effectively analyze large volumes of data in real-time. This allows them to quickly discover trends that may indicate cyber threats and rapidly take action to minimize any risks (Siau and Shen, 2021). Organizations can enhance the durability of sustainable supply chains, protect sensitive data, and uphold the integrity of environmental, social, and economic sustainability efforts in a more interconnected and digitalized world by incorporating strong threat detection mechanisms.

9.4.1 Overview of Common Cyber Threats Facing Sustainable Supply Chains

Sustainable supply chains face a myriad of cyber threats that can jeopardize their integrity, security, and resilience. By understanding these common cyber threats, organizations can implement proactive cybersecurity measures, enhance risk management strategies, and foster collaboration among stakeholders to safeguard the integrity, security, and resilience of sustainable supply chains. Some common cyber threats include:

- Phishing and social engineering: Cybercriminals frequently deploy phishing emails and social engineering tactics to deceive employees, suppliers, and other stakeholders into revealing critical information or engaging with harmful links (Bhambri et al., 2021). These assaults have the potential to result in data breaches, unauthorized entry into systems, and financial losses.
- Ransomware attacks: Ransomware assaults entail the utilization of malevolent software to encrypt data or restrict users from accessing their systems until a ransom is remitted. These attacks have the potential to cause disruptions in supply chain activities, result in the loss of data, and lead to financial extortion.
- Third-party vendor risks: Supply chains depend on external vendors and suppliers for a wide range of goods and services, which raises the

vulnerability to cyber threats due to the interconnected nature of networks. Inadequate cybersecurity protections implemented by vendors might act as vulnerabilities that allow attackers to get access to supply chain systems and compromise confidential information.

- Insider threats: Both purposeful and inadvertent insider threats provide substantial hazards to the security of the supply chain (Martin and Murphy, 2022). Malicious insiders have the potential to manipulate their authorized access to confidential information or computer systems for their own benefit, whereas unintentional actions by employees might lead to the exposure of data or other security breaches.
- IoT vulnerabilities: The increasing number of Internet of Things (IoT) devices in supply chains creates new opportunities for cyberattacks and weaknesses. Cybercriminals can exploit insecure IoT devices to illicitly access networks, disrupt operations, or pilfer critical data.
- Supply chain interdependencies: Supply chains are complex networks that involve several stakeholders, each with their own cybersecurity measures and weaknesses. Adversaries may focus their efforts on vulnerable points in the supply chain in order to infiltrate highly protected systems or cause disruptions throughout the entire network.
- Data breaches: Data breaches that result in the theft or exposure of sensitive information can have significant repercussions for sustainable supply chains. Compromised data may encompass confidential information, client data, or metrics pertaining to environmental and societal consequences, resulting in harm to one's reputation, penalties imposed by regulatory bodies, and legal responsibilities.
- Disruption of critical infrastructure: Targeting vital infrastructure, such as energy grids, transportation networks, or manufacturing sites, cyber assaults have the potential to disrupt supply chain activities and hinder the delivery of sustainable goods and services.

9.4.2 Techniques and Technologies for Detecting Cyber Threats in Supply Chain Networks

Efficiently detecting cyber threats in supply chain networks necessitates the utilization of a variety of approaches and technology to accurately identify and promptly respond to possible dangers. Organizations can improve their capability to detect and respond to cyber threats by implementing different strategies and technologies. This will help protect the integrity and security of their supply chain networks. Here are some commonly used techniques and technologies:

- Network traffic analysis: Examining the patterns and irregularities in network traffic might aid in identifying unusual behaviors that suggest cyber dangers such as illegal attempts to gain access, the unauthorized extraction of data, or the transfer of malicious software (Subramanian and Gunasekaran, 2023). Intrusion detection systems (IDS) and intrusion prevention systems

(IPS) are technological tools that continuously monitor network traffic and promptly notify users of any potential security breaches.

- Behavioral analytics: Behavioral analytics techniques utilize machine learning algorithms to study user activity and detect deviations from typical trends. Organizations can identify insider threats, compromised accounts, or other security breaches by creating benchmarks of normal user behavior and detecting any deviations from these benchmarks.
- Endpoint detection and response (EDR): EDR systems are designed to actively monitor and detect any indications of malicious behavior or illegal entry on endpoint devices, including laptops, desktops, and servers. These systems gather telemetry data from endpoints, evaluate behavior, and promptly respond to security problems, allowing enterprises to rapidly identify and control risks.
- Security information and event management (SIEM): SIEM solutions collect and analyze log data from several sources, such as network devices, servers, and apps, to offer a centralized perspective on security incidents. SIEM technologies utilize event correlation and pattern analysis to detect possible cyber threats and streamline incident response efforts.
- Threat intelligence feeds: By subscribing to threat intelligence feeds from credible sources, firms gain valuable information on developing cyber risks, attack patterns, and signs of compromise (Wong and Lai, 2020). Threat intelligence feeds enable firms to stay updated on the most recent threats and adapt their cybersecurity defenses accordingly.
- Machine learning and AI: Machine learning and AI technologies allow enterprises to automate the processes of detecting and responding to risks, analyze vast datasets to find patterns and anomalies, and identify hazards that were previously unknown. These technologies improve the efficacy and efficiency of detecting cyber threats in supply chain networks.
- Blockchain technology: Blockchain technology can bolster supply chain security by offering a decentralized and unalterable ledger for documenting transactions and monitoring items. Organizations may improve transparency, traceability, and integrity in their supply chains by utilizing blockchain-based solutions, thereby minimizing the possibility of tampering or fraud.
- Continuous monitoring: By implementing continuous monitoring capabilities, firms may effectively observe their supply chain networks in real time to detect any indications of suspicious activity or security breaches. Organizations can immediately detect and respond to cyber threats by continuously monitoring network traffic, user behavior, and system records.

9.5 MITIGATING CYBER THREATS IN SUPPLY CHAINS

To effectively address cyber threats in supply chains, it is necessary to adopt a holistic approach that combines proactive cybersecurity measures, risk management tactics, and collaborative efforts among stakeholders (Bhambri, 2024). Organizations can strengthen the ability of their supply chain to withstand and recover from disruptions by installing strong cybersecurity measures such as dividing the network into segments, using encryption, and requiring multiple forms of authentication. These measures are essential for safeguarding sensitive information and protecting key infrastructure (Berman and Li, 2020). In addition, it is crucial to regularly perform risk assessments, define explicit cybersecurity policies and procedures, and implement continuous training and awareness programs for employees and suppliers. These measures are necessary for recognizing and resolving potential vulnerabilities. Engaging in partnerships with external vendors, exchanging information about potential security issues, and establishing effective lines of communication can enhance the security measures of supply chains and reduce the vulnerabilities caused by cyber threats. By giving top priority to cybersecurity and promoting a culture of watchfulness and cooperation, firms can protect the honesty, security, and ability to recover their supply chains in a world that is becoming more digitalized and interconnected.

9.5.1 Proactive Measures for Mitigating Cyber Threats and Enhancing Supply Chain Security

Proactive measures for mitigating cyber threats and enhancing supply chain security involve a multilayered approach that encompasses various aspects of cybersecurity, risk management, and collaboration with partners and vendors (Kshetri, 2020). Organizations may bolster their cybersecurity stance, reduce potential risks, and improve the security of their supply chain by taking proactive measures. Continuously adapting and evolving these methods is crucial in order to successfully address emerging threats and weaknesses. Here are some key strategies:

- Risk assessment and management: Perform periodic risk assessments to detect weaknesses and potential dangers within your supply chain. Rank dangers according to their potential impact and probability of occurrence. Create customized risk mitigation methods based on the unique risks that have been identified.
- Vendor management: Define explicit security criteria for vendors and partners, encompassing contractual responsibilities pertaining to cybersecurity (Apte and Mason, 2020). Establish vendor risk management systems to evaluate the security stance of third-party providers. Consistently oversee and assess vendors' adherence to security protocols.
- Secure communication channels: Utilize encryption techniques to safeguard sensitive data during transmission and storage, preventing interception, or unauthorized entry. Deploy robust cryptographic technologies, such as HTTPS, Virtual Private Networks (VPNs), and encrypted email, to facilitate the secure exchange of critical information.
- Access control and authentication: Enforce robust access controls to limit access to important systems and data. Implement multfactor authentication (MFA) as a means of verifying the identity of users who are attempting to

access important resources (Rana and Bhambri, 2024). Implement a system to track and record user actions in order to identify and address any unauthorized attempts to get access.

- Cybersecurity training and awareness: Conduct routine cybersecurity training for employees, vendors, and partners to enhance their understanding of common threats and optimal strategies (Singh et al., 2021). Promote a security-oriented culture by actively encouraging staff to swiftly report any suspicious actions or potential security problems.
- Continuous monitoring and incident response: Implement security surveillance technologies to promptly identify abnormal activity and potential security breaches in real time. Create an incident response strategy that details the protocols for containing, investigating, and recovering from cybersecurity problems. Regularly do tabletop exercises and simulations to evaluate the efficacy of incident response protocols.
- Supply chain resilience: Diversify your supplier base to reduce reliance on a single vendor or region (Shanmuga and Bhambri, 2024). Establish redundancy and contingency plans to maintain operations in the event of disruptions to the supply chain such as cyberattacks or natural disasters. Collaborate with partners to share threat intelligence and coordinate response efforts.
- Regulatory compliance: Keep yourself updated on pertinent cybersecurity rules and industry standards, such as GDPR, California consumer privacy act (CCPA), national institute of standards and technology (NIST), international organization for standardization (ISO) 27001, and others. Ensure adherence to relevant requirements by conducting routine audits and assessments.

9.5.2 Integrating Cyber Threat Mitigation Strategies into Sustainability Practices

By intertwining cybersecurity with sustainability initiatives, businesses can cultivate a holistic approach to risk management that not only safeguards their digital assets but also contributes to long-term environmental and social well-being (Tang and He, 2020). One way to achieve this integration is by incorporating cybersecurity considerations into the design and implementation of sustainable technologies and processes. For instance, organizations can prioritize the development of secure and resilient infrastructure for renewable energy projects or smart city initiatives, ensuring that these innovations are protected from cyber threats while advancing environmental sustainability.

Embedding cyber threat mitigation strategies within sustainability practices fosters a culture of accountability and transparency, reinforcing organizational values of integrity and ethical conduct. Organizations can leverage sustainability reporting frameworks, such as the Global Reporting Initiative (GRI) or the Task Force on Climate-related Financial Disclosures (TCFD), to disclose their cybersecurity measures alongside environmental and social performance metrics (Zhang and Lim,

2023). By doing so, businesses demonstrate their commitment to managing cyber risks responsibly and building trust with stakeholders, including investors, customers, and communities. Ultimately, this integration promotes resilience across multiple dimensions, strengthening both cybersecurity defenses and sustainable business practices in tandem.

9.6 CASE STUDIES AND BEST PRACTICES

Sustainable supply chains are increasingly facing cyber threats as digitalization becomes more prevalent in the business world. By adopting proactive strategies and leveraging innovative technologies, organizations can overcome cyber threats and uphold sustainability principles in their operations. Following are some case studies, best practices, and real-world examples of sustainable supply chains overcoming cyber threats.

- **Case study:** Maersk Cyberattack Recovery–In 2017, Maersk, one of the largest shipping companies globally, fell victim to the NotPetya cyberattack. The attack disrupted their operations severely, causing significant financial losses. However, Maersk demonstrated resilience by implementing robust cybersecurity measures and recovery strategies. They improved their network segmentation, enhanced employee cybersecurity training, and invested in advanced threat detection systems (Li and Berman, 2023). By focusing on cybersecurity and leveraging sustainable practices in its supply chain, Maersk not only recovered but also strengthened its resilience to future cyber threats.

9.6.1 Best Practices

- Risk assessment and management: Conduct regular assessments of cybersecurity risks throughout the supply chain, including suppliers and partners. Develop risk management strategies to mitigate identified risks effectively.
- Cybersecurity training: Provide comprehensive cybersecurity training to employees, suppliers, and partners to enhance awareness and promote best practices for data protection and threat prevention.
- Supply chain visibility: Establish end-to-end visibility across the supply chain to monitor and detect potential cyber threats in real time. Implement advanced analytics and monitoring tools to identify anomalies and suspicious activities.
- Supplier engagement: Collaborate closely with suppliers and partners to ensure they adhere to cybersecurity standards and practices. Incorporate cybersecurity requirements into contracts and agreements.
- Incident response plan: Develop and regularly update an incident response plan to effectively respond to cyber incidents. Conduct regular drills and simulations to test the plan's efficacy and ensure swift and coordinated responses.

9.6.2 Real-world Examples

- IBM's blockchain solution: IBM implemented a blockchain-based supply chain solution to enhance transparency and security (Thomas and Chen, 2021). By utilizing blockchain technology, IBM provides immutable records of transactions and improved traceability, reducing the risk of cyber threats such as data tampering and counterfeiting.
- Toyota's cybersecurity initiatives: Toyota established a dedicated cybersecurity unit to address emerging cyber threats in its supply chain. The company collaborates with suppliers to enhance cybersecurity measures and conducts regular audits to ensure compliance with security standards.
- Schneider electric's supplier cybersecurity program: Schneider Electric implemented a comprehensive cybersecurity program for its suppliers, including assessment frameworks, training modules, and ongoing monitoring. By prioritizing cybersecurity across its supply chain, Schneider Electric aims to mitigate risks and safeguard sensitive data.
- Nestlé's supply chain resilience: Nestlé leverages digitalization and data analytics to enhance resilience in its supply chain. The company employs advanced cybersecurity measures to protect critical infrastructure and data, ensuring continuity of operations and mitigating potential cyber threats.

9.7 CONCLUSION

Safeguarding sustainable supply chains from cyber threats is paramount in the digital age. Through an examination of case studies, best practices, and real-world examples, several key findings emerge (Chatterjee, 2021). Firstly, proactive risk assessment and management are critical in identifying vulnerabilities and implementing effective mitigation strategies. Organizations must prioritize cybersecurity training, supply chain visibility, and supplier engagement to enhance resilience and protect against potential cyber threats. Furthermore, the act of collaborating and exchanging information across individuals or groups involved is crucial in order to provide a unified defense against cyber threats (Sharma and Thomas, 2020). Through the establishment of strong alliances and the adoption of cybersecurity protocols throughout the whole supply chain, companies can enhance their defensive capabilities and more efficiently reduce risks. Investing in emerging technologies like blockchain and advanced analytics can improve supply chain transparency and security, allowing firms to promptly identify and address cyber threats.

Entrepreneurs in the digital age must acknowledge the interdependence of contemporary supply chains and the inherent dangers presented by cyber threats. By adopting a proactive approach to cybersecurity and integrating it into sustainable supply chain management practices, entrepreneurs can protect their operations and maintain long-term viability (Oke and Srinivasan, 2020). Investing in employee training, cybersecurity partnerships, and innovative technologies can enhance resilience and mitigate the impact of cyber threats on business continuity and reputation. Moreover, entrepreneurs should prioritize transparency and collaboration with suppliers and partners to foster trust and facilitate information sharing, thereby strengthening the collective defense against cyber threats.

Looking ahead, future research and practice should focus on advancing cybersecurity strategies tailored specifically to sustainable supply chains. There is a need for further exploration of emerging technologies such as artificial intelligence and the Internet of Things (IoT) to bolster supply chain resilience and mitigate cyber risks (Wang and Cao, 2023). Additionally, understanding the socio-economic impacts of cyber threats on sustainability initiatives and stakeholder trust is crucial for developing holistic risk management strategies. Moreover, cross-sector collaboration and knowledge sharing can facilitate the development of standardized cybersecurity frameworks and best practices applicable to diverse industries and supply chain contexts. By embracing innovation and collaboration, organizations can effectively navigate the evolving threat landscape and protect sustainable supply chains from cyber threats in the digital era.

REFERENCES

Acquisti, A., & Friedman, A. (2021). Cybersecurity and Privacy in the Digital Age. *Journal of Economic Perspectives*, 35(1), 29–52.

Apte, A., & Mason, C. H. (2020). Cybersecurity Risks in Global Supply Chains. *Communications of the ACM*, 63(4), 60–69.

Berman, J., & Li, S. (2020). Strengthening Cybersecurity in Supply Chains: A Review and Research Agenda. *Journal of Management Information Systems*, 37(1), 268–309.

Bhambri, P. (2024). Wallets and Transactions. In D. Darwish (Ed.), *Decentralizing the Online Experience with Web3 Technologies* (pp. 90–106). IGI Global. https://doi.org/10.4018/979-8-3693-1532-3.ch004

Bhambri, P., Gupta, S., & Bhandari, A. (2005). Soft Computing Techniques. In UGC Sponsored National Conference (Abstract Book) (p. 100). Presented at the "Emerging Computing Technologies" conference at HRMMV, Jalandhar held during September 29–30, 2005.

Bhambri, P., & Khang, A. (2024). The Human-Machine Nexus With Art-Making Generative AIs. In S. Hai-Jew (Ed.), *Making Art with Generative AI Tools* (pp. 73–85). IGI Global. https://doi.org/10.4018/979-8-3693-1950-5.ch004

Bhambri, P., Singh, M., Jain, A., Dhanoa, I. S., Sinha, V. K., & Lal, S. (2021). Classification of the GENE Expression Data with the Aid of Optimized Feature Selection. *Turkish Journal of Physiotherapy and Rehabilitation*, 32(3), 1158–1167.

Chatterjee, R. (2021). Cybersecurity Threats and Mitigation Strategies in Sustainable Supply Chains. *International Journal of Production Economics*, 231, 107941.

Choo, K. K. R. (2022). Cybersecurity for Sustainable Supply Chains: Current Trends and Future Directions. *Sustainable Production and Consumption*, 32, 251–261.

de Haes, S., & van Grembergen, W. (2022). Cybersecurity Governance in Sustainable Supply Chains. *Journal of Information Technology*, 37(2), 195–212.

Finkelstein, M. (2020). The Role of Blockchain in Securing Sustainable Supply Chains from Cyber Threats. *Journal of Cleaner Production*, 259, 120883.

Ghosh, A. (2023). Assessing the Cybersecurity Risks in Sustainable Supply Chains: A Framework and Empirical Investigation. *Computers & Operations Research*, 139, 105270.

Hsiao, K. (2024). Cybersecurity Challenges and Solutions in Green Supply Chains. *Resources, Conservation and Recycling*, 181, 105349.

Kshetri, N. (2020). Cybersecurity and International Supply Chain Security: A Risk Governance Perspective. *Telecommunications Policy*, 44(1), 101896.

Lee, J. S., & Marciniak, R. (2021). Sustainable Supply Chain Security: Cyber Risk Assessment and Management. *Journal of Business Logistics*, 42(4), 307–319.

Li, S., & Berman, J. (2023). Leveraging Artificial Intelligence for Cybersecurity in Sustainable Supply Chains. *Decision Sciences*, 54(1), 143–169.

Martin, P., & Murphy, J. (2022). Enhancing Cyber Resilience in Sustainable Supply Chains: A Framework and Case Study. *International Journal of Production Research*, 60(5), 1483–1499.

McNeill, A., & Purvis, R. (2024). Digital Twins for Cybersecurity in Sustainable Supply Chains. *Computers & Security*, 111, 102562.

Oke, A., & Srinivasan, R. (2020). Cybersecurity in Sustainable Supply Chains: A Multi-Criteria Decision-Making Approach. *Annals of Operations Research*, 297(1–2), 277–301.

Pappas, I. O., & Mikalef, P. (2021). Protecting Sustainable Supply Chains: The Role of Blockchain Technology. *Journal of Enterprise Information Management*, 34(5), 1544–1566.

Qian, L., & Wang, L. (2022). Addressing Cybersecurity Risks in Sustainable Supply Chains: A Multi-Agent Reinforcement Learning Approach. *Transportation Research Part E: Logistics and Transportation Review*, 155, 102240.

Raghavan, V. (2023). Cyber Threat Intelligence Sharing in Sustainable Supply Chains: An Empirical Analysis. *Computers & Security*, 106, 102618.

Rana, R., & Bhambri, P. (2024). Healthcare Computational Intelligence and Blockchain: Real-Life Applications. In P. Bhambri, S. Rani, & M. Fahim (Eds.), *Computational Intelligence and Blockchain in Biomedical and Health Informatics* (pp. 155–168). CRC Press, Taylor & Francis Group. https://doi.org/10.1201/9781003459347

Rattan, M., Bhambri, P., & Shaifali. (2005). Institution for a Sustainable Civilization: Negotiating Change in a Technological Culture. In National Conference (Abstract Book) (p. 7). Presented at the "Technical Education in Globalized Environment-Knowledge, Technology & The Teacher" conference at SBBSIET, Padhiana, Jalandhar held during February 25–26, 2005.

Shanmuga, S. M., & Bhambri, P. (2024). Pulmonary and Lungs Nodule Classification using Deep Learning. In P. Bhambri, S. Rani, & M. Fahim (Eds.), *Computational Intelligence and Blockchain in Biomedical and Health Informatics* (pp. 320–332). CRC Press, Taylor & Francis Group. https://doi.org/10.1201/9781003459347

Sharma, R., Bhambri, P., & Sohal, A. K. (2020). Energy Bio-Inspired for MANET. *International Journal of Recent Technology and Engineering*, 8(6), 5580–5585. https://www.ijrte.org/wp-content/uploads/papers/v8i6/F8522038620.pdf. https://doi.org/10.35940/IJRTE.F8522.038620

Sharma, R., & Thomas, V. (2020). Cybersecurity Challenges in Sustainable Supply Chains: A Systematic Literature Review. *Information Technology & People*, 33(3), 760–786.

Siau, K., & Shen, Z. (2021). Blockchain Technology for Cybersecurity in Sustainable Supply Chains: A Conceptual Framework. *Industrial Management & Data Systems*, 121(7), 1467–1485.

Singh, M., Bhambri, P., Dhanoa, I. S., Jain, A., & Kaur, K. (2021). Data Mining Model for Predicting Diabetes. *Annals of the Romanian Society for Cell Biology*, 25(4), 6702–6712.

Singh, M., Singh, P., Kaur, K., & Bhambri, P. (2005). Database security. In Proceedings of National Conference (pp. 35–41). Presented at the "Future Trends in Information Technology" conference at SJPMLIET, Radaur, Yamuna Nagar held during March 6–7, 2005.

Srinivasan, R., & Oke, A. (2022). Managing Cybersecurity Risks in Sustainable Supply Chains: A Multi-Objective Optimization Approach. *Annals of Operations Research*, 308(1–2), 1–28.

Subramanian, N., & Gunasekaran, A. (2023). Cybersecurity in Sustainable Supply Chains: A Conceptual Framework. *International Journal of Production Research*, 61(5), 1450–1468.

Tang, C., & He, Q. (2020). Managing Cybersecurity Risks in Sustainable Supply Chains: A Hybrid Model of Fuzzy Analytic Hierarchy Process and Grey Relational Analysis. *International Journal of Production Economics*, 230, 107817.

Thomas, R., & Chen, L. (2021). Digital Twins for Cybersecurity in Sustainable Supply Chains: A Review and Outlook. *Journal of Manufacturing Systems*, 60, 632–644.

Trivedi, M., & Patel, V. (2022). Cybersecurity Threats and Vulnerabilities in Sustainable Supply Chains: A System Dynamics Approach. *Computers & Industrial Engineering*, 167, 107184.

Wang, C., & Cao, X. (2023). Blockchain Technology for Enhancing Cybersecurity in Sustainable Supply Chains: A Review and Future Directions. *Journal of Cleaner Production*, 326, 129590.

Wong, C. W. Y., & Lai, K. H. (2020). Cybersecurity Threats in Sustainable Supply Chains: A Systematic Review and Research Agenda. *International Journal of Production Research*, 58(6), 1682–1701.

Wu, T., & Ivanov, D. (2021). Cyber-Physical Security Risk Management in Sustainable Supply Chains: A Data-Driven Approach. *International Journal of Production Economics*, 239, 108096.

Xu, M., & Chen, H. (2022). Securing Sustainable Supply Chains: A Cyber Risk Assessment Framework. *Annals of Operations Research*, 306(1–2), 333–353.

Zhang, Y., & Lim, M. K. (2023). Enhancing Cyber Resilience in Sustainable Supply Chains: An Empirical Investigation. *Journal of Cleaner Production*, 318, 128476.

10 Decentralized Finance (DeFi) and Environmental Impact

Pankaj Bhambri and Marta Starostka-Patyk

10.1 INTRODUCTION

A variety of apps in the crypto-asset area referred to as "decentralized finance" (DeFi) rely on self-executing computer code, or "smart contracts," to disintermediate the offering of financial services. Thus far, DeFi has mostly been self-referential in the sense that it hasn't offered intermediation services to promote actual economic activity; instead, it has primarily made it easier to finance and trade crypto-assets. However, if tokenization of assets or the usage of applications based on DeFi by current financial institutions result in increased ties to conventional finance (TradFi), then this could alter in the future. We contend that many of the roles that TradFi performs and the risks associated with this intermediation are comparable to those of DeFi. Therefore, the same economic justification that has shaped financial regulation over decades can also be applied to the domains of cryptocurrency and DeFi. The intensity of failures in the market (externalities as well as data asymmetries) frequently increases the risks associated with DeFi. After contrasting the roles played by TradFi and DeFi, we demonstrate how DeFi is subject to regulations that safeguard consumers, uphold market integrity, and guarantee financial stability.

10.2 DECENTRALIZED FINANCE (DEFI) LITERATURE

Studies by Ozcan (2021), Katona (2021), Chohan (2021), Ozili (2022), Schar (2021), and others in the literature have shown how popular DeFi is because of its potential to replace traditional financial intermediaries, improve access to financial services, increase financial inclusion, speed up transactions, improve privacy and transparency, improve security, increase trust, and boost efficiency by reducing bank overhead costs. Yavin and Reardon (2021) demonstrated how DeFi might transform contemporary finance by increasing banking's accessibility and adaptability in society. Yavin and Reardon (2021) go on to suggest that banks enhance their own financial product and service offerings by utilizing DeFi solutions. DeFi, according to Meegan and Koens (2010), is an innovative framework that enables the delivery of decentralized financial services through the utilization of blockchain technology.

DOI: 10.1201/9781003484226-10

They demonstrated that while DeFi offers many advantages, there are also significant concerns. Meegan and Koens (2010) disagreed with the notion of using DeFi services in place of centralized finance (CeFi) services. Instead, they contended that in order for clients to profit from both systems, DeFi and CeFi must work together and coexist within the same financial system.

Research has assessed the challenges associated with DeFi. For instance, Zetzsche et al. (2020) offered a critical analysis of DeFi, arguing that it could potentially undermine traditional accountability mechanisms and undermine the effectiveness of traditional financial regulation as well as enforcement. The importance of regulating DeFi via "embedded regulation," which allows regulation to be seamlessly integrated into the design of DeFi, was underscored by Zetzsche et al. (2020). As Chen and Bellavitis (2020) illustrated, DeFi has the capacity to revolutionize the structure of contemporary finance, facilitate decentralized business models, and create fresh avenues for innovation and entrepreneurship. According to Chen and Bellavitis (2020), DeFi has the potential to increase decentralization in the financial system, but it currently faces some obstacles before realizing its full potential. According to Chohan's (2021) research, DeFi emphasizes disintermediation and decentralization in order to empower people in line with crypto-anarchist ideals. Further research by Chohan (2021) revealed that DeFi is frequently beset by a number of issues, such as market manipulation, distorted incentives, excessive short-termism, Ponzi schemes, and money-laundering difficulties, all of which seriously impede the widespread integration of DeFi's cutting-edge banking services into consumer experiences. As noted byChohan, in order for the transition to Emerging Finance (EmFi) to be effective, three important challenges must be resolved. The problems are: (1) organizational structure, operational model, and skills; (2) technical competency; and (3) leadership knowledge and commitment. EmFi, in accordance with Kore Fusion (2021), is capable of generating $7 trillion in revenues by 2030 and transforming the financial sector. The authors illustrated how the EmFi revolution will be propelled by the transition to e-commerce, the implementation of application programming interfaces (APIs), and the utilization of banking as a service (BaaS) providers. The authors decided that established companies and FinTech providers must establish their own EmFi strategies in order to capitalize on the EmFi potential. As stated by Teboul and Angelos Anastasiou (2021), the implementation of embedded financial services will be extremely expensive for businesses seeking to offer them to their customers. In addition, it necessitates substantial technological investment, specialized knowledge, and a profound understanding of regulatory, authorized, and compliance issues.

10.3 KEY COMPONENTS OF DECENTRALIZED FINANCE

DeFi is distinguished by a wide range of interrelated elements that collectively redefine conventional financial frameworks (Shanmuga & Bhambri, 2024). These essential elements constitute the fundamental components of a decentralized financial ecosystem, fostering transparency, connectivity, and innovation. This section offers a comprehensive examination of the key components that make up the essence of DeFi.

- Borrowing and lending platforms: DeFi borrowing and lending protocols provide the lending of digital assets with the opportunity to earn interest, as well as the borrowing of assets by offering security. Smart contracts streamline these procedures, eliminating the necessity for intermediaries (Singh et al., 2004). This component improves the availability of liquid assets, enabling users to generate profits from their cryptocurrency investments or obtain funds without depending on conventional banking institutions.
- Decentralized exchanges (DEXs): Decentralized Exchanges are crucial in the DeFi ecosystem as they enable direct trade of cryptocurrency between individuals without relying on centralized intermediaries. Decentralized exchanges (DEXs) leverage blockchain technology and employ smart contracts to carry out and finalize transactions. Users have the ability to engage in direct trading of digital assets from their wallets, which guarantees transparency and minimizes the risk associated with intermediaries.
- Decentralized autonomous organizations (DAOs): DAOs are governance frameworks in DeFi that enable participants to directly influence the decision-making procedures of decentralized protocols (Rana & Bhambri, 2024). Governance tokens confer voting privileges to users for suggestions pertaining to the development, enhancements, and modifications within the ecosystem. DAOs facilitate community involvement, enhance transparency, and enable decentralized governance.
- Liquidity mining and yield farming: It refers to the practice of users gaining rewards by supplying liquidity to decentralized protocols. Liquidity providers offer their digital goods to liquidity pools, receiving fees and governance tokens as compensation. This technique encourages users to actively engage in the expansion of DeFi platforms, hence improving liquidity and long-term viability.
- Insurance protocols: Insurance protocols in decentralized finance (DeFi) mitigate risks related to weaknesses in smart contracts and unforeseen circumstances. These protocols provide a decentralized form of insurance that enables users to safeguard their funds from potential hacks or failures of the protocol. The inclusion of insurance components plays a crucial role in reducing risk and fostering confidence in the decentralized finance ecosystem.
- Decentralized identity (DeID): Decentralized Identity solutions offer robust and privacy-enhancing techniques for user authentication in the context of DeFi (Singh & Bhambri, 2024). By utilizing blockchain technology, these solutions empower users to manage and distribute their identification information without depending on centralized authorities. DeID improves confidentiality, integrity, and user autonomy in decentralized financial applications.

10.4 SUSTAINABLE FINANCE

Sustainable finance (SuFi) was defined by Schumacher et al. (2020) as financial products and services that consider relevant environmental, social, and governance (ESG) considerations while funding and investing in the financial sector. According to Schoenmaker (2017), traditional finance only takes risk and financial return into account, whereas SuFi takes into account social, financial, and environmental returns in addition to long-term creation of value for the larger society. He maintained that insufficient private efforts and short-termism are the main barriers to SuFi. Fatemi and Fooladi (2013) argue that the conventional finance methodology for optimizing shareholder wealth is inadequate as a foundation for achieving sustainable wealth. This is due to its excessive focus on short-term results, which inadvertently compels numerous companies to shift the burden of environmental and social costs to external entities (Singh et al., 2005). Additionally, they contended that a great deal of undesirable results have been produced by an unflinching faith in the ability of markets to effectively identify long-term value repercussions of short-term results. According to Sandberg's (2018) argument, financial institutions have a social obligation to society to support sustainable financing. According to Kemfert and Schmalz (2019), a framework that pushes participants in the private financial market toward sustainability is necessary. Pimentel and Ramŭrez (2021) stressed the necessity of developing specific laws and rules that encourage participants in the financial sector to shift capital away from unsustainable investments and toward sustainable ones. According to Fullwiler (2016), blended-value investment, the realization that sustainability concerns can be linked to systemic risk, financial innovation to boost sustainability, and the development of SuFi-specific infrastructure are the four key trends driving the expansion of SuFi.

Hong et al. (2021) assert that the financial sector is under pressure to make a contribution towards constraining the rise in worldwide temperatures to 1.5–8°C above pre-industrial levels. The data indicates that financial institutions are facing pressure from governments and activists to allocate a portion of their portfolios to companies that are able to achieve net-zero emissions by 2050. According to Schoenmaker (2017), financial institutions are beginning to steer clear of unsustainable businesses due to risk concerns. Additionally, corporations that are leading the way in sustainable financing are increasingly making investments in sustainable businesses and initiatives in order to provide long-term value for the community. Nonetheless, insufficient private efforts and short-termism are the main barriers to SuFi.

According to Migliorelli (2021), there are too many inconsistent definitions, concepts, industry standards, and policy guidelines dominating the SuFi environment. According to the author, this kind of variability could impede the coherent advancement of the conceptual framework supporting SuFi and give birth to certain hazards that could undermine the agenda's legitimacy. Some of these risks include the rebranding of monetary flows, the disorderly adjustment of cost of capital ranges among industries, and the cleansing of products in an eco-friendly and sustainable manner. Ziolo et al. (2021) looked into the relationship between Sustainable Development Goals (SDGs) and SuFi. They conducted an analysis of Organization for Economic Cooperation and Development (OECD) member countries in Europe and contended that the SuFi model is essential to the implementation of SDGs and to guaranteeing that the goals take social and environmental sustainability into account. They discover that among the group of countries under analysis, the more sustainable the financial model, the more effectively the SDGs are achieved. Additionally, they discovered a substantial correlation between the SuFi model and the environmental (SDGs 11, 12, 13, and 15), social (SDGs 1, 3, 4, 5, 10, and 16), and economic (SDGs 8, 9, and 17) sustainability. Simultaneously develops several SuFi theories, such as the resource theory, priority theory, peer emulation theory, life span theory, positive signalling theory, and system disruption theory. According to Ozili (2022), these theories provide plausible justifications for how economic agents behave and respond toward SuFi.

10.5 BLOCKCHAIN AND SMART CONTRACTS

A new investment asset with few counterparts in history with regard to growth has emerged since Satoshi Nakamoto's 2008 release of the Bitcoin whitepaper, and particularly with the 2009 release of the Bitcoin software (Chan et al., 2019). The result of this crypto-punk ideal of a currency devoid of a centralized authority is an alluring investment vehicle that is risky (Grant & Hogan, 2015), requires a significant amount of energy (Corbet et al., 2021b), and has highly debatable financial characteristics. In fact, the COVID-19 epidemic brought large fluctuations in both the price of Bitcoin and the stock market, shedding light on issues related to safe havens, hedge funds, gold comparisons, market correlation, and other related topics (Mariana et al., 2021). Nakamoto's invention marked the start of a blockchain-based decentralization movement, which was later picked up by others looking to hone technical details. In order to achieve this, Buterin (2014) released the Ethereum whitepaper, which described a blockchain with fully computational features. The Ethereum program, which enables anyone to run code in a decentralized fashion and is also referred to as a smart contract, was developed shortly after (Ante, 2021). Ethereum uses smart contracts to democratize access to blockchain features. The following are some typical use cases for smart contracts:

- Developing online applications known as "web3" (the idea of a new age in the web;) or "dapps" (decentralized applications).
- The production, exchange, and custody of fresh tokens, commonly referred to as cryptocurrencies (Lee, 2019), as well as a few subtleties.
- Establishing and maintaining a self-sovereign identity in order to communicate with other users and blockchains (Liu et al., 2020).
- Supporting initiatives establishing initial coin offerings (ICOs), which are the virtual version of regulated and unregulated initial public offerings (IPOs) on the stock market (Bellavitis et al., 2021).
- Establishing entities known as decentralized autonomous organizations (DAOs) (Diallo et al., 2018), which enable the coordination of collectives through genuine voting, among other things.
- Registering and exchanging assets in non-fungible tokens (NFTs), with the ability to modify trades and uses in the future (Kong & Lin, 2021).

The primary characteristic of intelligent contracts is their ability to be carried out in any situation. Because of this, once an intelligent contract is implemented on a blockchain, its owner is powerless to halt it or alter its behavior. Because it fulfills the contract in this manner, no interaction with humans is required, and anyone can utilize it. This has three key implications: composability, permissionlessness, and decentralization. According to Catalini and Gans (2020), the absence of a central authority lowers the cost of transactions, monopoly costs, and gains from network effects. Additionally, it eliminates arbitrariness because anyone may utilize the contract with no authorization and because novel uses involving many smart contracts can be created by reason of the network's interoperability (Jain & Bhambri, 2006). Since then, the term "protocol" has gained popularity over "service" due to its distinctions from typical digital services. The service must be rendered by someone, but a protocol on a blockchain is just an open specification and an open-source implementation. However, there are hazards associated with the smart contract strategy (Grant & Hogan, 2015). Compared to standard digital items, there is a greater chance of an erroneous or intentional malfunction with this highly innovative technology. In actuality, the lack of regulations or their existence combined with anonymity serves as a draw for bad behavior. There is a flip side to smart contracts' unstoppable state, though. This is not feasible even in the event of new circumstances or if both parties decide to terminate the agreement. In conclusion, there are dangers and expenses associated with the elimination of a central authority, but they differ somewhat from the conventional ones. Decentralization, thus, creates a new paradigm for digital applications that facilitate human coordination of information and value.

10.6 FUTURE DIRECTIONS AND IMPLICATIONS OF DECENTRALIZED FINANCE

DeFi is leading the way in financial innovation and determining the next phase of global financial ecosystems. As the field of decentralized finance progresses, various potential paths and consequences arise, affecting not just the tech itself but also its wider societal and economic effects. This section examines the possible paths that DeFi could follow and the consequences for different parties involved.

- Trends and innovations: The future of DeFi is being shaped by continuing advances and developing trends, which will lead to constant innovation in the field. Promoting increased compatibility among various decentralized protocols, including Layer 2 scaling responses, and introducing innovative financial instruments are probable avenues of development. Moreover, the integration of DeFi with cutting-edge technologies such as non-fungible tokens (NFTs) as well as decentralized identity (DeID) has the potential to open up fresh opportunities for financial innovation and inclusivity.
- Scalability solutions: Tackling scalability difficulties is a crucial component of the future of DeFi. Implementing scalable alternatives, such as Layer 2 interfaces or sidechains, will be essential for improving transaction throughput and alleviating congestion on current blockchain networks. These innovations are intended to enhance user experience and meet the increasing need for decentralized monetary services.

- Regulatory evolution: Anticipated changes are likely to occur in the regulatory framework governing DeFi. Regulators worldwide are expected to modify their frameworks to tackle the distinct difficulties and opportunities posed by decentralized systems of finance. The regulatory discourse will focus on finding a balance between promoting innovation and safeguarding consumer protection.
- Decentralized governance maturity: Decentralized Autonomous Organizations (DAOs) along with governance tokens are expected to evolve and become more established in their respective roles (Rattan et al., 2005). Developing advanced governance models and procedures for decentralized decision-making is crucial for maintaining the long-term viability and flexibility of DeFi protocols. The incorporation of advanced DAO functionality can enable users to engage more actively in influencing the development of financial platforms.
- Cross-chain collaboration: The popularity of cross-chain interoperability is expected to increase, facilitating smooth collaboration and transfer of assets between diverse blockchain networks. The use of standard protocols allowing cross-chain communication has the potential to enhance connectivity and interoperability within the DeFi ecosystem. This, in turn, can diminish the existence of isolated systems and improve overall efficiency.
- Financial inclusion and accessibility: DeFi has the capacity to greatly influence financial inclusion by granting those who are unbanked, as well as underbanked, the opportunity to access financial services. With the increasing user-friendliness and accessibility of DeFi protocols, a wider range of people can now take advantage of decentralized lending, borrowing, and other kinds of financial activities, which promotes global financial inclusion.
- Environmental sustainability: The ecological ramifications of blockchain networks, specifically in proof-of-work consensus systems, have attracted considerable interest. Possible future developments may involve a greater acceptance of environmentally friendly consensus mechanisms such as proof-of-stake, together with continuous endeavors to reduce the environmental impact linked to decentralized financial activity.

10.7 CONCLUSION

DeFi is a powerful and revolutionary force that is reshaping the conventional structure of the financial industry. As we explore the complex aspects of blockchain technology, digital currencies, and the fundamental elements that make up the foundation of DeFi, it becomes clear that this shift in thinking goes beyond simple technological advancement. It represents a major reorganization of financial systems, promoting ideas of decentralization, accessibility, and inclusion. Exploring the technology underpinnings, operational complexities, and future directions of DeFi reveals its capacity to fundamentally transform the way people engage with financial services. Smart contracts, exchanges that are decentralized, DAOs, yield agriculture, and other elements jointly contribute to a decentralized ecosystem. In this ecosystem, users can engage in lending, borrowing, trading, and governance activities without relying on traditional middlemen. Anticipating the future, the trajectory of DeFi reveals itself through various auspicious pathways. The future of the industry will be influenced by advancements and current patterns, with a focus on improved compatibility, expandability solutions, and collaboration between different blockchain networks taking the lead. The development of decentralized governance, the progression of regulations, and a dedication towards ecological sustainability are crucial factors that will shape the path of decentralized systems of finance. As we anticipate the future, the social and economic consequences of DeFi become more evident. Financial inclusion is becoming increasingly important, as it provides anyone worldwide with the chance to receive services and engage in economic activity. The democratization of financial institutions and the ability to fundamentally change established economic institutions highlight the significant influence of DeFi on the wider socio-economic environment. Nevertheless, there are ongoing obstacles such as regulatory factors, security apprehensions, and the requirement for sustained technological advancement. Achieving a harmonious equilibrium between innovation as well as regulation will be crucial for ensuring the long-term and environmentally friendly expansion of DeFi. Decentralized financing is not solely a technological occurrence, but rather a driving force for a future of finance that is more comprehensive, transparent, and fair. As academics, programmers, regulators, and users actively participate in the advancement of DeFi, the ongoing story signifies the collaborative effort to reshape our perception and interaction with finance. The evolution of the decentralized financial system is characterized by its dynamic nature, great potential, and limitless possibilities, as it drives us towards a future where financial systems are decentralized and accessible to all.

REFERENCES

Ante, L. (2021). Non-fungible token (NFT) markets on the Ethereum blockchain: Temporal development, cointegration and interrelations. SSRN. https://ssrn.com/abstract=3904683 or http://dx.doi.org/10.2139/ssrn.3904683

Bellavitis, C., McNaughton, R. B., & Fisch, C. (2021). COVID-19 and the global venture capital landscape. Small Business Economics, 1–25. https://doi.org/10.1007/s11187-021-00547-9

Catalini, C., & Gans, J. S. (2020). Some simple economics of the blockchain. Communications of the ACM, 63(7), 80–90. https://doi.org/10.1145/3359552

Chan, E. K. F., Timmermann, A., Baldi, B. F., Moore, A. E., Lyons, R. J., Lee, S. S., Kalsbeek, A. M. F., Petersen, D. C., Rautenbach, H., Förtsch, H. E. A., Bornman, M. S. R., & Hayes, V. M. (2019). Human origins in a southern African palaeo-wetland and first migrations. *Nature*, 575(7781), 185–189. https://doi.org/10.1038/s41586-019-1714-1

Chen, Y., & Bellavitis, C. (2020), "Blockchain disruption and decentralized finance: The rise of decentralized business models", *Journal of Business Venturing Insights*, Vol. 13, C. https://doi.org/10.1016/j.jbvi.2019.e00151.

Chohan, U.W. (2021), "Decentralized Finance (DeFi): An emergent alternative financial architecture", Critical Blockchain Research Initiative (CBRI) Working Papers. https://doi.org/ 10.2139/ssrn.3791921.

Corbet, S., Hou, Y. (G.), Hu, Y., Larkin, C., Lucey, B., & Oxley, L. (2021b). Cryptocurrency liquidity and volatility interrelationships during the COVID-19 pandemic. Finance Research Letters, 45, 102137. https://doi.org/10.1016/j.frl.2021.102137

Fatemi, A. M., & Fooladi, I. J. (2013), "Sustainable finance: a new paradigm", *Global Finance Journal*, Vol. 24, No. 2, pp. 101–113.

Fullwiler, S. (2016), "Sustainable finance: Building a more general theory of finance", *Routledge Handbook of Social and Sustainable Finance*, Routledge, pp. 17–34.

Grant, G. H., & Hogan, R. T. (2015). Bitcoin: Risks and controls. *Journal of Corporate Accounting & Finance*, 26, 29–35. https://api.semanticscholar.org/CorpusID:154323815

Hong, H., Wang, N., & Yang, J. (2021), "Welfare consequences of sustainable finance", National Bureau of Economic Research, No. w28595, available at: https://www.nber.org/papers/w28595.

Jain, V. K., & Bhambri, P. (2006), *Fundamentals of Information Technology & Computer Programming* (4th reprint ed.). Katson Publishers. ISBN 81-88458-59-7.

Kane-Diallo, A., Srour, B., Sellem, L., Deschasaux, M., Latino-Martel, P., Hercberg, S., Galan, P., Fassier, P., Guéraud, F., Pierre, F. H., Kesse-Guyot, E., Allès, B., & Touvier, M. (2018). Association between a pro plant-based dietary score and cancer risk in the prospective NutriNet-Santé cohort. *International Journal of Cancer*, 143(9), 2168–2176. https://doi.org/10.1002/ijc.31593

Katona, T. (2021), "Decentralized finance: The possibilities of a blockchain 'money lego' system", *Financial and Economic Review*, Vol. 20, No. 1, pp. 74–102.

Kemfert, C., & Schmalz, S. (2019), "Sustainable finance: Political challenges of development and implementation of framework conditions", *Green Finance*, Vol. 1, No. 3, pp. 237–248.

Kong, D. R., & Lin, T. C. (2021). Alternative investments in the Fintech era: The risk and return of Non-Fungible Token (NFT). Available at SSRN 3914085.

Kore Fusion (2021), "Embedded finance: This decade's largest creator of value", A Kore Fusion Report, New York, available at: https://korefusion.com/embedded-finance-this-decades-largest-creator-of-valu/.

Lee, A. J. (2019). U-statistics: Theory and Practice. Routledge.

Liu, H., Gong, P., Wang, J., Clinton, N., Bai, Y., & Liang, S. (2020). Annual dynamics of global land cover and its long-term changes from 1982 to 2015 [Dataset]. PANGAEA. https://doi.org/10.1594/PANGAEA.913496

Mariana, C. D., Ekaputra, I. A., & Husodo, Z. A. (2021). Are Bitcoin and Ethereum safe-havens for stocks during the COVID-19 pandemic? Finance Research Letters, 38, 101798. https://doi.org/10.1016/j.frl.2020.101798

Meegan, X., & Koens, T. (2010), "Lessons learned from decentralised finance (DeFi)", ING, available at: https://new.ingwb.com/binaries/content/assets/insights/themes/distributed-ledger-technology/defi_white_paper_v2.0.pdf.

Migliorelli, M. (2021), "What do we mean by sustainable finance? Assessing existing frameworks and policy risks", *Sustainability*, Vol. 13, No. 2, p. 975.

Ozcan, R. (2021), "Decentralized finance", *Financial Ecosystem and Strategy in the Digital Era* (pp. 57–75). https://doi.org/10.1007/978-3-030-72624-9_4.

Ozili, P. K. (2022b), "Decentralized finance and cryptocurrency activity in Africa", Contemporary Studies in Economic and Financial Analysis, CSEF, Vol. 109, available at: https://ssrn.com/abstract54001707.

Pimentel, B. M., & Ramırez, G. (2021), "Sustainable finance and its sustainability", Working Paper. Plaid and Accenture (2021), "Embedded finance: financial services whenever and wherever customersneed them", A Plaid Report, California.

Rana, R., & Bhambri, P. (2024), "Artificial Intelligence and Internet of Things (IoT) facilitated digital twin for Industry 4.0 application domains", in S. Rani, P. Bhambri, S. Kumar, P. K. Pareek, & A. A. Elngar (Eds.), *AI-Driven Digital Twin and Industry 4.0: A Conceptual Framework with Applications* (1st ed., pp. 168–181). CRC Press. https://doi.org/10.1201/9781003395416.

Rattan, M., Bhambri, P., & Shaifali. (2005), "Information retrieval using soft computing techniques", In National Conference (pp. 58–60) on Bio-informatics Computing. Presented at TIET, Patiala held during March 18–19, 2005.

Sandberg, J. (2018), "Toward a theory of sustainable finance", *Designing a Sustainable Financial System* (pp. 329–346). Cham: Palgrave Macmillan.

Schar, F. (2021), "Decentralized finance: On blockchain-and smart contract-based financial markets", Second Quarter 2021, Federal Reserve Board of St. Louis Review, pp. 153–174. https://doi.org/10.20955/r.103.153-74.

Schoenmaker, D. (2017), *Investing for the Common Good: A Sustainable Finance Framework*. Brussels: Bruegel, p. 80.

Schumacher, K., Chenet, H., & Volz, U. (2020), "Sustainable finance in Japan", *Journal of Sustainable Finance and Investment*, Vol. 10, No. 2, pp. 213–246.

Shanmuga, S. M., & Bhambri, P. (2024), "Pulmonary and lungs nodule classification using deep learning", in P. Bhambri, S. Rani, & M. Fahim (Eds.), *Computational Intelligence and Blockchain in Biomedical and Health Informatics* (pp. 320–332). CRC Press, Taylor & Francis Group. https://doi.org/10.1201/9781003459347.

Singh, M. P., & Bhambri, P. (2024), "AI-driven digital twin conceptual framework and applications", in S. Rani, P. Bhambri, S. Kumar, P. K. Pareek, & A. A. Elngar (Eds.), *AI-Driven Digital Twin and Industry 4.0: A Conceptual Framework with Applications* (1st ed., pp. 127–141). CRC Press. https://doi.org/10.1201/9781003395416.

Singh, P., Singh, M., & Bhambri, P. (2004), "Interoperability: A problem of component reusability", In International Conference (Abstract Book) (p. 60). Presented at the "Emerging Technologies in IT Industry" at PCTE, Ludhiana on November 27, 2004.

Singh, P., Singh, M., & Bhambri, P. (2005), "Embedded systems", in U.G.C. Sponsored National Seminar (pp. 10–15). Presented at the "Embedded Systems" seminar at KMV, Jalandhar held during January 24–25, 2005.

Teboul, L., & Angelos Anastasiou, A. (2021), "The embedded finance journey: Innovation that differentiates the customer experience", A Goldman Sachs report on transaction banking, available at: https://www.goldmansachs.com/what-we-do/transaction-banking/insights/baas.pdf.

Yavin, O., & Reardon, A. J. (2021), "What digital banks can learn from decentralised finance", *Journal of Digital Banking*, Vol. 5, No. 3, pp. 255–263.

Zetzsche, D. A., Arner, D. W., & Buckley, R. P. (2020), "Decentralized finance", *Journal of Financial Regulation*, Vol. 6, No. 2, pp. 172–203.

Ziolo, M., Bak, I., & Cheba, K. (2021), "The role of sustainable finance in achieving Sustainable Development Goals: does it work?", *Technological and Economic Development of Economy*, Vol. 27, No. 1, pp. 45–70.

11 Green Data Centers and Renewable Energy Sources

Kurbanov Khayrillo Abdurasulovich, Karlibayeva Raya Khodjabaevna, and Karlibayeva Gulshat

11.1 INTRODUCTION TO GREEN DATA CENTERS

In Uzbekistan, computer technology is advancing, formed on the basis of client-server architecture, which has attracted interest from many users towards new, promising working conditions. These users include individuals, medium-sized companies, and large corporations. To ensure scalability and reliability of operation, the server-side requirements encompassed the entire service infrastructure. Currently, the term "data center" or "data processing center" (DPC) became established, referring to comprehensive solutions that encompassed purpose-built facilities for powerful computers, conditions for uninterrupted power supply, and infrastructure development. With the development of outsourcing of business processes to external resources, data center technologies began to evolve, geared toward maximizing resource utilization while tightening service quality standards. Both organizations and governmental bodies require flexible access to external resources and have the task of ensuring continuous business process continuity, necessitating the planning of specialized facilities that guarantee enhanced security for data storage, processing, and transmission. There are certain benchmarks building up necessities for the plan of information centers; right now, to decide the unwavering quality of a center, the American standard created for information centers by the broadcast communications industry is utilized, presenting layouts for designing information centers and categorizing them by an unwavering quality level. In Uzbekistan, there are no specific standards with which data centers are equipped. If a company wants to offer international services from their data centres, they need to adhere to local communication facility standards and follow the foreign TIA942 standard. Uzbekistan faces a complex set of issues that require solutions.

Inside the system of the "Digital Uzbekistan-2030" strategy, the primary high-performance information capacity and handling center has been built up in Uzbekistan.

Within the center of Uzbekistan, in collaboration with the Chinese company Huawei, a framework with a capacity of 20 petabytes has been introduced; 580 servers and diesel generators with a capacity of 5 megawatts guarantees the consistent

 DOI: 10.1201/9781003484226-11

operation of the framework, while building cooling and fire protection systems to meet cutting-edge prerequisites. Overall, the information center is certified as TIER III by the global company Uptime Institute.

However, at present, most Uzbek companies and organizations use their own server rooms and small data centers. Moreover, few wish to engage in the installation and operation of engineering systems, partly due to the lack of necessary expertise. Therefore, it is not surprising that there is significant interest in all-in-one solutions, including all necessary engineering components of data centers: power distribution units, cooling facilities, racks, cabinets, as well as fire safety systems. Such data centers can vary in size from microdata centers with a single rack to containerized and modular ones with several tens or even hundreds of racks. Solutions of this class were presented by several companies at the Eurasia Data Center & Cloud Forum, including globally renowned companies such as Delta Electronics and Schneider Electric, as well as manufacturers with Russian roots – Astergo, ITK, and NAG.

The topic of uninterruptible power supply (UPS) systems and energy storage has become one of the key issues today. Such solutions allow for uninterrupted supply of quality electricity to IT equipment, thereby ensuring continuous provision of IT services. For Uzbekistan, this issue is doubly important, as the country has a relatively low quality of their energy system and faces many power shortages. Power outages and voltage fluctuations often occur, and customers encounter difficulties in obtaining technical specifications for connecting to the technological networks.

Schneider Electric and Delta Electronics are among the global leaders in the field of UPS, and their products are characterized by some of the highest energy efficiency and power density indicators. However, even when the efficiency of modern UPS systems seems to have reached its maximum, leaders continue to work on improving it. Delta Electronics' announcement in Tashkent about the release of a new UPS with a double conversion efficiency of 97.5 percent at the beginning of the next year was impressive.

Another important element of data centers is the cable system (Chakrvarthy, 2001). In Uzbekistan, solutions in this area have been presented by Siemon, a company that holds leading positions worldwide, as well as the aforementioned Astergo and ITK. High-density and high-performance optical systems, which are part of the arsenal of these companies, are predominantly used in data centers. Intelligent management systems are efficient for automating structured cabling system (SCS) operation procedures. Siemon offers such a system, while ITK is developing a similar solution.

It is understood that UPS, SCS, and other "engineering" components are necessary to ensure the main goal – the operation of IT services. Increasingly, and even for critical applications, cloud services are being used (Bhambri & Mangat, 2005). The CEO of the cloud provider Cloupard reported that clouds in Uzbekistan are already being utilized by two-thirds of companies.

Huawei 3Com (H3C) presented its comprehensive solutions for building cloud platforms at the Eurasia Data Center & Cloud Forum. Having completed a number of major projects in China, the company "possesses a full stack of solutions for building data centers and clouds." This includes switches (including 400G core switches), servers, software-defined storage (SDS), and virtualization systems. The company's developments in the field of liquid cooling for servers and switches are of great interest.

The research center, Data Center Dynamics, conducted an analysis of the operation of modern data centers and concluded that the majority of data center operators are firms from the information technology sector (37.6 percent of the total number of racks), the financial sector (23.9 percent of the total number of racks), and telecommunications operators (19.7 percent of the total number of racks), with construction companies owning 12.2 percent of the total number of racks. Government organizations account for 6.6 percent of the total number of racks.

Taking computational speed as a qualitative characteristic, it is necessary to identify a certain number of data centers from the total amount that surpass others in this indicator. The term "hyperscale" alludes not just to the measure of an information center, but moreover to its arrange design and hardware approach. As it stands, a couple of worldwide computerized companies are able to make such offices, for example, AWS, Microsoft, Yahoo, Twitter, Baidu, and Oracle. The research organization, Synergy Research Group, inspected the 24 biggest cloud and web benefit companies within the world. Most of them are within the United States, with America driving the positioning in terms of the number of hyperscale information centers with 135 offices. China and Japan follow with 24 and 21 hyperscale centers, respectively. Filling out the best five are the UK with 15 offices and Australia with 12. The development of hyperscale data centers is steadily increasing; it is forecasted that there will be an additional 100 such centers added to the existing 300 worldwide, surpassing the 400 mark. On average, major administrators such as Amazon, Google, or Microsoft have around 13 capable information centers dispersed around the world, but as we have seen, most of these offices are found within the United States. In terms of the rate of the entire world, America possesses 45 percent of computational capacity.

Overall, according to forecasts by the research agency Gartner, global IT expenditures will reach $3.5 trillion in 2022, which is 1.4 percent higher than in 2016. However, the development rate has moderated compared to the past quarter's estimate by 2.7 percent. It is expected that in 2022, the data center systems segment will grow by only 0.3 percent. There are several reasons for this. Many corporations are abandoning their own projects of building corporate data centers, preferring to lease resources from established brands. Moreover, preference is given to operators who strive to use energy-efficient technologies. Another reason is the increasing demands for the stability and security of new data centers, which is a natural consequence as the value of data significantly increases. Thus, it can be concluded that the data storage and processing market has a positive dynamic, indicating a constant need for this type of activity.Data centers are high-energy consumers and at the same time, powerful sources of heat. These two indicators are a challenge for developers of modern data centers and are the main criteria for assessing energy efficiency. Let's start with the heat dissipation problems. The physical environment of the data center is strictly regulated and closely monitored. The regulating report, "Thermal Guidelines for Data Processing Environments," prescribes a temperature run of 18–27°C. The temperature within the information center rises as the electrical energy utilized warms the discuss around the gear.

The larger part of information centers utilize chillers or discuss conditioners to cool gear, devouring 30–70 percent of all energy utilized. Typical data centers can increase cooling costs from 50 cents to $1.50 or more for every dollar of electricity they use. The second energy efficiency indicator is the data center's electricity consumption. Here, it is necessary to mention very important energy usage effectiveness (PUE) coefficient proposed by the international organization Green Grid in 2007. The term refers to the ratio of the total power consumed by the entire data centre to the power consumed specifically by the IT equipment used for processing and storing data within that data centre (Singh et al., 2005). The coefficient value can range from 1.0 to infinity. When the coefficient value approaches 1.0, it indicates 100 percent efficiency – all energy is used only by IT equipment, i.e., server racks, and is not used elsewhere in the facility, including lighting, air conditioning, and server rack cooling. However, practically achieving this indicator is difficult. Green Grid provides a table that allows for an assessment of a data center's energy consumption organization based on the figures obtained when evaluating PUE parameters (Table 11.1).

Monitoring the environmental state of modern data centers: An increasing number of data center clients, when choosing a service provider, are interested in environmental solutions implemented by operators at their facilities, in addition to economic indicators. And although many operators actively advertise their achievements in "greening" data processing processes, obtaining objective and up-to-date information from independent sources remains a challenge. Greenpeace annually publishes the results of its "Clean Clicks" research, presented in the form of a ranking of the world's largest environmental polluters among high-tech companies, with the aim of motivating the transition of production processes to energy from renewable sources. As an example of such a ranking, one can consider the indicators of internet search engine operators in terms of their energy consumption (Table 11.2). Clients of operators whose resources are hosted in corresponding data centers demand that their digital infrastructure be connected to clean sources of electricity. These clients will then utilize this data to gain advantages.

TABLE 11.1
Evaluating PUE Parameters

Assessment	PUE
Platinum – perfect organization of the information center	<1.25
Gold – ideal organization of the information center	1.25–1.43
Silver – great organization of the information center	1.43–1.67
Bronze – palatable organization of the information center	1.67–2
Recognize – improvements are possible	2–2,.5
Not recognized – measures to improve energy efficiency are not accepted	>2.5

Source: Developed by the authors

TABLE 11.2
The Rating of Internet Search Engine Operators Based on their Energy Consumption

	Final Grade	Clean Energy (%)	Natural Gas (%)	Caul (%)	Unclear (%)
Baidu	F	24	3	67	3
Bing	B	32	23	31	10
Daum.net	C	2	19	39	31
Google.com	A	56	15	14	10
Nate	F	2	19	39	31
Naver	D	2	19	39	31
Yahoo	B	74	5	12	6
Zum	F	2	19	39	31

Source: Developed by the authors

The competitiveness of renewable energy sources, compared to those using fossil fuels, increases due to price security in long-term contracts.

The competitiveness among IT companies and the affiliation of brand character with renewable energy sources are taken into consideration, given the developing concern about climate change among representatives and clients.

These data are partially confirmed by a recent survey of retail clients of colocation providers conducted by Data Center Knowledge. The survey showed that 70 percent of corporate clients pay increased attention to environmental issues when choosing commercial data center providers. About one-third of respondents expressed a desire for providers to offer the option of operating energy infrastructure with electricity from renewable sources. Therefore, the efforts of data center operators to make their data centers "greener" are in line with modern trends. According to Greenpeace, about two dozen major American commercial data center operators and internet companies have announced commitments to transition to 100 percent renewable energy.

Innovative solutions to data center problems: A key focus of modern innovation is their energy efficiency, as "smart things," "smart homes," and "smart cities" technologies not only significantly increase the demand for new computing power but also greatly increase energy consumption. Already, for example, in Switzerland, 8 percent of the country's total energy consumption is related to the functioning of the internet. And this share will only grow. As unused information centers are required to back the foundation of creating advances, administrators in this field are progressively centering on creating innovations that offer assistance balancing the rise in energy utilization with a "green" point of view for productive information centers of tomorrow.

Based on distributed reports and industry improvement plans inside the "Industry 4.0" concept, the number of ventures utilizing cloud computing is consistently expanding. Increasingly, gadgets and sensors are being associated with the web, permitting trade proprietors to always screen, degree, and alter their generation, executing advanced coordination of topographically conveyed offices. According to Gartner's estimate, there are, around 4.9 billion associated gadgets; and Cisco specialists anticipate up to 50 billion associated gadgets by 2025, Anticipating an increase in demand for this course in the next three years. Customers, particularly, will utilize these organized gadgets within the "smart home" portion. For example, a "smart" fridge will be able to make a modern shopping list or offer to reorder well-known items direct from the store's administration server.

Geographically, the majority of German information centres are mostly located in Frankfurt am Main. This concentration is mainly attributed to the city's aspiration to be a global financial hub, vying with London, which declared its departure from the European Union as a result of Brexit. The expanding drift of digitizing the economy and society leads to a major request for computing control. Modern advancements, such as "Industry 4.0," machine learning, or expanded reality, do not only refer to a new kind of technology but also represent a growing demand for reliable and efficient IT infrastructure. Germany currently holds the top position in supporting and establishing IT frameworks, positioning itself as the largest manufacturer of data centres in Europe and the third largest globally, thanks to its continuous growth and development in this field. . Geographically, the majority of German information centres are mostly located in Frankfurt am Main. This concentration is mainly attributed to the city's aspiration to be a global financial hub, vying with London, which declared its departure from the European Union as a result of Brexit. One such urban center is the fifth largest information center within the world, covering a region of roughly 700,000 square feet. On a worldwide scale, it has generally been the case that most expansive data centers are found within the United States, supporting the advancement of one of the world's most capable economics. Applying different parameters, information centers expend from three to five percent of the world's control.

Data center operators who have already focused on measuring and optimizing energy consumption in their centers have made significant progress in energy cost savings. In particular, the issues and technologies of energy-efficient and economical server cooling are at the forefront of operators' attention to reduce their significant operating expenses. Energy investment funds not only save money but also make a significant commitment to climate protection. To this conclusion, cooling strategies are getting to be more developed and versatile. Another technique that information center administrators utilize to become more energy productive is the utilization of characteristic assets. Typically, to dissipate heat and thus cool data center premises, so-called chillers are used – powerful, energy-intensive devices that circulate and supply air processed in a central air conditioner to the premises, as well as heat and/or cool the air (Bhambri & Singh, 2006). Recently, the idea of using free cooling technology – free cooling – has been gaining popularity. Free cooling has tremendous energy-saving potential, allowing the elimination of the most energy-intensive links in the technological cooling and heat dissipation chain.

In one of its data centers operating in Finland, Google uses seawater for cooling, a practice considered today as an energy-efficient alternative to expensive chillers. The Swedish company Interxion has further developed this idea and now utilizes the same seawater across multiple data centers, resulting in additional savings. This is achieved through the modularity of data centers. After cooling the first module, the water remains cold enough to be used for cooling the next module. The water temperature supplied to the first module is 6°C and it reaches 12°C at the outlet. Passing through the second module, the water heats up to 18°C and to 24°C through the third. The heat absorbed by the water is then released to heat local residential homes and offices. Interxion recouped its initial investments in approximately one year: the cost of electricity using seawater cooling equipment amounts to $0.03 per kWh. Upon full-scale project implementation, the data center will occupy a technical space of 2,200 square meters (23,680 sq. ft.) in full operation, with a power capacity of up to 3.5 MW.

Facebook has announced plans to create infrastructure for using the heat generated by its data center equipment in Odense, Denmark, to heat nearby homes. It is reported that after heat recovery, the thermal energy will be transferred through the city's centralized heating network. Facebook engineers expect the new system to heat 6.9 thousand residential properties. Developers believe that such heat recirculation processes will allow Facebook to position its data centers as beneficial neighbors. Thus, it can be confidently stated that a new business model for data center operations near centralized heating networks is emerging, which is much more attractive than operating in areas with only cheap land. This model makes the data center operation project more financially viable and attractive, both for data center operators and for municipal authorities seeking to reduce the environmental impact and urban budget burden.

It is clear that the presented successful interaction will not work for every city with powerful data centers. The model will only work in places that are cold enough and where living conditions necessitate an interest in obtaining heat, and data centers must be powerful enough to have significant heat surpluses. Despite its limited applicability, this model has already sparked interest. The Swedish government entered the IT technology market to implement a concept aimed at expanding the scale of centralized heating and attracting more data centers to Sweden. Sweden has announced tax incentives for data centers, reducing the electricity tax rate by 97 percent, effectively reducing electricity costs by approximately 40 percent, exempting them from the electricity tax in Sweden. Considering that Sweden currently has the lowest electricity prices in Europe and centralized heating in its cities, and given its cold climate, it becomes evident that this country will soon have the most favorable conditions for the development of data centers such as those operated by Facebook, meeting local computing power needs. Additionally, it is worth mentioning another Swedish innovation: the development of Stockholm "data parks," where city authorities prepare a site within the city limits for a data center. The site includes connection to the city's fiber optic infrastructure and centralized heating system. The park provides cooling for tenant equipment, providing heat pumps in exchange for the use of their waste heat – an innovative solution: "cooling as a service." At full capacity, the data center located in the park will be able to transfer up to 30 million kWh of thermal energy per year to the central heating network. This is enough to heat radiators in 10,000 residential properties. City authorities hope that by using this technology, they can satisfy 10 percent of the city's thermal energy needs.

Thus, it becomes evident that the demand for "green" solutions for data centers will be driven by factors such as the need to reduce operational expenses, tighter government regulations aimed at reducing carbon footprint, and the desire of corporate clients to enhance their image. Overall, the initiatives of leading internet companies to "green" high-cost industries such as data centers based on energy-efficient principles serve as a stimulus for further development of green technologies, and the transition to 100 percent renewable energy sources has become a catalyst for involving a broader range of corporations in this process. Moreover, investments in renewable energy sources for data centers help stimulate other industries unrelated to data processing processes.

Within the advanced world, where innovation is a necessary portion of our day-to-day lives, the request for information centers has altogether expanded. Be that as it may, information centers devour a considerable sum of energy, which raises operational costs but also hurts the environment. To address this issue, numerous companies have started to actualize green information centers, which are pointed at lessening energy utilization and expanding maintainability. Green information centres employ energy-efficient technologies, renewable energy sources, and other sustainable practices to minimise carbon footprints and reduce environmental impact. Here are a few key points regarding the concept of green information centers:

1. Energy-efficient advances: Green information centers utilize energy-efficient advances such as virtualization and cloud computing to diminish energy utilization and raise efficiency. By giving a solidified see of assets, virtualization permits information centers to utilize fewer servers, driving to lower energy utilization (Dumitrescu,2018). Cloud computing, on the other hand, empowers companies to get to computing assets on request, decreasing the requirement for exclusive information centers and minimizing energy utilization.
2. Renewable energy sources: Green information centers utilize renewable energy sources such as sun power and wind control to play down their carbon footprint. For illustration, Facebook's information center in Sweden is

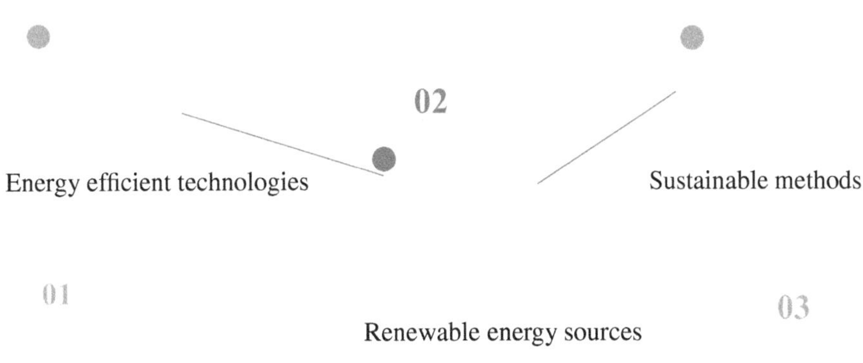

FIGURE 11.1 Getting to know eco-friendly data centers. Source: Developed by the author.

totally based on renewable energy sources, counting hydroelectric control and wind energy.

3. Eco-houses: Green information centres implement practical techniques such as minimising resource waste, conserving water, and using environmentally-friendly products to reduce their ecological impact.For illustration, Google's information center in Finland employs seawater from the Baltic Ocean for cooling and minimizes the utilization of freshwater assets.

Green information centers speak to an exceptional arrangement for lessening energy utilization and improving supportability. By implementing energy-efficient technologies, utilising renewable energy sources, and adopting feasible practices, corporations can significantly reduce their carbon footprint and make a positive contribution to environmental conservation.

11.2 THE SIGNIFICANCE OF ENERGY PRODUCTIVITY IN INFORMATION CENTERS

Information centers have ended up playing a role in an essentail portion of our advanced life, serving as the spine of the computerized era. They store and prepare endless sums of information each day, making them one of the foremost energy-intensive offices within the world. This towering energy utilization has pushed the world to critical natural issues, counting carbon emissions and global warming. Hence, it is pivotal to upgrade the energy proficiency of information centers by receiving maintainable arrangements to diminish their carbon footprint.

Here are a few reasons why energy effectiveness is vital in information centers:

1. Reduce costs: Energy utilization is the foremost critical operational cost for information centers, accounting for up to 40 percent of the whole budget. By expanding energy effectiveness, information centers can reduce energy utilization and save on their energy bills.
2. Reduce carbon footprint: The carbon footprint of information centers may be a critical natural issue. Information centers within the United States alone expend roughly 91 billion kilowatt hours of power each year, emitting over 100 million metric tons of carbon contamination. By receiving energy-efficient systems, information centers can diminish their carbon footprint and contribute to a maintainable future.
3. Improve quality: Over-utilizing energy can lead to overheating and framework failures in information centers. By expanding energy proficiency, information centers can diminish the hazard of framework failures, guaranteeing dependable and continuous operations.
4. Compliance with administrative prerequisites: Governments and administrative bodies are progressively forcing strict energy proficiency directions on information centers. By complying with these rules, information centers can dodge fines and punishments, guaranteeing smooth operations.

Financial productivity is significant for the economical operation of information centers. By receiving energy-efficient systems such as virtualization, effective cooling frameworks, and renewable energy sources, information centers can decrease energy utilization, carbon footprints, and operational costs thereby giving solid and continuous administrations to their clients.

11.3 DCCI'S COMMITMENT TO SUPPORTABILITY

Within the advanced world, where environmental issues are at record highs, maintainability has ended up a buzzword in each industry. The information center industry is no exception. As information centers devour a critical sum of energy, the industry's carbon footprint is considerable (Bhambri et al., 2020). Subsequently, it is pivotal to address the issue of energy effectiveness and maintainability in information centers. DCCI (Information Center Company Inc.) is taking steps toward a more feasible future. From utilizing renewable energy sources to diminishing their carbon footprint, DCCI is committed to supportability. Here are a few ways in which DCCI advances energy proficiency and maintainability:

1. Renewable energy: DCCI points to utilizing renewable energy sources such as sun-based and wind control to decrease their carbon footprint. They have introduced sun-oriented panels on their housetops and have taken steps to buy renewable energy credits to counterbalance their energy utilization.
2. Energy-efficient hardware: DCCI utilizes energy-efficient gear in their information centers, making a difference to diminish energy utilization. For instance, they utilize low-energy servers and utilize progressed cooling frameworks that use less energy.
3. Carbon offsetting: DCCI looks to diminish their carbon footprint by offsetting their outflows through carbon credits. They collaborate with different organizations that help them in offsetting their carbon emissions.
4. Reusing: DCCI reuses their IT hardware such as servers, switches, and capacity gadgets, to diminish waste and advance maintainability. Moreover, they reuse other materials utilized within the information center such as paper and cardboard.
5. Maintainable operations: DCCI enhances economic operations by encouraging its employees to adopt cost-effective practices, such as reducing paper and energy use. Additionally, they have a dedicated team focused on promoting sustainable practices inside the company.

DCCI endeavors to upgrade energy proficiency and supportability in their information centers. They have implemented many sustainable practices, such as utilising renewable energy sources and promoting sustainable operations, in order to reduce their carbon footprint. With these endeavors, they are not only making a difference to work toward a more economical future but also driving the information center industry.

Information centers expend a noteworthy sum of energy, and it is vital to decrease their carbon footprint to relieve environmental harm. Information Center Company

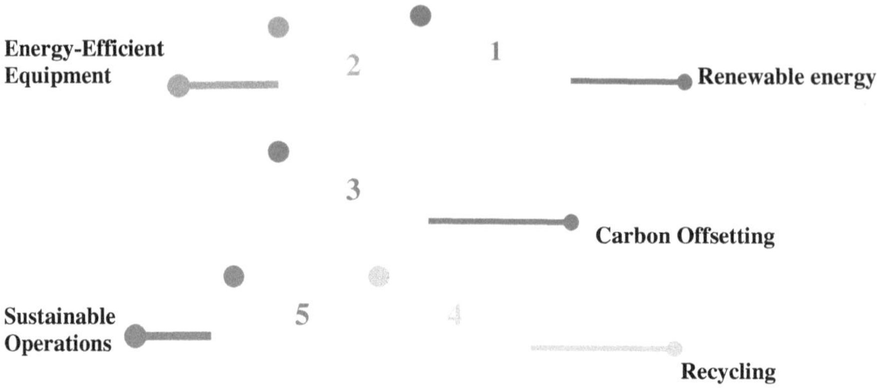

FIGURE 11.2 Energy-saving innovations actualized by DCCI. Source: Developed by the author.

Inc. (DCCI) has actualized different energy advances to extend energy productivity and supportability. These advances have not diminished energy utilization but, moreover, improved the general proficiency of the information center. From a commerce owner's viewpoint, contributing in energy innovations can lead to critical investment funds in energy costs, and it can upgrade the business's branding as a socially mindful company. From a nature viewpoint, accomplishing such innovations can essentially diminish the information center's carbon footprint. Underneath are the energy-saving innovations actualized by DCCI:

1. Virtualization: DCCI has achieved server virtualization, permitting different virtual servers to run on one physical server. This innovation has not diminished the number of required physical servers but moved forward asset utilization, thus attainingenergy reserve funds.
2. Cooling advances: DCCI has executed different cooling innovations such as hot and cold path control, which isolates hot and cold wind stream to decrease energy utilization. Furthermore, DCCI utilizes free cooling frameworks that utilize outside discuss to cool the information center rather than depending on mechanical cooling frameworks.
3. Renewable energy sources: DCCI has contributed to renewable energy sources such as solar power and wind control. By creating clean energy on site, DCCI can diminish its reliance on fossil fuels and diminish its carbon footprint.
4. DCCI uses Energy Star certified hardware that meets strict energy proficiency proposals set by the Natural Assurance Organization.
5. Power administration innovations: DCCI utilizes control administration advances such as shrewdly control dispersion units, which screen and control the sum of control utilized by each server.

This innovation guarantees that servers don't devour more control than vital, ending up with noteworthy energy reserve funds. DCCI has actualized various energy

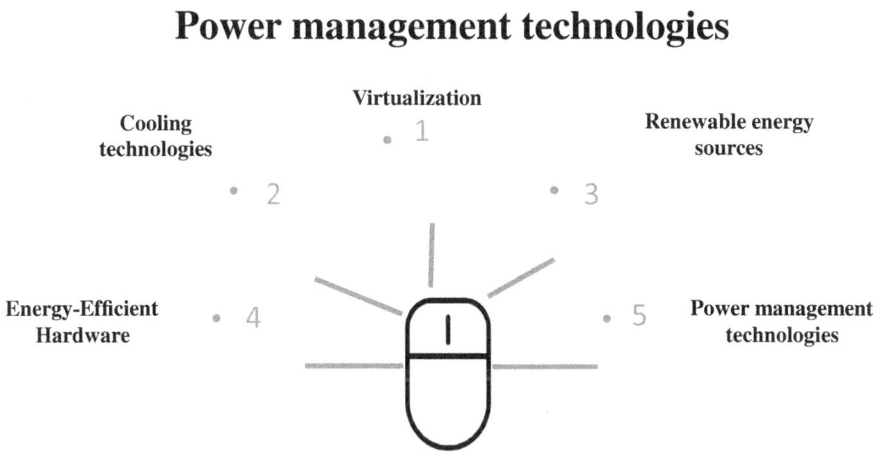

FIGURE 11.3 Energy-saving technologies implemented by DCCI.

advances to extend energy productivity and supportability. These innovations not only decrease energy utilization but upgrade the general productivity of the information center. Through participating in energy-saving innovations, organisations can significantly reduce their carbon emissions and make progress towards their advantage.

11.4 GREEN IT PRACTICES IN DCCI INFORMATION CENTERS

Within the present-day world, where innovation has ended up consuming a fundamental portion of our lives, we must consider the effect of our technical advances on the environment. Consequently, firms are implementing strategies to reduce their carbon emissions and promote sustainable practices. DCCI is one such company that has executed green IT practices in its information centers, guaranteeing that they contribute to worldwide endeavors towards energy, proficiency, and supportability.

One of the foremost vital green IT practices received by DCCI is the utilization of energy-efficient gear. DCCI information centers are prepared with cutting edge, energy-efficient servers that expend less control than conventional gear, lessening the amount of energy required to operate the information center. Moreover, DCCI has executed virtualization advances that permit different virtual servers to share one physical server, reducing the general number of servers required to manage the information center. This not only conserves energy but also reduces the amount of physical space required, saving money for investment funds that have taken a toll.

Another green IT practice executed by DCCI is the utilization of renewable energy sources. DCCI has introduced solar-powered panels on the rooftops of its information centers, permitting them to create their own power and decrease their reliance on conventional energy sources. This not only aids in decreasing their carbon footprint, but makes a significant difference on energy costs.

DCCI has executed proficient cooling frameworks in its information centers. Conventional cooling frameworks devour a noteworthy sum of energy, which can be a major burden to the information center's carbon footprint. DCCI has executed advanced cooling frameworks that utilize less energy and are more proficient, decreasing the sum of energy required to cool the information centers.DCCI strives to promote sustainable practices and reduce their carbon emissions. By executing green IT practices in their information centers, they contribute to worldwide endeavors toward energy effectiveness and supportability. Through the utilization of energy-efficient equipment, renewable energy sources, and proficient cooling frameworks, DCCI sets an example for other companies.

11.5 FOCAL POINTS OF GREEN INFORMATION CENTERS

Green information centers are getting to be progressively well known, and there are compelling reasons for this viewpoint. They not only offer assistance in diminishing energy utilization and carbon outflows, but they also offer a range of other benefits that are frequently ignored. Green data centres can provide significant benefits to both businesses and consumers, including reduced costs and greater reliability.

Here are a few key points of interest of green information centers:

1. Economic productivity: Green information centers are planned to be more energy-efficient than conventional information centers. They utilize different strategies to decrease energy utilization, such as utilizing renewable energy sources, optimizing cooling frameworks, and utilizing energy-efficient hardware. By decreasing energy utilization, green information centers can offer assistance on lowering energy bills and carbon outflows.
2. Cost investment funds: One of the primary advantages of green information centers is cost savings. By lessening energy utilization, green information centers can offer assistance on lowering power bills and operational costs. Also, green information centers regularly require less upkeep and have a

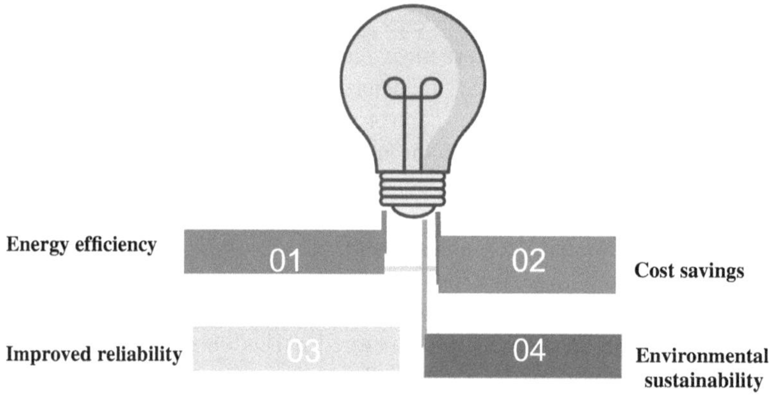

FIGURE 11.4 The benefits of green data centers.

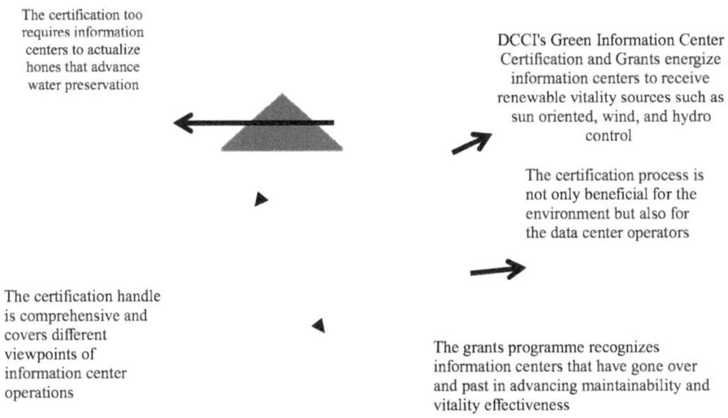

FIGURE 11.5 DDCIs Green Information Center Certification and Grants.

longer life expectancy than conventional information centers, which can offer assistance to diminish costs over time.

3. Enhanced unwavering quality: Green information centers are outlined to be more solid than conventional information centers. They utilize excess control and cooling frameworks to guarantee that basic frameworks stay operational within the occurrence of control blackouts or other types of failures. This will offer assistance to minimize downtime and guarantee that data is continuously accessible when required.
4. Environmental supportability: Maybe the foremost noteworthy advantage of green information centers is their natural maintainability. By utilizing renewable energy sources and diminishing energy utilization, green data centers can offer assistance to diminish carbon emissions and minimize their effect on the environment. This is often a basic thought for companies and consumers who are looking to decrease their carbon footprint and advance supportability.

By and large, green information centers offer a gamut of focal points that make them an appealing choice for both businesses and consumers. Green data centers can provide a solution by offering expanded reliability and sustainability, while also reducing the impact on reserve money. As more businesses and buyers have become mindful of these benefits, we anticipate seeing a forward move towards more energy-efficient and economical information center settings.

11.6 DCCIS GREEN CERTIFICATION AND GRANTS

DCCI has been a pioneer in advancing supportability and energy proficiency within the information center industry. One way they have accomplished this is by executing information center certification and grants. This activity points to recognize and advance information centers that are outlined and managed in an ecologically capable way. Certification is granted to information centers that meet a set of strict

FIGURE 11.6 Future plans for DCCIs sustainability efforts.

criteria, which consider components such as energy proficiency, water preservation, and the utilization of renewable energy sources.

1. The certification requirement is comprehensive and covers different viewpoints of information centers. It looks at components such as the utilization of energy-efficient equipment and cooling frameworks, as well as the usage of virtualization and cloud computing advances to diminish energy utilization.
2. Certification requires information centers to actualize water-saving measures and water supply frameworks (Pärssinen et al., 2019).
3. DCCI's Green Information Center certification and grants energize information centers to embrace renewable energy sources such as sun-based, wind, and hydroelectric control. This not only makes a difference in diminishing carbon emissions but alsodiminishes reliance on conventional energy sources.
4. The certification handle offers benefits not just for the environment but also for information center administrators. By adopting energy-efficient homes, information centers can effectively reduce their operational expenses.This is often since energy utilization is one of the biggest costs for information centers.
5. The grants program recognizes information centers that have outperformed and surpassed maintainability and energy effectiveness guidelines. For illustration, the Greenest Information Center grant recognizes information centers that have accomplished the most elevated level of energy productivity and maintainability.

Certification and grants from DCCI for Green Information Center practices are a great activity that contribute to maintainability and energy effectiveness within the information center industry. By adopting energy-efficient homes and reducing carbon emissions, information centers can reduce their operational expenses and also contribute to a more sustainable future.

11.7 FUTURE PLANS FOR SUSTAINABLE DEVELOPMENT AT DCCI

As a driving supplier of information center systems, DCCI is committed to expanding energy effectiveness and maintainability. We understand the significance of diminishing our carbon footprint and moderating the effect of our operations on the environment. That is why we continuously strive to make progress in our sustainability efforts, using the latest technologies and best practices to achieve our objectives. In this area, we are going to examine DCCI's future plans for guaranteeing maintainability, understanding diverse points of view, and presenting a list that gives point-by-point data for our plans (Borodko & Pantyukhin, 2019).

1. Attaining Carbon Neutrality: DCCI acknowledges the importance of carbon neutrality in reducing detrimental emissions and mitigating the impact of our activities on the environment. Our objective is to achieve carbon neutrality by 2030, and our approach encompasses the following steps:
   - Increasing the utilization of renewable energy sources such as sunpower and wind control.
   - Improving energy effectiveness and decreasing energy utilization through the execution of energy-efficient innovations and best practices.
   - Teaming with partners to create and actualize economical arrangements.

2. Water conservation: DCCI acknowledges that water could be a valuable asset, and we endeavor to moderate it. Our plan includes the followingr methodologies:
   - Reducing water utilization through the execution of water supply and best practices.
   - Utilizing and reusing water where conceivable.
   - Implementing water gathering frameworks to gather and store water for future utilization.

3. Waste diminishment: DCCI recognizes that waste decrease is key to accomplishing maintainability. Our plan includes the following procedures:

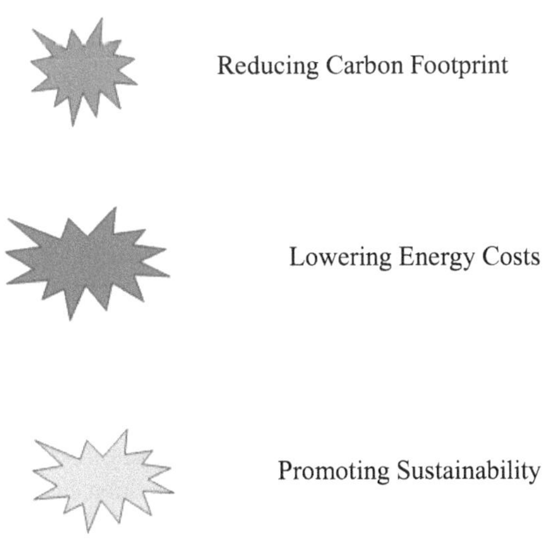

FIGURE 11.7 The part of green information centers in a economical future.

- Reducing waste through the execution of waste diminishment innovations and best practices.
- Implementing reusing programs for waste such as paper, plastic, and metals.
- Partnering with organizations specializing in waste decrease and reusing to create feasible arrangements.

DCCI points to upgrade energy productivity and supportability, and our future plans reflect this commitment. Through the execution of renewable energy sources, water preservation procedures, and waste control endeavors, we strive to realize our objectives in maintainable advancement and carbon footprint reduction. We accept that maintainability isn't just an obligation but an opportunity to enact positive change and make a distinction within the world.

11.8 THE PART OF GREEN INFORMATION CENTERS IN A FEASIBLE FUTURE

As information centers continue to expand and evolve, the need for environmentally friendly, cost-effective, and energy-efficient practices becomes increasingly important. Green information centers aim to reduce the harmful impact of information centers on the environment by implementing sustainable practices. The part of green information centers in a maintainable future is significant as they offer different benefits for the environment and the economy. From lessening carbon footprint to expanding energy productivity, green information centers clear the way for a feasible future. In this section, we are going talk about the part of green information centers in an economical future and give point-by-point data on how they contribute to a more feasible world.

1. Carbon footprint diminishment: Green information centers are planned to decrease carbon emissions and energy utilization (Kaur et al., 2019). By utilizing renewable energy sources such as wind and solar-energy control, green information centers can totally diminish their carbon footprint. For illustration, Facebook's information center in Sweden utilizes a hydroelectric control plant, which gives a renewable energy source, diminishes carbon outflows, and makes a difference in relieving climate change.
2. Energy Reduction: Green information centers are designed to decrease energy use, resulting in decreased energy expenses. By utilizing energy-efficient innovations such as virtualization and energy-efficient cooling frameworks, green information centers can reduce their energy bills and save money. Green Mountain's information center in Norway utilizes cold discuss technology to cool its servers, resulting in a 50-60% reduction in energy expenses compared to traditional information centers.
3. Encouraging maintainability: Green information centers promote sustainability by reducing energy consumption, minimizing carbon emissions, and employing renewable energy sources. These practices offer assistance in

promoting a more maintainable future by moderating assets and diminishing natural effect. For illustration, Google's information center in Belgium employs reused water for cooling frameworks, lessening water utilization by 40 percent.

The energy sector is the fundamental basis of any nation. It gives options for financial improvement, comfortable living conditions for individuals, and the creation of conditions for the ordinary working of such basic social circles as instruction, healthcare, social security, utilities, and so on. However, maintaining a continuous energy supply in normal working conditions is a challenging task, let alone increasing power generation. It is now unable to achieve this without transferring the sector to the framework of a market-based economy. When the foremost advanced accomplishments are utilized in energy development and operation, each of which has it possess and exceptionally mindful part, will the set objectives be accomplished. energy improvement based on advertise laws makes conditions for competition, giving an opportunity to draw in the foremost qualified financial specialists to extend usage, who have not, as it were, the implies but moreover advanced innovations and encounter, permitting them to present into the energy segment of Uzbekistan innovation and strategy that will stay significant for decades.

11.9 ENERGY TRANSITION – ONE OF THE MOST POWERFUL GLOBAL TRENDS

Energy transition is a decades-long strategy aimed at changing the structure of energy production and consumption by humanity. It has both devoted supporters and vehement opponents, but what does it entail? A global transformation of the world's energy system, encompassing four directions: energy efficiency, decarbonization, decentralization, and digitalization (Baudry, 2015).

This transformation is currently an officially accepted strategy of global humanity, as enshrined in the Paris Climate Agreement, signed by 180 countries in 2015.

The choice to move to "green energy" is based on the speculation that current human energy exercises are contributing to climate change, the so-called "worldwide warming." Not all researchers concur with this hypothesis, but it characterizes the global view. Inside its system, it is accepted that warming is caused by an increment in CO_2 outflows.

The undeniable fact is that for over 60 years, there has been a consistent and growing annual increase in the release of carbon dioxide by humans.For example, yearly estimations have been made since 1958 at the observatory on Mauna Loa well of lava (Hawaii). In any case, not everybody concurs with its anthropogenic nature – a few researchers point out that humankind transmits up to 50 billion tons of carbon dioxide amid its economic activities. This can be 10 percent of the entire planetary CO_2 generation, scarcely discernible against the foundation of the 450-550 billion tons radiated yearly by the world's seas. In addition, CO_2 itself accounts for 5 percent to 26 percent of the nursery impact (more than 70 percent of the nursery impact is due to water vapor).

Nevertheless, the Paris Agreement has been adopted and has begun to take effect. Under its framework, participants have agreed to reduce carbon emissions to "net zero" by 2050.

Green energy refers to energy generated from renewable sources. It is recharged at a more prominent rate than it is expended. Humankind has been creating carbon dioxide since they started using fire. Warm generation through the combustion of carbon powers (from dry wood to liquefied natural gas (LNG)) is the essential strategy of getting energy. Oxidation of carbon produces its oxide–CO_2, and there's nothing to be done about it. Subsequently, "green energy" basically includes abandoning fossil fuels (coal, oil, and common gas).

In the event that we do not burn carbon, where will the energy come from? By extending the utilization of "green" (renewable) sources (RES). Green energy within the world owes its title to alternative energy sources (it is additionally called renewable or regenerative). This implies that energy resources are always renewed by changes happening within the environment. Renewable energy is obtained from sources such as hydroenergy, wind energy, sun-powered energy, geothermal energy, biomass, and tidal energy. It is accepted that they, not at all like fossil fuels, are not drained (Armbrust et al., 2009).

Supporters of RES logically argue that we are literally surrounded by an unimaginable amount of energy. For example, the sun alone pours about 173 petawatts (or 173 million gigawatts) of energy onto the planet annually – this is more than four orders of magnitude above global needs.

1. Solar energy holds a driving position within the improvement of green energy, and in 2019, and after that in 2021, it set another record: the entire capacity of sun-based establishments around the world come to 760 GW. This is twice as much as nuclear power plants produce. There are two types of solar energy generation. The first generates energy through the internal photo effect using photovoltaic elements, which are a series of solar panels. The second involves using solar energy to convert water into steam and convert kinetic energy into electricity. By the way, both types of solar energy are recognized as the most accessible and environmentally friendly (Sviridov, 2022).
2. Wind energy is the change of wind energy into power employing a wind turbine. Wind turbines can be inland or introduced at the ocean in coastal zones. In Denmark, wind control fulfills 47 percent of power need, in Ireland – over 30 percent, and in Portugal and Spain – over 20 percent. The International Energy Agency (IEA) estimates that the potential of wind power generation exceeds current electricity demand by 40 times. But only if all necessary technological barriers are overcome, which is still far from being achieved. However, wind turbines can easily meet humanity's electricity needs.
3. Hydroelectric power is based on renewable sources, in this case, water, that is, the potential energy of the water flow is used. Hydropower plants have been built for a long time; this is the most developed renewable energy

technology. Unfortunately, despite all the advantages of hydroelectric power plants, they can be built far from everywhere (Barham et al., 2003).
4. Bioenergy is the production of energy from biofuels obtained by processing biological waste. It sounds impressive, but "burning firewood in the stove" is also a form of bioenergy. Wood is a renewable fuel, but still not the most environmentally friendly.
5. Geothermal energy uses the heat energy of the earth's crust. Potentially, this is a sea of energy literally underfoot. Practically, there are not so many places where this heat is close enough to the surface of the planet, and these places are usually quite remote.
6. Tidal energy is the use of tidal energy, that is, essentially, the attraction of the Moon. The Moon is highly sustainable, as it does not experience any environmental degradation. However, establishing infrastructure on the Moon is a hard and costly endeavor.

Global investments in renewable energy have increased from $33 billion to over $300 billion in a span of 20 years. Obviously, humanity is seriously considering, at least, trying to do without hydrocarbons. Currently, the primary method of generating profits from renewable energy sources (RES) is through investment. There are several strategies here. Those who are used to working for the future should consider companies involved in future technologies – for example, solar energy generation. It is unlikely that quick dividends should be expected; however, given the development of the industry and the reduction in the cost of RES, the prospects are very good. Perhaps, generating and selling electricity companies could become a more reliable way to invest funds.

Renewable green energy has numerous drawbacks. Despite the promising prospects, efforts, and investments, why do we constantly hear more about the challenges of energy transition than about its successes? The European energy crisis, Californian and Texan lockdowns, Australian energy collapse, freezing homes in England, grid outages in China, and the staggering rise in energy and fuel prices – where is the high-tech RES at this moment?

Unfortunately, any new technology does not scale up immediately. What works on a small scale cannot always be easily increased by simply adding resources and funds because effects occur that were not observed on a small scale.

What are the unresolved problems of green energy?

1. "Energy production-consumption gap." The production and consumption of "green" energy do not coincide throughout the day. In the morning and evening, when consumption is at its peak, the sun either does not shine yet or no longer shines, and the wind picks up after the sun heats the earth. The gap widens even further during unfavorable weather conditions. This requires the substitution of generating capacities, the creation of global energy transfers (imports), or its accumulation. Substituting capacities means the same traditional energy sources that we want to move away from, transfers require colossal infrastructure (try pumping enough electricity across half

a continent to meet the needs of a whole country), and accumulating technologies await the invention of something more capacious and cheaper than current batteries.
2. Not all places on the planet are well illuminated by the sun, have stable winds, or geothermal sources. This leads to the same problem of global transfers – you can cover the entire Sahara with solar panels but no one needs so much electricity there. And where it is needed, for example, for smelting aluminum and heating homes, the sun does not shine for three months a year.
3. Green energy sources require a multitude of high-tech materials and rare minerals, the extraction and production of which also require a lot of energy. Due to energy shortages in China this year, many enterprises paused operations, including suppliers of components for RES. According to calculations by the analytical company Russell Group, overall energy consumption restrictions in China disrupted global commodity flows by $120 million – and this is related to the sharp reduction in the traditional energy sector.

The costs of transporting and transmitting green energy are several times higher compared to other types of electricity. For example, the cost of transmitting "wind" energy is three times higher than transporting coal-generated electricity. Typically, this is also due to the fact that the production of green electricity occurs in places far removed from the territories where it is consumed. Therefore, during the transportation of green energy, the load and operational costs on transmission lines increase many times over due to long distances.

Additionally, green energy sources require significant investments in infrastructure (Anand & Bhambri, 2018). For example, widespread construction and installation of charging stations. Green energy, such as solar energy, also requires smart distribution: pricing based on time of day, electricity cutoffs to certain companies depending on the amount of energy they consume.

There are other issues as well. For instance, strong storms can disrupt the consistent supply of energy in coastal countries. Consequently, backup sources of electricity are needed – for example, batteries with a minimum three-day reserve of power. These are seasonal problems, the solutions to which are quite costly and can undermine the economy.

Despite all the problems, the share of green energy use is growing and has great prospects. Despite all the difficulties, the world will move towards RES. Regardless of whether the global warming is anthropogenic or not, humans will still have to respond to it. Difficulties of the transition period await us in the near future, but technology scaling brings not only problems but also solutions. Technologies that were too expensive on a small scale become profitable on a large scale. For example, the cost of a hydrogen fuel cell was recently $1700 per kilowatt of power, but with mass production it has dropped to $300 per kilowatt, and the day is not far off when a hydrogen bus will be economically superior to a diesel one.

Carbon-free energy is not as distant or impossible as it may seem. For instance, in Uzbekistan, a country with not the best natural conditions for RES and yet rich in hydrocarbons, as of today, 40 percent of the generation comes from nuclear, wind, solar, and hydroelectric power plants.

Agreeing to experts' estimates, the taken-a-toll of renewable energy sources, fundamentally sun power and wind power, will altogether diminish by 2040;in most districts, its generation will begin to get cheaper compared to gas extraction by 2030. This will happen due to the quickened improvement of advances and, thus, the diminishment within the onset of clean energy. This, without a doubt, makes RES commercially practical.

Most likely, the difficulties of rapid RES growth are surmountable. Perhaps skeptics are right, and even if humanity manages to sharply reduce greenhouse gas emissions, the concentration of CO_2 in the atmosphere will not decrease for another century. But the technological leap that we will receive as a result of mass investments in green energy has significant value in itself.

In any case, there is no doubt that forthcoming changes, however complex they may seem, are inevitable. Operators have always prioritized data center availability and sought to avoid risks, and they face a new risk – falling behind competitors amid rising electricity costs and compliance with environmental standards. The only way to survive the impending storm is to act preemptively.

Currently, a total of 13 methods for acquiring power have been identified, representing a significant milestone in creating a competitive environment for the energy industry. Some of the firms involved in the production of free energy are Add up to Eren SA (France), Abu Dhabi Future Energy Company - Masdar (UAE), Aksa Enerji Üretim A.Ş. (Turkey), ACWA Control (Saudi Arabia), and others.

Thanks to the establishment of a conducive investment environment in Uzbekistan, there is no need to attract government-backed loans.

Pursuant to the presidential decree No. PP-4249 dated March 27, 2019, titled "On the strategy for further development and reform of the electricity industry of the Republic of Uzbekistan," a Venture Office has been established under the Ministry's supervision, which also collaborates with international financial institutions.

The Venture Office specialists have created a concept, which incorporates a staged move to a discount competitive power showcase, with three stages arranged for the period 2021–2025.

The main objective involves the deregulation of power utilities and the granting of licenses to independent producers who are willing to sell the electricity they generate.

During the process of reorganisation, a sole power distribution system administrator will be established, and responsibilities linked to the sale of electricity to consumers will gradually be transferred to suppliers.

The third stage involves conducting intraday transactions on an hourly basis. According to this arrangement, surplus or deficit amounts of hourly power generation and consumption will be traded online on a scheduled basis on the trading platform.

Endeavors to move the administrative system of the energy division forward will proceed. In specific, dynamic work is being carried out to look at the created ventures: the Law of the Republic of Uzbekistan "On Electricity" (together with specialists from the European Bank for Reconstruction and Development and the World

Bank), the Declare of the President of the nation "On additional measures to reform the power industry" and the Government Resolution "On approval of the Code of main electric networks."

REFERENCES

Anand, A., & Bhambri, P. (2018). Orientation, scale and location invariant character recognition system using neural networks. *International Journal of Theoretical & Applied Sciences*, 10(1), 106–109. ISSN No. (Print): 0975-1718. ISSN No. (Online): 2249–3247. Published by Research Trend. www.researchtrend.net.

Armbrust, M., Fox, A., Griffith, R., et al. (2009) Above the clouds: A Berkeley view of cloud computing. Department Electrical Engineering and Computer Sciences, University of California, Berkeley, CA, Report No. UCB/EECS-2009-28.

Barham, P., Dragovic, B., Fraser, K., et al. (2003). Xen and the art of virtualization. *ACM SIGOPS Operating Systems Review*, 37(5), 164–177.

Baudry, K. (2015). Data center site search and selection. In H. Geng (Ed.), *Data Center Handbook* (pp. 89–102). New Jersey: Wiley.

Bhambri, P., & Mangat, A. S. (2005). Wireless security. In Abstract Book of UGC Sponsored National Conference on Emerging Computing Technologies (p. 101). Jalandhar: HRMMV.

Bhambri, P., & Singh, M. (2006). Artificial intelligence. In National Seminar on E-Governance - Pathway to Progress (p. 14). Derabassi: SSIET.

Bhambri, P., Sinha, V. K., & Dhanoa, I. S. (2020). Development of cost effective PMS with efficient utilization of resources. *Journal of Critical Reviews*, 7(19), 781–786. http://www.jcreview.com/index.php?fulltxt=102893&fulltxtj=197&fulltxtp=197-1594892754.pdf.

Borodko, A. V., & Pantyukhin, O. I. (2019). Principles and basic stages of designing data centers for the system's interests "Safe City" // VI International Scientific and Practical Conference "Modern Scientific Research: Current Issues, Achievements and Innovations" (pp. 24–26).

Chakrvarthy, S. (2001). The batch markovian arrival process: A review and future work. *Advances in Probability Theory and Stochastic Processes* (pp. 21–39).

Dumitrescu, C. (2018). Assessment of data center energy efficiency. Methods and metrics. Conference: 2018 International Conference and Exposition on Electrical and Power Engineering (EPE) October 2018. https://doi.org/10.1109/ICEPE.2018.8559745.

Kaur, J., Bhambri, P., & Kaur, S. (2019). SVM classifier based method for software defect prediction. *International Journal of Analytical and Experimental Model Analysis*, 11(10), 2772–2776. http://ijaema.com/Volume-11-Issue-10-October-2019-1/.

Pärssinen, M., Wahlroos, M., Syri, S., & Manner, J. (2019). Waste heat from data centers: An investment analysis. *Sustainable Cities and Society*, 44, 428–444.

Sviridov, A. N. (2022). Analysis of increasing the data centers energy efficiency methods. *Journal Modern High Rechnologies,* 2, 110–115.

Singh, M., Bhambri, P., & Kaur, K. (2005). Network security. In Proceedings of the National Conference on Future Trends in Information Technology (pp. 155–161). SJPMLIET, Radaur, Yamuna Nagar.

12 Energy-Efficient Data Center Design

Harpreet Kaur Channi and Ramandeep Sandhu

12.1 INTRODUCTION

The pursuit of energy efficiency in data center architecture has become an urgent need in our age, characterized by the incessant growth of digital infrastructure. Data centers are essential nodes that drive our networked world and the digital economy as a whole, but they also present considerable energy consumption issues. Data center design that minimizes energy consumption has recently been a hot topic in the IT sector, due to rising concerns about the environment and the ever-increasing need for processing capacity. With an eye on energy efficiency, this chapter walks readers through the maze of data center design and operation [1]. There is a need to help businesses construct and oversee data centers that can handle the demands of the modern digital world while also reducing their impact on the environment and paving the way for a greener tomorrow by delving into cutting-edge tech, important strategic factors, and industry best practices. Embark with the authors on a quest for energy-efficient data center architecture, where sustainability, creativity, and operational excellence come together to mold the digital infrastructure of tomorrow. Data centers are the unsung heroes of today's lightning-fast digital environment, the unseen hands that keep everything linked. Massive data centers store the servers, storage arrays, and networking gear that make modern computing possible and allow us to go about our everyday lives without a hitch [2]. But the energy usage and environmental consequences of this digital infrastructure are substantial. Data centers are now in the limelight as big energy users, responsible for a significant amount of the world's electrical consumption, thanks to the exponential development of data needs. An energy-efficient design for data centers is more important than ever before due to growing concerns about sustainability and climate change. Striking a balance between meeting the constant need for computer power and minimizing environmental effect calls for creative solutions and careful planning [3].

Amidst the fast growth of digital infrastructure, energy efficiency in data centers is a crucial challenge. These buildings represent the lifeblood of contemporary computing, and as such, they have very high energy needs considering the continual running of servers, storage systems, and networking gear. Data center energy efficiency is crucial for several important reasons. A large amount of operating expenditures are related to energy, thus energy-efficient data centers are improving energy utilization and immediately results in cost reductions [4]. Additionally, it is crucial to

DOI: 10.1201/9781003484226-12

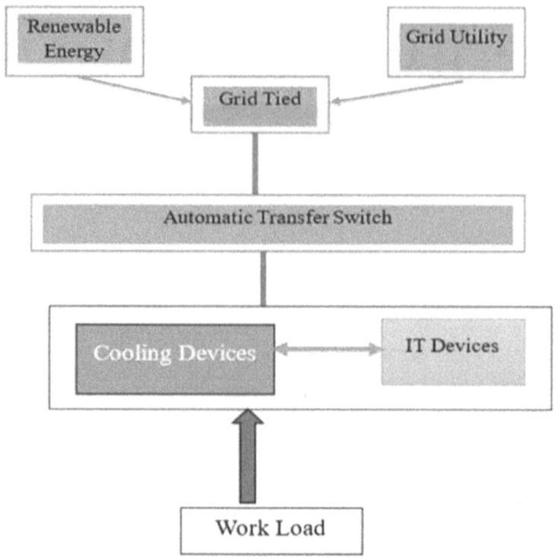

FIGURE 12.1 Green data center.

consider the environmental consequences of data centers' energy use, since the need for more environmentally friendly operations is being propelled by concerns about carbon emissions and sustainability. Energy efficiency standards are becoming more important due to regulatory restrictions. High-energy density, changing workloads, and legacy infrastructure are some of the obstacles that data centers must overcome in order to achieve energy efficiency as shown in Figure 12.1. Improvements are possible, nevertheless, via methods such as virtualization, effective cooling systems, hardware optimization, and the use of renewable energy sources. If data center operators use these measures, they may lessen their impact on the environment, save money on operations, and make sure their facilities can adapt to a changing digital world [5,6].

12.2 BACKGROUND

The idea of a data center has its roots in the middle of the 20th century when huge mainframe computers needed specific places to live and work. When compared to today's state-of-the-art data centers, their antiquated cooling systems and room- or building-occupying layouts were clear indicators of their primitive origins. Data center infrastructure became more complex in response to the exponential growth in data processing and storage needs caused by advances in computer technology [2]. There have been numerous significant turning points in the history of data centers:

Mainframe era: Data centers were mainly utilized to host the centralized systems of mainframe computers, which were dominant in the early days of computing.

Client-server architecture: Smaller servers placed in data centers proliferated with the development of client-server architecture, which decentralized computing.

Internet boom: The fast growth of data centers was driven by the increasing demand for hosting websites, online services, and e-commerce platforms, which in turn was spurred by the advent of the internet in the late 20th century.

Cloud computing: By allowing on-demand scaling and virtualizing resources, cloud computing radically altered the design of data centers in the 21st century.

Edge computing: Recent years have seen the emergence of edge computing, which involves decentralizing data processing and storage to the network edge, in response to the growth of Internet of Things (IoT) devices and the demand for low-latency processing [3–5].

12.2.1 Factors Driving Energy Efficiency

Several factors have contributed to the growing importance of energy efficiency in data centers:

- **Rising energy costs:** Data center operators are looking for strategies to improve energy use and minimize operational expenditures due to the rising cost of power.
- **Environmental concerns:** A larger focus on sustainability and lowering the carbon footprint of data center operations has resulted from the increasing awareness of environmental challenges such as climate change and carbon emissions.
- **Regulatory requirements:** Data center operators are under intense pressure to conform to new energy efficiency standards and regulations imposed by governments and regulatory agencies throughout the globe.
- **Corporate social responsibility:** Sustainable techniques, such as energy-efficient data center architecture, are becoming more important to many firms as they strive to fulfill their corporate social responsibility goals.
- **Technological advances:** Data center management software, cooling technology, and hardware advancements have all contributed to greater resource and energy efficiency [6].

12.3 LITERATURE SURVEY

G. et al. proposed the heterogeneous-cloud radio access network (H-CRAN) energy-efficient methods. This energy-efficient algorithm incorporates an energy-efficient resource allocation management design to deal with H-CRANs in 5G [7]. Fan et al. investigated a nodal importance assessment method based on the data center dynamic model for perturbation testing [8]. Chen et al. presented an overview of the data center power supply system covering the power delivery path from the grid edge to onboard point-of-load (PoL) conversion. They provided an overview for high-performance data center power supply system design. Reducing the energy consumption of data centers decreases operational expenses, as well as their carbon footprint [9]. Grigoryan et al. designed a technique to reduce data center power consumption by leveraging software-defined networking (SDN) and programmable data plane concepts. Coherent optical links are becoming increasingly attractive for intra-data center applications as data rates scale [10]. Maharry et al. described the optimal design spaces for low-cost and energy-efficient coherent links [11]. Abderrahim et al. proposed a data center-enabled high altitude platform (HAP) system where a flying data center supports the operation of terrestrial data centers. The findings demonstrate that a data center-enabled HAP is more energy efficient than a traditional terrestrial data center, owing to the naturally low temperature in the stratosphere and the ability to harvest solar energy [12]. Andrade et al. explored the transformative impact of custom ASICs on data center video processing, exploring their pivotal role in meeting the ever-evolving demands of this dynamic landscape [13]. Naug et al. presented PyDCM, a customizable data center model implemented in Python, that allows users to create unique configurations of IT equipment with custom server specifications and geometric arrangements of IT cabinets. PyDCM enables the use of deep reinforcement learning via the Gymnasium wrapper to optimize data center cooling and offers a user-friendly platform for testing various data center design prototypes [14]. Dutta et al. attempted to quantify the potential for gains in energy efficiency in software, based on a set of examples drawn from common, everyday decisions made by software developers and enterprise architects. Understanding the relationship between power, temperature, consolidation, and performance is crucial to enable an energy-efficient management at the data center level [15]. Arroba et al. proposed novel power and thermal-aware strategies and models to provide joint cooling and computing optimizations from a local perspective based on the global energy consumption of metaheuristic-based optimizations [16].

The objectives of energy-efficient data center design revolve around optimizing resource usage, reducing environmental impact, and ensuring operational effectiveness. Following are some key objectives:

- The main goal of designing an energy-efficient data center is to reduce energy usage while still satisfying the facility's operating demands.
- Reducing emissions of greenhouse gases and other pollutants linked to energy consumption is another goal of energy-efficient data center design, which tries to lessen the environmental effect of data center operations.
- Data centers that prioritize energy efficiency construct their infrastructure with the goal of making the most effective use of resources including servers, storage, and networking gear.

12.4 FOUNDATIONS OF ENERGY-EFFICIENT DESIGN

To minimize energy consumption and maximize operational efficiency, data center energy-efficient architecture is based on both theoretical ideas and practical tactics. Hardware choices like servers, storage devices, and networking gear that favor performance over power consumption are key to this strategy. By increasing server utilization rates, decreasing the total number of servers required, and decreasing energy consumption, virtualization and workload consolidation further improve resource use. Data centers rely on efficient cooling solutions like in-row cooling and hot/cold aisle containment to keep temperatures ideal while reducing energy use. Further reductions in cooling-related energy needs may be achieved by smart layout design and airflow control, which guarantee optimal cooling distribution and eliminate hotspots. By keeping tabs on energy use in real time, comprehensive monitoring and management systems make it easy to spot inefficiencies and maximize resource allocation, therefore reducing energy waste [17]. To achieve sustainability objectives, it is necessary to integrate renewable energy sources like solar or wind power in order to decrease dependency on conventional fossil fuels. Data centers may keep their environmental effect to a minimum by following proper lifecycle management procedures such as recycling and appropriate disposal of old equipment. Lastly, initiatives that raise awareness and educate personnel foster an energy-efficiency culture, which in turn encourages actions that cut down on energy use and maximize operational efficiency. Data center operators may build sustainable facilities with minimal energy use and environmental effect by adhering to these basic concepts and practices. A number of important ideas and methods for reducing energy consumption and increasing operational efficiency form the basis of data center energy-efficient architecture as shown in Figure 12.2 [18].

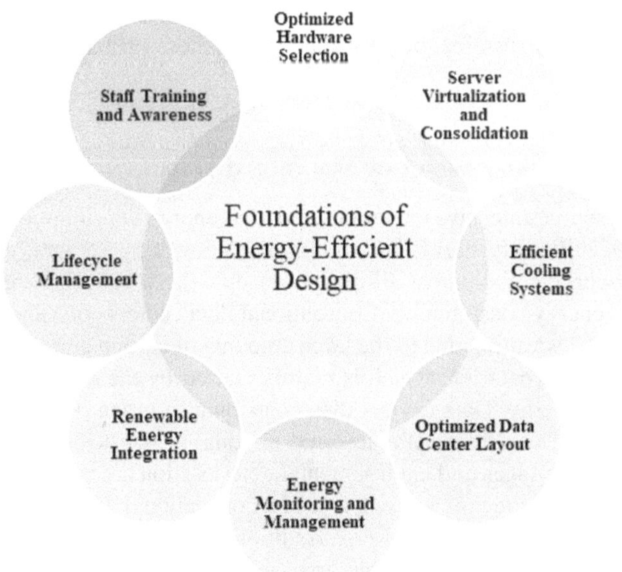

FIGURE 12.2 Foundations of energy-efficient design.

12.4.1 Overview of Energy-Efficient Data Center Principles

An all-encompassing strategy to lessening power usage, lessening environmental effect, and maximizing operational efficiency are the tenets of energy-efficient data center architecture. In order to build data center facilities that are both efficient and ecologically friendly, these concepts rely on strategic planning, new technology, and sustainable practices. A key premise is to maximize performance while limiting power consumption by strategically selecting energy-efficient hardware and infrastructure components [19]. This includes things like servers, storage devices, and cooling systems. Further optimization of resource use is achieved by virtualization and workload consolidation, which maximize server utilization rates while minimizing the total number of servers required. Optimization of temperature control in the data center environment while limiting energy use is achieved by efficient cooling systems such as hot/cold aisle confinement and innovative cooling technology. Solar and wind power are examples of renewable energy sources that may be integrated to assist in meeting sustainability objectives while also reducing dependency on conventional fossil fuels. By keeping tabs on energy use in real time, comprehensive monitoring and management systems may help find inefficiencies and put optimization strategies into action [20]. The data center's environmental effect is minimized throughout its existence by the use of proper lifecycle management procedures such as the appropriate disposal and recycling of old equipment. Through education and training, businesses may raise awareness about the importance of energy efficiency and inspire their employees to adopt practices that cut down on power use and boost productivity. In addition to meeting the increasing needs of contemporary computing, data center operators may decrease energy use, reduce environmental impact, and assure long-term sustainability by adhering to these principles while creating their facilities [21].

12.4.2 Environmental Impact of Traditional Data Centers

Due to their resource-intensive operations and high energy consumption, traditional data centers significantly harm the environment. Powering servers, cooling systems, and networking equipment all day, every day, these facilities are known for their ravenous energy consumption. Conventional data centers contribute to climate change and global warming due to the large amounts of carbon emissions produced by their continuous operation, which is mainly caused by the combustion of fossil fuels to produce energy. Data centers also contribute to urban heat islands, which have an impact on local microclimates and air quality due to the heat they produce [22]. Water shortages and environmental deterioration are consequences of the water-intensive cooling techniques employed by conventional data centers, including evaporative cooling and cooling towers. Traditional data centers also cause habitat loss and resource depletion due to the large quantities of land, raw materials, and energy needed for their construction and operation. The disposal of old IT equipment produces electronic trash, which adds to pollution and the release of harmful elements into the environment. There is an immediate need to use energy-efficient design methods and sustainable technology to lessen the environmental impact of conventional data centers and to foster digital sustainability [23]. High energy usage, carbon emissions, and resource exploitation are some of the ways in which traditional data centers significantly harm the environment. Among the most significant negative effects that conventional data centers have on the environment are:

- **Energy consumption:** Powering computers, cooling systems, networking gear, and other infrastructure around the clock, traditional data centers are electrical hogs. Running these plants nonstop uses a lot of power, which means more fossil fuels are needed and more greenhouse gas is released into the atmosphere.
- **Carbon emissions:** Carbon emissions are high in conventional data centers because of their high energy consumption, which is derived mostly from the combustion of fossil fuels. Extreme weather, higher sea levels, and other environmental hazards are a result of these emissions' contribution to climate change and global warming.
- **Heat generation:** When computers and cooling systems are running, data centers produce a lot of heat. In highly populated regions next to data centers, this surplus of heat may exacerbate urban heat island effects, which in turn affect local microclimates and air quality.
- **Water consumption:** When it comes to cooling, data centers, especially in hotter countries, need a lot of water. Evaporative cooling, cooling towers, and other water-intensive cooling technologies may contribute to water shortages and environmental deterioration by putting a burden on local water supplies and ecosystems.
- **Waste generation:** The disposal of old or broken IT equipment is the main source of electronic trash (e-waste) in conventional data centers. The discharge of harmful compounds into the environment and contamination of landfills and aquatic bodies are potential outcomes of improper e-waste disposal.
- **Resource depletion:** Traditional data centers need a lot of energy, land, and raw materials during construction and operation. Environmental degradation, including the loss of habitat and forests, may result from the harvest and exploitation of natural resources.
- **Air pollution:** Emissions from cooling towers, diesel backup generators, and other infrastructure components of traditional data centers contribute to air pollution. Particulate matter, nitrogen oxides (NO_x), and sulfur dioxide (SO_2) are some of the pollutants found in these emissions. These may harm both humans and the environment.

To encourage environmental stewardship in the digital age and offset the environmental impacts of conventional data centers, it is crucial to embrace energy-efficient design methods and sustainable technology [24–26].

12.4.3 Economic and Operational Drivers for Efficiency

Aside from ecological considerations, operational and financial considerations also play a role in motivating data center efficiency. From a monetary standpoint, data center operators may save money by reducing energy use. Since power is a major operational expenditure, cutting costs by increasing efficiency reduces utility bills and total operating expenses, which boosts profitability. Data centers that are efficient with energy use may better manage the growing demand for digital services without breaking the bank on electricity. Additionally, data center operational efficiency is often associated with energy efficiency [27]. Data center operators may enhance the performance and reliability of essential infrastructure components by managing energy use. For instance, having an efficient cooling system in place does double duty: reducing energy usage and assisting in maintaining appropriate working temperatures, which in turn increases the longevity of equipment like servers. Data centers may operate more efficiently due to energy-efficient design concepts like server virtualization and workload consolidation, which increase resource usage. Data center operators are, therefore, better able to respond to fluctuating workloads and evolving business needs as a result of increased scalability and flexibility. Aligning financial objectives with environmental sustainability and operational excellence is crucial for data center operations, as it reflects the relevance of energy-efficient design methods and technology. This is driven by both operational and economic considerations for efficiency [28].

12.5 ARCHITECTURAL CONSIDERATIONS

Data centers that need less energy to run are the result of careful architectural planning that takes into account factors like airflow control, insulation, and building architecture. The data center's physical architecture is an important factor to consider since it may have a major influence on energy efficiency as shown in Figure 12.3. Operators of data centers may lessen the strain on cooling systems and save energy usage by organizing server racks and equipment in a way that maximizes airflow and minimizes hotspots. In addition to enhancing energy efficiency,

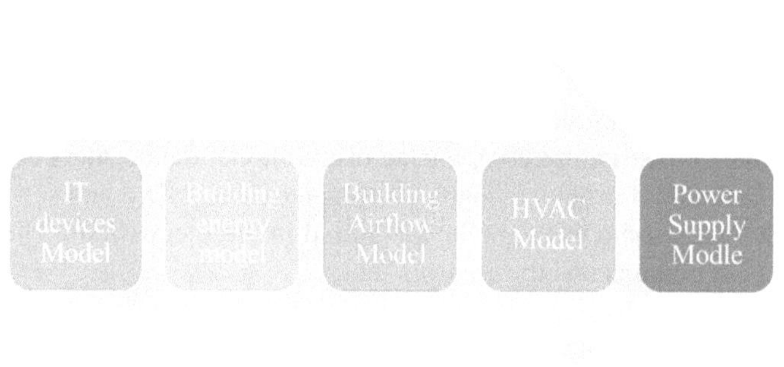

FIGURE 12.3 Data center architecture.

stable interior temperatures are achieved by adequate insulation and sealing of the building envelope.

The cooling infrastructure design of the data center is another architectural factor to think about. To keep operational temperatures ideal while limiting energy use, efficient cooling technologies like hot/cold aisle confinement and in-row cooling are important. Improve cooling efficiency and minimize burden on mechanical cooling equipment with these systems, which segregate hot and cold airflows and distribute cool air exactly where it is required. By using ambient air temperatures to augment or replace mechanical cooling when circumstances permit, energy efficiency may be further enhanced via the combination of outdoor air economization and free cooling systems [29]. The choice of building materials and techniques of construction are other aspects of architecture to think about. To lessen their influence on the environment and make the most of the energy they save, energy-efficient data centers often use sustainable construction materials and design concepts. Reduced heating and cooling loads, energy-efficient windows, and reflective roofing materials are a few examples of how to lessen the impact on the environment and save money on utility bills. Natural lighting and ventilation are examples of passive design ideas that may improve the data center's comfort and usefulness while also increasing energy efficiency [30]. In conclusion, several facets of a data center's architectural design, including its insulation, cooling infrastructure, and overall layout, are crucial to the design and management of energy-efficient data centers. Data center operators may limit energy consumption, decrease environmental impact, and enhance operating efficiency by using sustainable construction materials, smart layout design, and efficient cooling systems [31].

- **Building layout:** The diagram would illustrate the layout of the data center facility, including the arrangement of server racks, cooling equipment, and other infrastructure components. It would highlight the strategic placement of equipment to optimize airflow and minimize hotspots, as well as the use of containment systems to segregate hot and cold airflows.
- **Insulation and sealing:** The insulation and sealing techniques used in the building envelope to preserve consistent interior temperatures and reduce heating and cooling loads are shown in this section of Figure 12.3. Details on window treatments, insulation materials, and air leak prevention sealing methods may be included.
- **Cooling infrastructure:** The cooling infrastructure of the data center, including the arrangement of the airflow control systems, ducts, and cooling units, would be shown in Figure 12.3. It would showcase effective cooling techniques such as outside air economization, in-row cooling, and hot/cold aisle confinement.
- **Building materials:** The use of sustainable construction materials and design concepts to improve energy efficiency and reduce environmental effect will be covered in depth in this section. It could include details on the energy-efficient windows, reflective roofing, and high-performance insulation that were employed in the building of the data center.

- **Passive design strategies:** Incorporated within the architectural design of the data center, passive design techniques like natural lighting and ventilation would be shown in Figure 12.3. To minimize artificial lighting and mechanical ventilation, it would demonstrate how to optimize natural light and airflow via the use of daylighting methods, skylights, and moveable windows.

Overall, by integrating sustainable construction practices and effective cooling solutions, the architectural considerations diagram would provide a visual depiction of the many design aspects and tactics used to produce an energy-efficient data center facility [32–35].

12.5.1 Sustainable Building Design

An all-encompassing strategy, sustainable building design places human well-being, resource economy, and environmental responsibility at the forefront of a building's lifetime. Sustainable building design in the context of data centers refers to a range of tactics and ideas that are meant to minimize the negative effects on the environment, maximize interior space use, and lower energy usage. In order to optimize energy efficiency, reduce waste, and foster environmental stewardship, this strategy carefully evaluates the building materials, construction techniques, energy systems, and site placement [36]. Energy-saving technology, such as high-performance insulation, reflective roofing materials, energy-efficient lighting, and energy-efficient HVAC systems, are often included into data center architecture to minimize energy use and operating expenses. Moreover, to further lessen dependency on fossil fuels and cut carbon emissions, sustainable building design places a strong emphasis on the utilization of renewable energy sources such as wind and solar power. In addition, to create a healthy and productive work environment, sustainable building design places a high priority on indoor air quality, water conservation, and occupant comfort. Businesses may lessen their environmental impact, save operating expenses, and show their dedication to environmental sustainability by incorporating sustainable building design concepts into the development and management of data centers [37].

As a crucial component of designing an energy-efficient data center, layout optimization aims to minimize energy consumption and maximize resource usage throughout the whole building. By strategically arranging and designing the infrastructure's parts, this method may eliminate hotspots, improve cooling efficiency, and maximize airflow – all of which can save a substantial amount of energy. Strategically placing server racks and equipment inside the data center space is a crucial component of layout optimization. Data center administrators may guarantee good heat dissipation and reduce the strain on mechanical cooling equipment by setting up server racks to minimize airflow blockages and enhance the performance of cooling systems. By separating hot and cold airflows, avoiding air stream mixing, and boosting cooling efficiency, hot/cold aisle containment systems are another way to improve layout optimization [38]. Through the elimination of superfluous cable lines and the optimization of power supply to equipment, improving the architecture of power distribution systems and cabling infrastructure may also assist to decrease energy losses and reduce power usage. By improving airflow, boosting cooling efficiency, and lowering energy consumption throughout the facility, layout optimization, taken as a whole, is a key factor in increasing energy efficiency and cutting operating costs in data centers [39].

Layout optimization for energy savings in data center architecture is a multimodal technique at Chandigarh University that aims to maximize efficiency and minimize environmental effect, as shown in Figure 12.4 [40]. The university's data center layout optimization techniques concentrate on a few crucial areas in order to save energy.

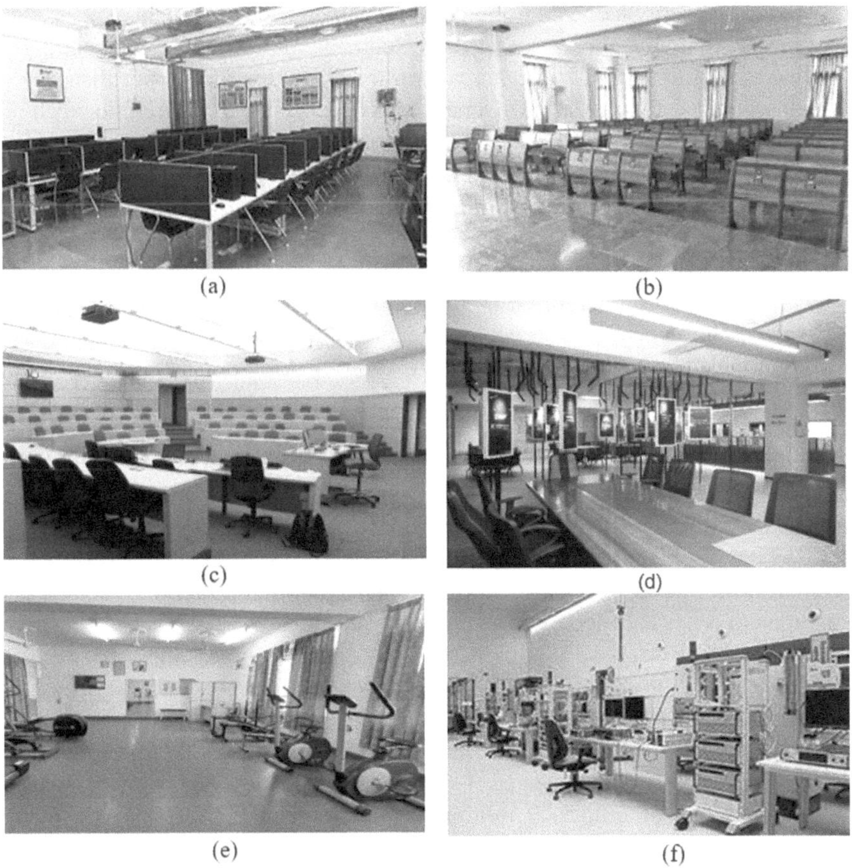

FIGURE 12.4 Pictures of some classrooms in the sample: (a) Computer lab; (b) Lecture hall; (c) Seminar hall; (d) Meeting hall; (e) Health practice room; (f) Engineering practice room.

- **Server rack placement:** It is essential to arrange server racks strategically in the data center to maximize ventilation and reduce energy use. Chandigarh University guarantees good heat dissipation and lessens the strain on mechanical cooling equipment by setting up server racks to minimize obstacles and enhance the performance of cooling systems.
- **Hot/cold aisle containment:** By separating hot and cold airflows, hot/cold aisle containment systems assist in avoiding air stream mixing and increase cooling efficiency. This method is used by Chandigarh University to separate hot and cold aisles, ensuring that cool air is sent exactly where it is required and cutting down on energy waste.
- **Efficient cooling systems:** To maintain ideal temperatures in the data center environment, Chandigarh University uses effective cooling systems including evaporative cooling technologies and precision air conditioning units. The institution ensures dependable functioning of IT equipment while reducing energy usage by using these cutting-edge cooling technologies.
- **Power distribution optimization:** Reducing power usage and energy losses requires optimizing the cabling infrastructure and power distribution systems. Chandigarh University uses effective power distribution techniques to reduce the need for pointless cable runs, enhance power delivery to devices, and increase energy efficiency.
- **Renewable energy integration:** Chandigarh University investigates possibilities for incorporating renewable energy sources, such as solar power, into the data center's energy supply in order to further improve energy savings and sustainability. The institution lowers its carbon footprint and lessens its dependency on conventional fossil fuels by using clean and renewable energy.

Ultimately, Chandigarh University's strategy for data center architecture optimization for energy savings shows a dedication to increasing productivity, minimizing environmental effect, and adopting sustainable methods in keeping with its overarching sustainability objectives. The institution achieves significant energy savings while guaranteeing dependable and sustainable operation of its data center infrastructure via smart layout design, effective cooling systems, optimal power distribution, and integration of renewable energy [41–43].

12.5.1.1 Methodology

Numerous important techniques and factors need to be taken into account while designing an energy-efficient data center. Chandigarh University's technique for designing energy-efficient data centers uses a methodical approach with the goals of enhancing operational efficiency, minimizing environmental effect, and optimizing energy use. As shown in Figure 12.5, the process starts with a comprehensive planning and evaluation phase where the current data center infrastructure is assessed to find opportunities for improvement. The evaluation serves as a basis for creating a comprehensive design plan that specifies objectives, tactics, and performance indicators for attaining energy savings. Gear optimization, which involves carefully

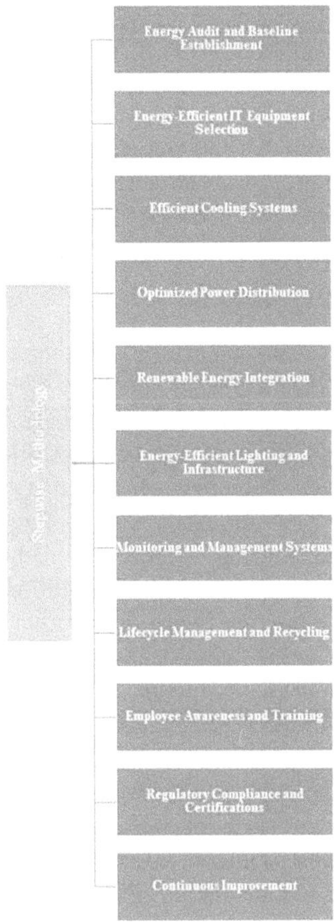

FIGURE 12.5 Stepwise methodology.

choosing energy-efficient servers, storage units, and networking gear to reduce power usage while preserving performance, is a crucial component of the process. In order to optimize resource use and minimize the total number of servers required, server virtualization and consolidation are also employed. In order to maximize cooling system performance and maintain ideal temperatures inside the data center environment, cutting-edge cooling technologies and airflow control techniques are also used [44]. A key component of the concept is the incorporation of renewable energy sources. Chandigarh University, for example, integrates solar power into the data center's energy supply to lessen its dependency on fossil fuels and cut down on carbon emissions. Energy losses are further reduced and efficiency is increased via the use of effective power distribution techniques such as zoned power distribution and intelligent power management software. Real-time application of optimization measures and detection of inefficiencies are made possible by the continuous monitoring of energy use and performance parameters. To teach data center workers energy-saving techniques and encourage an energy-efficient culture, training courses and awareness campaigns are held. In order to achieve significant energy savings and guarantee the sustainable and effective operation of the data center infrastructure, Chandigarh University's methodology for designing energy-efficient data centers emphasizes a holistic approach that integrates hardware optimization, cutting-edge cooling technologies, integration of renewable energy, efficient power distribution, and continuous monitoring and optimization [45].

12.5.2 Cooling Systems Optimization

Optimizing the cooling system in data centers is essential for improving energy efficiency and preserving ideal operating conditions for IT hardware. Heat mixing may be avoided by using techniques such as hot/cold aisle confinement, which efficiently separates air streams. Utilizing devices like airflow baffles and perforated floor tiles to regulate airflow maximizes the supply of cold air while reducing recirculation. By dynamically adjusting airflow in response to cooling needs, variable speed fans and pumps lower energy usage during times of low load. When practical, using free cooling systems allows for the use of ambient conditions for cooling, while liquid cooling technologies provide highly efficient computing at high densities. Ensuring equipment durability and lowering energy consumption may be achieved by precise temperature and humidity management within approved limits [46]. Optimizing the delivery of cold water and implementing effective heat rejection systems enhance the overall efficiency of cooling. Sustained efficacy in cooling system optimization requires thermal confinement, optimized data center structure, and routine monitoring and maintenance. In many types of facilities, including data centers, energy-efficient HVAC (heating, ventilation, and air conditioning) systems are essential for cutting down on energy use and advancing sustainability. These systems maximize ventilation, heating, and cooling while reducing energy waste by combining cutting-edge technology and architectural concepts. Using energy-efficient equipment, such as heat pumps, energy recovery ventilation (ERV) systems, and variable refrigerant flow (VRF) systems, is one of the key techniques. Furthermore, sophisticated control systems make it possible to precisely monitor and modify airflow and temperature, guaranteeing the best possible conditions for both residents and equipment while using the least amount of energy [47]. Reduced heat input and loss from air leaks, air sealing, and building orientation all help to improve HVAC efficiency. Adding renewable energy sources, such as solar thermal or geothermal systems, may also make HVAC operations more sustainable. All things considered, installing energy-efficient HVAC systems cuts carbon emissions and operating expenses, paving the way for a more sustainable and environmentally friendly future [48].

12.5.3 Hardware Efficiency Strategies

A variety of methods are included in hardware efficiency strategies with the goal of maximizing the energy and performance of computer hardware in data centers [49, 50]. The efficiency of servers, storage systems, networking devices, and other hardware components is the main goal of these tactics. The use of server virtualization to combine tasks; the selection of low-power, energy-efficient hardware; the implementation of dynamic power management to modify power use in response to workload needs; and the routine replacement of hardware with newer, more energy-efficient models are important strategies. Hardware efficiency is further influenced by effective cooling systems, data compression and deduplication methods, appropriate workload allocation, and energy monitoring tools. Furthermore, including renewable energy sources into the architecture of data centers improves overall energy efficiency and lessens the effect on the environment [51]. Data center operators may maintain high performance levels and save a lot of money and protect the environment by putting these techniques into practice. Two significant certifications that assist customers in identifying IT equipment that are both environmentally friendly and energy-efficient are Energy Star and Electronic Product Environmental Assessment Tool (EPEAT) ratings. The US Environmental Protection Agency (EPA) launched the Energy Star program, which establishes energy-efficiency criteria for a range of items, including servers, displays, and PCs. When a product has the Energy Star certification, it means that it uses less energy in standby and operation than conventional models and that it meets or exceeds these criteria [52]. In contrast, EPEAT is a worldwide grading system that assesses how environmentally friendly electronic items are, taking into account things as energy efficiency, recyclable nature, and usage of hazardous components. Bronze, Silver, and Gold are the EPEAT grades, with Gold-rated items fulfilling the strictest environmental requirements. Customers may help with sustainability initiatives, cut down on environmental impact, and limit energy usage by selecting IT equipment that has earned the Energy Star and EPEAT certifications. In addition, these ratings provide transparency and direction to companies looking to fulfill sustainability objectives and adopt green buying practices [53].

12.5.4 Monitoring and Management Systems

By giving real-time visibility and control over many infrastructure components, monitoring and management systems are essential to the smooth running of data centers [54–57]. These systems include a variety of hardware and software programs that are intended to monitor and control important variables including energy use, humidity, temperature, and equipment performance. Through the process of gathering and evaluating information from sensors and monitoring devices positioned around the data center, these systems help administrators spot any problems, maximize the use of available resources, and apply energy-saving strategies [58]. Proactive maintenance and predictive modeling are made possible by advanced features like machine learning algorithms and predictive analytics, which also assist to minimize downtime and maximize productivity. Furthermore, management systems provide operators centralized control over workload orchestration, power distribution, and cooling systems, enabling them to dynamically modify parameters and assign resources in response to demand. In general, data center operations' dependability, effectiveness, and sustainability depend on monitoring and management systems [59].

12.5.4.1 Real-time Energy Monitoring

In order to identify and assess patterns in energy use in real time, modern data center management requires real-time energy monitoring. Servers, cooling systems, power distribution units (PDUs), and uninterruptible power supplies (UPSs) are just a few of the components of a data center's design that are continuously monitored. Using real-time energy monitoring solutions, data center administrators may swiftly get insight into the energy use of different systems and equipment. They are able to detect anomalies, identify inefficiencies, and adapt energy use to changing conditions, thanks to this information. In most cases, these systems provide detailed information on power use, including metrics like kWh, carbon emissions, and power usage effectiveness (PUE) [60].

Operators may immediately implement energy-saving measures, such as adjusting cooling setpoints, reducing server loads, or redistributing resources to equipment with better energy efficiency, thanks to real-time energy monitoring. Data collected in real time may also be used for future capacity planning, trend research, and performance benchmarking purposes. In addition, real-time energy monitoring supports sustainability initiatives by helping data center operators reduce their environmental effect and accomplish energy efficiency goals. By identifying optimization opportunities and implementing targeted steps, organizations may lessen their negative influence on the environment, boost operational efficiency, and save a ton of money. When it comes down to it, modern data center management couldn't function without real-time energy monitoring. It provides invaluable insights that drive continuous improvements in sustainability and energy efficiency [61].

12.5.4.2 Automation and Artificial Intelligence Applications

In order to improve efficiency, effectiveness, and performance, data center operations are increasingly using automation and AI technology. For many parts of data center administration, these technologies provide several benefits, including:

- **Resource provisioning and scaling:** Automated analysis of workload patterns and demand forecasting by AI allows for proactive provisioning and scalability of storage and servers to meet demand changes.
- **Predictive maintenance:** Machine learning models may proactively reduce downtime and maximize equipment longevity by analyzing data from sensors and monitoring systems to forecast equipment problems in advance.
- **Energy management:** Utilizing AI algorithms, automation systems have the ability to optimize energy consumption. By continuously altering cooling systems, lighting, and other infrastructure components in response to real-time data and predictive analytics, substantial energy savings may be achieved.
- **Workload optimization:** Workload orchestration solutions powered by AI can optimize resource usage, distribute workloads intelligently across data center infrastructure, and ensure that key applications get the resources they need while reducing waste.

- **Security enhancement:** Improving overall cybersecurity and lowering the risk of data breaches are both made possible by automation systems that are AI-enabled and can continually monitor network traffic, identify abnormalities, and react to security threats in real time.
- **Troubleshooting and diagnostics:** Analyzing log files, performance measurements, and historical data, AI-powered automation may help diagnose and troubleshoot problems inside the data center environment by identifying underlying causes and suggesting remedial steps.
- **Optimized cooling and airflow:** By evaluating airflow patterns, temperature data, and equipment loads, AI algorithms may optimize cooling systems. This optimization process involves dynamically adjusting cooling settings and airflow management tactics, resulting in improved efficiency and reduced energy usage.
- **Data center automation:** In order to reduce human error and free up IT workers for more strategic efforts, automation solutions may automate typical administrative operations like server provisioning, software upgrades, and configuration management.
- **Capacity planning:** In order to make capacity planning choices, AI-driven analytics may look at past data and predict how demand will change in the future. This way, the data center can manage expansion without over-provisioning.
- **Service level management:** Automation solutions driven by AI can keep tabs on service levels in real time, identify performance issues, and automatically reallocate resources to satisfy service level agreement (SLAs) and improve user experience.

Organizations may gain a competitive edge in today's digital world and boost efficiency, dependability, and agility in data center operations by using AI and automation [62].

12.5.5 Data Analytics for Performance Optimization

Data analytics is essential for optimizing performance in many different areas since it uses data-driven insights to make things better. Organizations may get important insights from massive datasets to drive changes and enhance decision-making by using advanced algorithms and statistical methodologies. The optimization of company processes is a key area that data analytics may help with. Businesses may find inefficiencies, improve workflows, and allocate resources more effectively by examining operational data including production outputs, resource usage, and supply chain measurements. Better overall performance, increased productivity, and reduced costs are the results.

With the use of data analytics, businesses may learn more about consumer habits, tastes, and industry tendencies in the field of sales and marketing. Businesses may enhance sales processes, customize customer experiences, and adjust marketing tactics based on consumer data like demographics and purchase history. This helps to boost revenue and customer happiness. When it comes to healthcare performance improvement, data analytics is equally vital. Healthcare practitioners may improve treatment approaches, anticipate patient requirements, and spot trends by studying clinical results, medical records, and patient data. Better use of resources, more effective healthcare delivery, and better patient outcomes are the results of this. Sports, transportation, and logistics are just a few other areas where data analytics plays a crucial role in performance optimization. Take sports as an example: Teams use data analytics to study individual stats, plan games, and increase their odds of winning. Data analytics are used by transportation and logistics organizations to enhance delivery schedules, optimize routes, and decrease fuel use. In sum, data analytics is a potent instrument for optimizing performance in a wide range of companies and fields. In today's data-driven world, companies may gain a competitive edge, improved results, and more efficiency by using data to discover important insights, drive informed decision-making, and constantly improve operations [63].

12.5.6 Renewable Energy Integration

The term "renewable energy integration" describes the steps taken to guarantee the efficiency, stability, and dependability of energy systems while using renewable energy sources as shown in Figure 12.6. To lessen our impact on the environment and improve our energy independence, this integration is vital as we move away from fossil fuels and toward a more sustainable energy mix. The unpredictability and intermittency of renewable power sources, including wind and solar, is a major obstacle to their widespread use. Renewable energy production is time- and weather-dependent; in contrast to conventional fossil fuel-based power, which can be dispatched on demand. To tackle this issue, energy management systems and sophisticated forecasting methods are used to maximize the integration of renewable energy into the

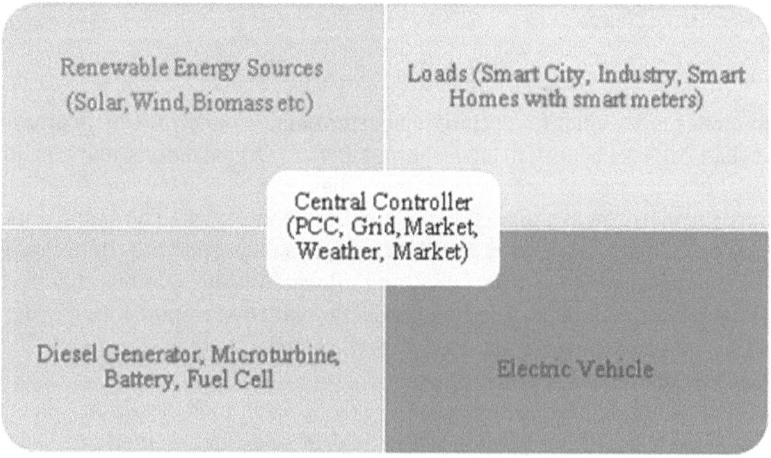

FIGURE 12.6 Renewable energy integration.

grid and anticipate when it will be available. When integrating renewable energy sources, grid stability is an additional important factor to consider [64]. Voltage and frequency swings are two consequences of renewable energy's unpredictability that threaten grid stability. Energy storage systems, smart grid infrastructure, and grid-scale power electronics are some of the technologies that grid operators use to balance supply and demand in real time and assure grid stability [65].

In addition, energy storage is critical for renewable energy integration since it allows for the discharge of surplus energy during low generation periods and the storage of excess energy during high production periods. Grid stability is achieved by the use of novel storage technologies such as battery storage and pumped hydro storage, which assist to mitigate variations in renewable energy supply. Renewable energy integration is also greatly aided by well-thought-out policy and regulatory frameworks. Various incentives are in place to stimulate investment in renewable energy infrastructure and its integration into the grid. These include feed-in tariffs, renewable energy credits, and net metering. Renewable energy resources may be easily integrated into the current energy infrastructure via grid modernization programs and connectivity standards. In sum, integrating renewable energy sources calls for a comprehensive strategy that blends technology advancement, governmental backing, and grid modernization initiatives. Cleaner, more robust, and longer-lasting energy systems are possible when renewable energy sources are properly integrated into the energy mix [66].

12.5.7 Operational Best Practices

Organizations use operational best practices to manage their day-to-day operations efficiently and effectively. These best practices include a broad diversity of methods and procedures. The tenets of sustainability, risk management, and optimization form the basis of these procedures. Methodologies such as Lean and Six Sigma aim to reduce process unpredictability and waste, while continuous improvement programs provide workers the tools they need to spot and fix inefficiencies. To improve consistency and automate mundane operations, standardizing processes and integrating technology are essential, while cross-functional cooperation encourages departments to work together and come up with new ideas. In addition, by putting the customer first, operational choices may be made with the goal of providing value and going above and beyond what the customer expects. Organizations may boost operational efficiency, promote growth, and respond to changing market dynamics by implementing these best practices in a fast-paced business environment [67].

12.6 CASE STUDIES

12.6.1 Google Data Center – The Dalles, Oregon

The operations of Google's data centers are dedicated to sustainability and energy efficiency. One of Google's most cutting-edge energy-efficient data centers is located in a city in Oregon named The Dalles. The "evaporative cooling" technology is an innovative cooling system that Google uses at its data center in The Dalles. By drawing on the area's naturally cold weather and relatively low humidity levels, this system may chill buildings without relying on costly and inefficient mechanical cooling systems. A hot aisle containment system is used in the data center, where servers are lined up in rows with their hot air exhausts facing each other in covered lanes. A more efficient cooling system with less energy usage is achieved by this design, which prevents hot and cold air from mingling. Google has pledged to use renewable energy to power its data centers. Wind and solar power are among the renewable energy sources used by The Dalles data center, which helps to lessen the facility's impact on the environment and its need on fossil fuels. The data center's modular architecture enables both efficient space use and flexible expansion. Google is able to optimize its infrastructure and decrease energy use by using this modular strategy, which allows for fluctuations in the demand for computer resources. When compared to older, less efficient data centers, Google's facility in The Dalles is light years ahead. Cooling energy usage has been drastically cut, thanks to the novel cooling system and hot aisle containment design. This has resulted in lower operating costs and less environmental effect. Using renewable energy sources also shows that Google is serious about cutting carbon emissions from its data center operations, which is in line with its environmental aims [68].

12.6.2 Microsoft Data Center – Dublin, Ireland

Data center designs at Microsoft and other major companies also place particular emphasis on minimization of energy use and environmental impact. One example of Microsoft's dedication to environmentally friendly data center techniques is the Dublin data center. In order to make the most of the chilly weather in the area, the data center in Dublin uses a free air conditioning system. Significant energy savings are achieved by circulating outside air throughout the building, which minimizes the need for mechanical cooling systems, which are energy expensive. At its data center in Dublin, Microsoft has installed a water recycling system that collects and treats rainwater and wastewater for later use in landscaping and cooling plants. This method lessens the facility's influence on the environment by decreasing its water use. Microsoft has also pledged to use renewable energy to power its data centers, much like Google. Renewable energy installations in the area provide the wind and solar electricity that the Dublin data center uses. To keep data center power consumption low and processing performance high, Microsoft uses energy-efficient servers and storage devices. This encompasses the use of server virtualization technologies, energy-efficient CPUs, and solid-state storage. There have been major gains in energy efficiency and positive effects on the environment at Microsoft's data center in Dublin. Free air cooling, water recycling, and the purchase of renewable energy have all contributed to a decrease in the facility's operational expenses and carbon impact. Data center operations also benefit from reduced energy consumption and better utilization of available resources when IT gear is designed with efficiency in mind [69].

12.6.3 Facebook Data Center – Luleå, Sweden

Energy efficiency and environmental responsibility are priorities for Facebook's data center operations. Facebook has one of its most cutting-edge and energy-efficient data centers in Sweden, the Luleå data center. Located in an area of Sweden with a frigid environment, the Luleå data center has consistently low temperatures all year 'round. To take advantage of this inherent benefit, Facebook uses free cooling methods, which include drawing air from outside the data center to keep it cold instead of relying on mechanical cooling systems. Hydroelectric power, in particular, accounts for a significant chunk of Sweden's renewable energy output. Facebook's Luleå data center is even more environmentally friendly since it uses only renewable energy, which is generated entirely by hydroelectric power stations in the area. Facebook has created a revolutionary server rack design that maximizes cooling efficiency by optimizing airflow. With the Open Compute Project's (OCP) rack architecture, air can circulate more freely, which improves server performance and reliability while decreasing cooling energy consumption. In order to recycle the excess heat from the computers, the data center uses heat recovery equipment. The facility's energy efficiency and environmental effect are both enhanced by heating office areas and surrounding buildings with this collected heat. Facebook's data center in Luleå has accomplished its environmental and energy-efficiency targets with great success. There have been considerable reductions in energy use and carbon emissions as a consequence of the integration of free cooling, renewable energy procurement, smart server rack design, and heat recovery technologies. One way Facebook is showing its concern for the environment is via this facility, which is a model for sustainable data center design and management [70].

12.6.4 Apple Data Center – Maiden, North Carolina, USA

Apple has pledged to reduce the environmental effect of its data centers by powering them with renewable energy. The North Carolina-based Maiden data center exemplifies Apple's commitment to sustainability and energy efficiency. Energy from solar panels and fuel cells on-site are in use. A large solar array and fuel cells fueled by biogas are among the on-site renewable energy sources that power Apple's Maiden data center. While fuel cells provide dependable backup power, the data center's solar array – which spans over 100 acres – generates a substantial amount of the energy required. Modern insulation, energy-saving LED lighting, and a highly efficient HVAC system are all part of the data center's high-efficiency design. All things considered, these architectural features aid in reducing HVAC, lighting, and total energy use. At its Maiden data center, Apple has installed water-efficient landscaping and rainwater collection systems to reduce water use. The facility's water consumption and environmental effect are reduced via these methods, which is especially important in an area that often experiences water shortages. The Maiden data center has attained Leadership in Energy and Environmental Design (LEED) accreditation, which is a world-renowned benchmark for environmentally conscious construction practices. The achievement of this accreditation highlights Apple's dedication to running its data centers in an environmentally responsible manner. There have been major gains in energy efficiency and positive effects on the environment at Apple's Maiden data center. In its data center operations, Apple prioritizes sustainability and environmental stewardship. This is evident via the use of on-site renewable energy sources, high-efficiency building design, water conservation measures, and LEED certification [71].

12.6.5 Switch Data Centers – Las Vegas, Nevada, USA

Renowned for its dedication to sustainability and energy efficiency, Switch is a front-runner in the industry of colocation data center solutions. Among the world's biggest and most energy-efficient data center buildings, the Switch Las Vegas facility stands out. Modern methods of cooling for precise temperature regulation in its data center buildings: Switch makes use of cutting-edge cooling technology such as indirect evaporative cooling and adiabatic cooling systems. When compared to more conventional mechanical cooling techniques, these devices significantly reduce energy usage. All of Switch's data centers will be run by sustainable energy sources. For the purpose of offsetting its power use and reducing its carbon footprint, the corporation has made investments in large-scale renewable energy projects, such as wind and solar farms. In order to reduce power usage and optimize efficiency in its data centers, Switch uses energy-efficient infrastructure components including high-density server racks, energy-efficient UPS systems, and LED lighting. LEED accreditation is one of the stringent environmental criteria that Switch data centers are constructed and run to achieve. Data centers that have achieved LEED certification showcase Switch's dedication to eco-friendly construction methods and protecting the environment. There have been significant advances in energy efficiency and environmental advantages at Switch's data center site in Las Vegas. Reduced energy consumption and carbon emissions have been achieved via the use of LEED certification, energy-efficient infrastructure, renewable energy procurement, and sophisticated cooling technology. Switch's dedication to sustainability highlights the significance of conducting operations in an eco-conscious manner and serves as an example for other data center companies [72].

12.6.6 Equinix Data Center – AM3 Science Park, Amsterdam, Netherlands

Sustainable practices and energy efficiency are key to Equinix's business model. The firm operates data centers and interconnections across the world. Among Equinix's most prominent facilities is the AM3 Science Park data center in Amsterdam, which has cutting-edge cooling technology and an energy-efficient layout. Adiabatic cooling systems are used to control temperatures at Equinix's AM3 data center. These systems work using evaporative cooling principles. Data centers may reduce their reliance on energy-intensive mechanical cooling systems by installing these systems, which use outside air to cool the equipment. The modular architecture of the AM3 data center enables optimal space use and flexible expansion. With Equinix, you can optimize resource usage and minimize energy consumption by deploying new infrastructure modules as required. To collect and reuse the thermal energy that data center equipment produces, Equinix installs heat recovery systems. Nearby buildings and facilities are heated using the trapped heat, which helps with energy efficiency and lessens the total environmental effect. Renewable energy sources will power Equinix's data centers. In order to power its data centers with sustainable energy, the firm works with local renewable energy suppliers to acquire wind and solar electricity. There have been major gains in energy efficiency and positive effects on the environment at Equinix's data center at AM3 Science Park. Significant reductions in energy consumption and carbon emissions have been achieved via the use of adiabatic cooling systems, modular design concepts, heat recovery technologies, and sources of renewable energy. Sustainable data center operations are within reach, as shown by Equinix's dedication to sustainability, which further solidifies the company's position as a market leader [73].

These case studies show how big giants like Apple and Facebook are running their data centers in an eco-friendly and energy-efficient way by using renewable energy and creative design. These businesses are leading the way in sustainability and helping the fight against climate change on a worldwide scale.

12.7 FUTURE TRENDS AND INNOVATIONS

By enhancing their sustainability, cost-effectiveness, and resilience, future developments in energy-efficient data center architecture have the potential to completely transform the data center industry. Powering data center operations using renewable energy sources is becoming more common. For one, we need to put more money into renewable energy sources like solar, wind, hydro, and geothermal power so we can cut down on our use of fossil fuels and our impact on the environment. In order to lessen the negative effects on the environment and increase energy efficiency, data center cooling systems are increasingly using cutting-edge cooling technologies, including liquid cooling and direct-to-chip cooling. Modular and prefabricated data center modules are two examples of the novel architectural concepts that are gaining popularity; these designs enable quicker deployment, greater scalability, and better resource utilization. Additionally, it is anticipated that more and more data centers will utilize AI and ML algorithms to improve operations, forecast equipment breakdowns, and dynamically modify energy use according to demand patterns. Improved data processing and less reliance on energy-intensive long-distance data transfers are two additional benefits of the emerging edge computing infrastructure. This technology moves computing resources closer to the people who will be using them. All things considered, these new developments in energy-efficient data center architecture will be pivotal in reducing the negative impact of data center operations on the environment, while simultaneously satisfying the ever-increasing demand for digital services [74].

Organizations may now optimize performance and reliability while also meeting sustainability targets, thanks to new technologies that are reshaping energy-efficient data center architecture. Fluid immersion cooling and direct-to-chip cooling are two examples of such cutting-edge cooling technologies. These novel methods allow for higher-density computing while decreasing energy usage compared to conventional air-based cooling systems. Using algorithms for machine learning and artificial intelligence to improve data center operations is another new technology. By utilising AI-powered predictive analytics, it is possible to optimise energy efficiency without compromising performance. This technology can anticipate fluctuations in workload demand and adjust cooling and power usage in real-time to achieve the required levels. Renewable energy sources, including solar, wind, and geothermal power, are also becoming more practical and affordable to include. In an effort to lessen their impact on the environment, data centers are using renewable energy technology [75]. An additional sustainable and adaptable option is the rise of prefabricated and modular data center designs. The fast scalability and effective use of space and resources are made possible by these modular architectures, which may be erected rapidly. Finally, edge computing is changing the way data centers are designed by moving processing closer to end users and decentralizing power. Edge computing reduces the need for data to travel great distances, which in turn increases efficiency and reduces energy use. In the realm of energy-efficient data center architecture, there are new technologies that are revolutionizing sustainability, performance, and cost-effectiveness. These innovations are setting the stage for a digital infrastructure that is both durable and environmentally conscious [76].

12.8 CONCLUSION

Finally, cutting-edge technology and environmental concerns are coming together to propel data center architecture toward greater energy efficiency. Reduced energy usage and operating expenses are on the horizon, thanks to new innovative cooling techniques including direct-to-chip cooling and liquid immersion. These innovations enhance temperature management, which in turn allows for more efficient and dense processing. In addition, data centers can now adapt their cooling and power consumption in real time to meet fluctuating demand patterns, thanks to the integration of AI and ML algorithms. Improving energy efficiency without sacrificing performance is possible with the help of predictive analytics, which optimize the allocation of resources. Reduced energy usage, increased equipment longevity, and decreased maintenance expenses are all results of this intelligent optimization. On top of that, renewable energy sources, including solar, wind, and geothermal power, are quickly gaining popularity. These renewable energy sources are helping data centers cut down on their use of fossil fuels and their carbon footprint. By incorporating renewable energy sources, data centers may demonstrate their commitment to sustainability and meet regulatory obligations.

Modular and prefabricated data center designs provide scalability and flexibility, along with technology developments. Rapid deployment and effective use of resources are made possible by these modular buildings, which save construction time and costs and respond to shifting company demands. In addition, computer capacity is decentralized using edge computing technologies, which improves overall efficiency and decreases energy consumption related to long-distance data transmission. When taken as a whole, these tendencies indicate a sea change in the direction of data center operations that are both efficient and ecologically friendly. Organizations may maximize resource use, minimize environmental effect, and achieve financial savings in the long run by adopting sustainable practices and energy-efficient technology. Collaboration among data center operators, technology providers, lawmakers, and environmental campaigners is necessary for a successful deployment, however. Finally, cutting-edge innovation, eco-friendly procedures, and teamwork are the keys to future-proof data center architecture that save energy. To fulfill increasing demand while minimizing environmental impact and optimizing operational efficiency, data centers must prioritize energy efficiency as they expand into crucial infrastructure underpinning the digital economy.

REFERENCES

1. Liang, Y., Yang, H., Wang, H., Bao, X., & Cui, H. (2024). Enhancing energy efficiency of air conditioning system through optimization of PCM-based cold energy storage tank: A data center case study. *Energy, 286*, 129641.
2. Aili, A., Long, W., Cao, Z., & Wen, Y. (2024). Radiative free cooling for energy and water saving in data centers. *Applied Energy, 359*, 122672.
3. Lin, L., Wang, Z., Tian, L., Wu, J., & Wu, W. (2024). A PSO-based energy-efficient data collection optimization algorithm for UAV mission planning. *Plos One, 19*(1), e0297066.
4. Kang, M., & Jeon, S. W. (2024). Energy-efficient data aggregation and collection for multi-UAV-enabled IoT networks. *IEEE Wireless Communications Letters, 13*(4), 178–190.
5. Duan, Z., & Wang, T. (2024). Towards learning-based energy-efficient online coordinated virtual network embedding framework. *Computer Networks, 239*, 110139.
6. Li, Y., Wang, X., Luo, P., & Pan, Q. (2019). Thermal-aware hybrid workload management in a green datacenter towards renewable energy utilization. *Energies, 12*(8), 1494.
7. G, M., Deshmukh, P., N. L., U. K., Macedo, V. D. J., K B, V., N, A. P., & Tiwari, A. K. (2023). Resource allocation energy efficient algorithm for H-CRAN in 5G. *International Journal on Recent and Innovation Trends in Computing and Communication, 11*(3s), 118–126.
8. Fan, W., Xiao, F., Lv, M., Han, L., Wang, J., & He, X. (2023). Node essentiality assessment and distributed collaborative virtual network embedding in datacenters. *IEEE Transactions on Parallel and Distributed Systems, 34*(4), 1265–1280.
9. Chen, Y., Shi, K., Chen, M., & Xu, D. (2022). Data center power supply systems: From grid edge to point-of-load. *IEEE Journal of Emerging and Selected Topics in Power Electronics, 13*(5), 3851–3856.
10. Grigoryan, G., & Kwon, M. (2023, June). Towards greener datacenters via programmable data plane. In *2023 IEEE 24th International Conference on High Performance Switching and Routing (HPSR)* (pp. 62–67). IEEE.
11. Maharry, A., Andrade, H., Misak, S., Liu, J., Xia, Y., Wissing, A., ... Schow, C. L. (2023). Integrated SOAs enable energy-efficient intra-data center coherent links. *Optics Express, 31*(11), 17480–17493.
12. Abderrahim, W., Amin, O., & Shihada, B. (2023). Data center-enabled high altitude platforms: A green computing alternative. *IEEE Transactions on Mobile Computing, 23*(5), 6149–6162. https://doi.org/10.1109/TMC.2023.331620.
13. Andrade, E., Ndachena, A., Macz, B., Tang, C., Wang, M., Yen, S., ... Tao, B. (2023, October). Custom ASICs for data center video processing: Advancements in AI-ML integrated, high-performance VPUs for hyper-scaled platforms. In *Applications of Digital Image Processing XLVI* (Vol. 12674, pp. 294–306). SPIE.

14. Naug, A., Guillen, A., Luna Gutiérrez, R., Gundecha, V., Ghorbanpour, S., Dheeraj Kashyap, L., ... Sarkar, S. (2023, November). PyDCM: Custom data center models with reinforcement learning for sustainability. In *Proceedings of the 10th ACM International Conference on Systems for Energy-Efficient Buildings, Cities, and Transportation* (pp. 232–235).
15. Dutta, K., & Vandermeer, D. (2023). A case for sustainability and environment friendliness in software development and architecture decisions by taking energy-efficient design decisions. *arXiv preprint arXiv:2311.01680.*
16. Arroba, P., Risco-Martín, J. L., Moya, J. M., & Ayala, J. L. (2018). Heuristics and meta-heuristics for dynamic management of computing and cooling energy in cloud data centers. *Software: Practice and Experience*, *48*(10), 1775–1804.
17. Lu, T., Lü, X., Välisuo, P., Zhang, Q., & Clements-Croome, D. (2024). Innovative approaches for deep decarbonization of data centers and building space heating networks: Modeling and comparison of novel waste heat recovery systems for liquid cooling systems. *Applied Energy*, *357*, 122473.
18. Kim, J., Han, S., Ko, G., Kim, J. H., Lee, C., Kim, T., ... Kim, J. Y. (2024). EPU: An energy-efficient explainable AI accelerator with sparsity-free computation and heat map compression/pruning. *IEEE Journal of Solid-State Circuits*, *59*(3), 830–841. https://doi.org/10.1109/JSSC.2023.334691.
19. He, W., Xu, Q., Liu, S., Wang, T., Wang, F., Wu, X., ... Li, H. (2024). Analysis on data center power supply system based on multiple renewable power configurations and multi-objective optimization. *Renewable Energy*, *222*, 119865.
20. Bai, X., Xu, X., Mi, R., & Li, Z. Climate zones for the application of water-side economizer in a data center cooling system. SSRN. https://ssrn.com/abstract=4687775 or http://dx.doi.org/10.2139/ssrn.468777.
21. Koo, M., & Park, J. (2024). MEC server sleep strategy for energy efficient operation of an MEC system. *Applied Sciences*, *14*(2), 605.
22. Satyanarayan, R. (2024). A novel approach for energy-efficient container migration by using gradient descent namib beetle optimization. *Contemporary Mathematics*, *5*(3), 3462–3483. https://ojs.wiserpub.com/index.php/CM/article/view/308
23. Ni, W., Hu, X., Du, H., Kang, Y., Ju, Y., & Wang, Q. (2024). CO2 emission-mitigation pathways for China's data centers. *Resources, Conservation and Recycling*, *202*, 107383.
24. Sun, Y., Xing, Z., & Liu, G. (2024). Achieving resilient cities using data-driven energy transition: A statistical examination of energy policy effectiveness and community engagement. *Sustainable Cities and Society*, *101*, 105155.
25. Devred, A., Ballarino, A., Bourcey, N., Mangiarotti, F., Milanese, A., & Petrone, C. (2024). Proof-of-principle of an energy-efficient, iron-dominated electromagnet for physics experiments. *IEEE Transactions on Applied Superconductivity*, *34*(5), 1–7. https://doi.org/10.1109/TASC.2024.335587.
26. Simaiya, S., Lilhore, U. K., Sharma, Y. K., Rao, K. B., Maheswara Rao, V. V. R., Baliyan, A., ... Alroobaea, R. (2024). A hybrid cloud load balancing and host utilization prediction method using deep learning and optimization techniques. *Scientific Reports*, *14*(1), 1337.
27. Sarkar, A., Dasmahapatra, S., Chaudhuri, S., Saha, R., Mookherjee, S., & Sanyal, D. (2024). A novel order-separated generalized feedforward design for motion control in energy-efficient electrohydraulic system with proportional and integral feedback. *ISA Transactions.*
28. Agbajor, F. D., & Mewomo, M. C. (2024). Green building research in South Africa: A scoping review and future roadmaps. *Energy and Built Environment*, *5*(2), 316–335.
29. Ebekozien, A., Aigbavboa, C., Thwala, W. D., Amadi, G. C., Aigbedion, M., & Ogbaini, I. F. (2024). A systematic review of green building practices implementation in Africa. *Journal of Facilities Management*, *22*(1), 91–107.

30. Sharanya, B. K., Ghoshb, S., Bhandari, S. S., Poudel, S., Padikkal, G. M., & Kumarc, M. S. (2039). Green and sustainable building practices for museums. In *AIP Conference Proceedings* (Vol. 20010, No. 2018).
31. Pesch, D., Rea, S., Torrens, J. I., Zavrel, V., Hensen, J. L. M., Grimes, D., ... Tsachouridis, V. (2017). Globally optimised energy efficient data centers. *ICT-Energy Concepts for Energy Efficiency and Sustainability*, 187–214.
32. Sharanya, B. K., Ghoshb, S., Bhandari, S. S., Poudel, S., Padikkal, G. M., & Kumarc, M. S. (2039). Green and sustainable building practices for museums. In *AIP Conference Proceedings* (Vol. 20010, No. 2018).
33. Wang, J., Chen, D., Yang, Z., Feng, Y., Geng, L., Zhang, Y., ... Mu, H. (2024, January). Analyzing the existing problems in the operation and management of green public buildings. In *2023 2nd International Conference on Public Service, Economic Management and Sustainable Development (PESD 2023)* (pp. 103–115). Atlantis Press.
34. Kaur, H., & Kaur, K. (2012). Energy conservation: An effective way of energy utilization. *International Journal of Managment, IT and Engineering*, *2*(5), 623–637.
35. Efosa, I. O., James, A. O., & Joseph, D. I. E. (2023). Impact of green finance on green infrastructure development in Nigeria (1995–2022). *Strategic Management Practices & Sustainable Development in a Global Economy*, 273.
36. Stefanatou, A., Lagkadas, M., Petousi, I., Schiza, S., Stasinakis, A. S., Fyllas, N., & Fountoulakis, M. S. (2024). Vertical flow constructed wetlands as green facades and gardens for on-site greywater treatment in buildings: Two-year mesocosm study on removal performance. *Science of The Total Environment, 906*, 167362.
37. Pu, J., Shen, C., Lu, L., Long, Y., Zhang, C., Shuai, Y., & Kalogirou, S. A. (2024). Ammonia powered thermal-responsive smart window with spectral regulation of Cu2+ and sodium copper chlorophyllin. *Energy Conversion and Management, 299*, 117815.
38. Jain, M., Agarwal, M. R., & Jain, A. S. (2030). Impact of green Economy-A performance analysis of India's transition to green energy, JK Business School Journal Special Issue, *1*(2), 8–22. ISSN 2581-7337. https://www.jkbschool.org/research-journals.
39. Mamajonova, N., Oydin, M., Usmonali, T., Olimjon, A., Madina, A., & Marg'uba, M. (2024). The role of green spaces in urban planning enhancing sustainability and quality of life. *Holders of Reason*, *2*(1), 346–358.
40. Ali, A., Khan, L., Javaid, N., Aslam, M., Aldegheishem, A., & Alrajeh, N. (2024). Exploiting machine learning to tackle peculiar consumption of electricity in power grids: A step towards building green smart cities. *IET Generation, Transmission & Distribution*, *18*(3), 413–445.
41. Pan, J., Bi, Q., & Wu, X. (2024). Spatial accessibility analysis of the prefab building suppliers using Gaussian based 2-step floating catchment area method: A case study of Fuzhou city. In *Constructional Engineering and Ecological Environment* (pp. 81–89). CRC Press.
42. Feng, C., Ye, X., Li, J., & Yang, J. (2024). How does artificial intelligence affect the transformation of China's green economic growth? An analysis from internal-structure perspective. *Journal of Environmental Management, 351*, 119923.
43. Hassan, Q., Algburi, S., Sameen, A. Z., Salman, H. M., & Jaszczur, M. (2024). Green hydrogen: A pathway to a sustainable energy future. *International Journal of Hydrogen Energy, 50*, 310–333.
44. MacIvor, J. S., Williams, N. S., & Lundholm, J. (2024). The role of engineered green infrastructure in supporting urban biodiversity. In *Routledge Handbook of Urban Biodiversity* (pp. 333–345). Routledge.
45. Hines, N. (2024). Corporate climate litigation and environmental justice: How green amendments can be used to advance accountability and equity. *Indiana Journal of Law and Social Equality*, *12*(1), 3.

46. Babalola, A., & Harinarain, N. (2024). Policy barriers to sustainable construction practice in the Nigerian construction industry: An exploratory factor analysis. *Journal of Engineering, Design and Technology*, *22*(1), 214–234.
47. Rashid, A., Rasheed, R., & Ngah, A. H. (2024). Achieving sustainability through multifaceted green functions in manufacturing. *Journal of Global Operations and Strategic Sourcing*, *17*(2), 402–428. https://doi.org/10.1108/JGOSS-06-2023-005.
48. Cai, S., & Gou, Z. (2024). Impact of COVID-19 on the energy consumption of commercial buildings: A case study in Singapore. *Energy and Built Environment*, *5*(3), 364–373.
49. Adams, S., & Kaffo Fotio, H. (2024). Economic integration and environmental quality: Accounting for the roles of financial development, industrialization, urbanization and renewable energy. *Journal of Environmental Planning and Management*, *67*(3), 688–713.
50. Chidolue, O., Fafure, A. V., Illojianya, V. I., Ngozichukwu, B., Daudu, C. D., & Ibekwe, K. I. (2024). Integration of renewable energy in industrial operations: Experiences from Canada, USA, and Africa, GSC Advanced Research and Reviews, *18*(01),213–221. https://doi.org/10.30574/gscarr.2024.18.1.001.
51. Hassan, Q., Viktor, P., Al-Musawi, T. J., Ali, B. M., Algburi, S., Alzoubi, H. M., ... Jaszczur, M. (2024). The renewable energy role in the global energy transition. *Renewable Energy Focus*, 100545.
52. Rajapandiyan, A., & Thangavel, S. (2024). Energy management in distribution system due to the integration of renewable energy powered EV charging infrastructures. In *Power Systems Operation with 100% Renewable Energy Sources* (pp. 281–297). Elsevier.
53. Shahzad, S., Abbasi, M. A., Ali, H., Iqbal, M., Munir, R., & Kilic, H. (2023). Possibilities, challenges, and future opportunities of microgrids: A review. *Sustainability*, *15*(8), 6366.
54. Aljarrah, R., Fawaz, B. B., Salem, Q., Karimi, M., Marzooghi, H., & Azizipanah-Abarghooee, R. (2024). Issues and challenges of grid-following converters interfacing renewable energy sources in low inertia systems: A review. *IEEE Access*, *12*, 5534–5561. https://doi.org/10.1109/ACCESS.2024.334963.
55. Kaur Channi, H., Gupta, S., & Dhingra, A. (2020). Optimization and simulation of a solar–wind hybrid system using HOMER for Rural Electrification. *International Journal of Advanced Science and Technology*, *29*, 2108–2116.
56. Sunaina, H. K. C., & Gupta, S. (2020). Optimization and simulation of solar PV based hybrid system using Homer software. *International Journal of Advanced Science and Technology*, *29*(12s), 715–728.
57. Kushwah, B., & Channi, H. K. (2023, June). Net zero energy building: A case study of jaisalmer. In *International Conference on Intelligent Manufacturing and Energy Sustainability* (pp. 423–435). Singapore: Springer Nature Singapore.
58. Liang, Y., Yang, H., Wang, H., Bao, X., & Cui, H. (2024). Enhancing energy efficiency of air conditioning system through optimization of PCM-based cold energy storage tank: A data center case study. *Energy*, *286*, 129641.
59. Kang, M., & Jeon, S. W. (2024). Energy-efficient data aggregation and collection for multi-UAV-enabled IoT networks. *IEEE Wireless Communications Letters*, *13*(4), 1004–1008. https://doi.org/10.1109/LWC.2024.335593.
60. Lu, T., Lü, X., Välisuo, P., Zhang, Q., & Clements-Croome, D. (2024). Innovative approaches for deep decarbonization of data centers and building space heating networks: Modeling and comparison of novel waste heat recovery systems for liquid cooling systems. *Applied Energy*, *357*, 122473.
61. Bai, X., Xu, X., Mi, R., & Li, Z. Climate zones for the application of water-side economizer in a data center cooling system, SSRN. https://ssrn.com/abstract=4687775 or http://dx.doi.org/10.2139/ssrn.468777.

62. Sun, Y., Xing, Z., & Liu, G. (2024). Achieving resilient cities using data-driven energy transition: A statistical examination of energy policy effectiveness and community engagement. *Sustainable Cities and Society*, *101*, 105155.
63. Jung, J., Kim, S., Jang, W., Seo, B., & Lee, K. J. (2024). An energy-efficient, unified CNN accelerator for real-time multi-object semantic segmentation for autonomous vehicle. *IEEE Transactions on Circuits and Systems I: Regular Papers.*
64. Chen, X., Krishnakumar, A., Ogras, U., & Chakrabarti, C. (2024). PED: Probabilistic energy-efficient deadline-aware scheduler for heterogeneous SoCs. *Journal of Systems Architecture*, *147*, 103051.
65. Bhambri, P., Sinha, V. K., Dhanoa, I. S., & Kaur, J. (2019). Genome DNA sequence matching using HBM algorithm. *International Journal of Control and Automation*, 12(5), 531–539. http://sersc.org/journals/index.php/IJCA/article/view/2686/1887.
66. Tondon, N., & Bhambri, P. (2017). Novel approach for drug discovery. *International Journal of Research in Engineering and Applied Sciences*, *7*(6), 28–46. www.euroasiapub.org.
67. Rana, R., Chhabta, Y., & Bhambri, P. (2019). A review on development and challenges in wireless sensor network. Paper presented at the International Multidisciplinary Academic Research Conference at CT University, Ludhiana (pp. 184–188) held during September 6–7, 2019.
68. Shanmuga, S. M., & Bhambri, P. (2024). Pulmonary and lungs nodule classification using deep learning. In P. Bhambri, S. Rani, & M. Fahim (Eds.), *Computational Intelligence and Blockchain in Biomedical and Health Informatics* (1st ed., pp. 320–332). CRC Press. https://doi.org/10.1201/9781003459347.
69. Singh, M. P., & Bhambri, P. (2024). AI-driven digital twin conceptual framework and applications. In S. Rani, P. Bhambri, S. Kumar, P. K. Pareek, & A. A. Elngar (Eds.), *AI-Driven Digital Twin and Industry 4.0: A Conceptual Framework with Applications* (1st ed., pp. 127–141). CRC Press. https://doi.org/10.1201/9781003395416.
70. Rana, R., & Bhambri, P. (2024). Artificial Intelligence and Internet of Things (IoT) facilitated digital twin for Industry 4.0 application domains. In S. Rani, P. Bhambri, S. Kumar, P. K. Pareek, & A. A. Elngar (Eds.), *AI-Driven Digital Twin and Industry 4.0: A Conceptual Framework with Applications* (1st ed., pp. 168–181). CRC Press. https://doi.org/10.1201/9781003395416.
71. Wilberforce, T., Olabi, A. G., Sayed, E. T., Mahmoud, M., Al-Alami, A. H., & Abdelkareem, M. A. (2024). The state of renewable energy source envelopes in urban areas. *International Journal of Thermofluids*, 100581.
72. Laimon, M., & Yusaf, T. (2024). Towards energy freedom: Exploring sustainable solutions for energy independence and self-sufficiency using integrated renewable energy-driven hydrogen system. *Renewable Energy*, 119948.
73. Rozhkov, A. (2024). Decentralized renewable energy integration in the urban energy markets: A system dynamics approach (Preprint, Version 1). Research Square. https://doi.org/10.21203/rs.3.rs-3835543/v1.
74. Hassan, A. A., & El-Rayes, K. (2024). Optimal use of renewable energy technologies during building schematic design phase. *Applied Energy*, *353*, 122006.
75. Arévalo, P., Cano, A., & Jurado, F. (2024). Large-scale integration of renewable energies by 2050 through demand prediction with ANFIS, Ecuador case study. *Energy*, *286*, 129446.
76. Niu, Q., Tang, L., Yu, L., Wang, H., & Yang, Z. (2024). Unit commitment considering electric vehicles and renewable energy integration—A CMAES approach. *Sustainability*, *16*(3), 1019.

13 AI for Environmental Monitoring and Conservation

Leszek Ziora

INTRODUCTION

Contemporary intelligent technologies play significant roles in the process of environment monitoring and conservation. Artificial Intelligence solutions embrace machine learning; big data analytics systems; different methods, techniques, and tools of data science and mining; natural language processing with large language models or sentiment analysis; and generative AI with GAN networks form an ecosystem of technological and organizational solutions which become indispensable in supporting decision-makers or policymakers involved in environment concerns (Filipsson, 2024; Florkin, 2024). These AI tools are important for decision-making support while also positively influencing the quality of decisions as well as in the facilitation and improvement of their daily work. The chapter aims to pose the following research questions: 1) what is the current state of AI deployment in the environment monitoring and conservation? 2) what are the advantages and disadvantages resulting from implementation of these solutions in the mentioned research area? and 3) what are the future trends in development of AI applied in environment monitoring and conservation? (Harnessing the Power of AI for Biodiversity Preservation, 2023; How artificial intelligence is helping tackle environmental challenges, 2022)

13.1 THE CHARACTERISTICS OF ENVIRONMENT MONITORING AND CONSERVATION

Jones et al. mention the interest in monitoring conservation by breaking the topic down into knowledge- and action-focused issues. The following stages embrace such reasons as to learn about the system, which is valuable in enhancing our understanding of dynamic interconnections in biological systems (Fan et al., 2023). Especially in terms of long-lasting monitoring, this allows us to pinpoint trends and further prepare some significiant forecasts, and to detect unexpected change. Also, long-term monitoring contributes to knowledge enhancements by raising awareness among the public policymakers through organization of various public events focused on environment protection, leading to implementation and upholding national laws. to

DOI: 10.1201/9781003484226-13

audit management actions, thereby having a positive influence on the motivation of authorities towards pro-ecological measurements; and to inform management decisions where the practice of adaptive management enables learning from monitoring outcomes to improve management effectiveness (Jelonek et al., 2020; Jelonek et al., 2018; Jones et al., 2013). Jones et al. also describe the systems of monitoring for conservation such as statistical approaches, applying statistical methods and tools, e.g., time series, hyptheses testing, Bayesian modeling; accounting for detectability with multiple methods such as species richness and community diversification; the remote sensing deploys, e.g., satellite data for the purpose of image analysis and ecosystems mapping like animal habitats in different parts of the world with additional application of airborne LiDarsor high-fidelity imaging spectroskopy (HiFIS) (Jones et al., 2013).

Stem et al. state that different approaches and tools applied in monitoring and conservation require different needs, and there is no single methodology that is suitable for all conservation efforts and monitoring, only quantitative biological information is insufficient (Stem et al., 2005). In case of environmental sciences, AI might be applied in climate change prediction enabling modeling and predicting climate patterns as well as shaping policy decisions. The process involves deployment of AI models, evaluating extensive datasets obtained from satellite images, ocean buoys, and weather stations, bringing advantages such as improved comprehension of climate change or better environment policy implementation (Filipsson, 2024; Konya and Nematzadeh, 2024; Lamba et al., 2018; Nadella, 2023; Pentapalli, 2023; Perucica and Andjelković, 2022; Popescu et al., 2024). Deep learning enables automation of large datasets (Lamba et al., 2019). Florkin perceives AI as a game- changer in environmental monitoring, helping to optimize the use of resources by predicting the most effective strategies for conservation and sustainability and facilitating compliance and reporting. As well as streamlining operational efficiency, creating new revenue streams, enhancing decision-making with data driven insight with optimized resource allocation, and proper risk management (Florkin, 2024). Artificial Intelligence may serve as an enabler of environmental sustainability (Perucica & Andjelković, 2022). Note that the amount of power consumption and the time required to train an AI model can greatly affect the carbon emissions it produces, exacerbating the challenges posed by climate change (Konya, Nematzadeh, 2024).

13.2 THE ROLE OF ARTIFICIAL INTELLIGENCE IN THE ENVIRONMENT MONITORING AND CONSERVATION

Chisom et al. present the role of AI in environmental monitoring and conservation utilizing ecosystem management, including wildlife tracking, habitat assessment, and resource conservation (with image analysis allowing for forest health assessment, detection of deforrestation state, identification of lands which should be restored), biodiversity analysis, natural disaster prediction, and species identification (analysis of acoustic recordings, environmental DNA or camera-trap footage) (Rattan et al., 2005; Bhambri et al., 2005). What is more, AI may provide and support flood early- warning systems, but Chisom et al. also concentrate on challenges where they mention data quality and availability, algorithmic bias, and infrastructure limitations (Chisom et al., 2024). AI is applied in biodiversity conservation where predictive analytics is applied in habitat loss prediction, analysis of illegal wildlife trade, and assessing climate change. AI is used in wildlife monitoring such as image analysis from camera traps, acoustic monitoring, anti-poaching efforts resulting in drone patrols of protected areas (Dziembek and Ziora, 2023). In ecological restoration, AI solutions enable simulation of an ecosystem's behavior, testing restoration scenarios, invasive species control, soil health monitoring. Support in resource allocation for policymakers, analysis of legal documents with suggested amendments in environmental laws, deployment in sentiment analysis are also used in AI (What Is AI in Biodiversity Conservation?, 2023). Pentapalli mentions the usage of drones for automated image recognition to evaluate population densities, tracking movement of endangered species, allowing more effective decision-making, analyzing big data to predict future ecological changes or real-time detection of illegal logging activities in protected forests (Pentapalli, 2023; Bhambri and Singh, 2005). Predictive analytics can be used to predict the impact of conservation strategies (Harnessing the Power of AI for Biodiversity Preservation, 2023).

Fan et al. claim that AI and deep learning are deployed to improve the prediction of harmful algal blooms in rivers out performing traditional regression analysis and also allows conservationists to derive information on bird population movement trends (Zen et al., 2023). Popescu et al. indicate the application of AI in detecting hazardous substances in the environment for the purpose of protecting human well-being and ecosystems where AI-driven sensor systems and IoT are applied in detecting air pollutants, water contaminants, and soil toxins (Popescu et al., 2024). Ditria et al. state that AI implementation provides managers with more information to exchange and develop knowledge about the target ecosystem, and the utilization of machine learning methods aids in better understanding of the ecosystem (Ditria et al., 2022) and this leads to taking preventive measures in advance.

On the basis of the references noted, the experts cover a general framework indicating artificial intelligence technologies applied in environment and conservation while also embracing machine learning with deep learning and reinforcement learning (Bhambri and Gupta, 2005; Rana and Bhambri, 2024; Rattan et al., 2005; Shaji et al., 2023; Singh and Bhambri, 2024; Stem et al., 2005).

It is worth mentioning that qualitative or quantitative data, which are sent and collected by various sensors using Internet of Things solutions, as well as images captured by ground and satellite cameras (Ziora, 2019; Singh and Bhambri, 2024; Dunay, 2023) in order to be processed and analyzed, are, in most cases, stored in the cloud (Dziembek Ziora, 2023). This allows for real-time analysis of such data and conducting various analyses offering multiple benefits. These benefits can be visualized and presented in managerial dashboards. Machine learning models (Ziora, 2022b) and natural language processing (NLP) solutions also increase efficiency of the decision-making process and positively influence the acceleration. The automation is also possible by the application of robotic process automation solutions (RPA) (Xu, 2019; What is AI in Biodiversity Conservation?, 2023; Ziora, 2019; Ziora, 2022a; Ziora, 2022b).

13.3 THE REVIEW OF SELECTED CASE STUDIES AND PRACTICAL EXAMPLES

One of the case studies related to AI applicability in the environment monitoring and conservation domain is the collaboration between World Wildlife Fund and Intel Corporation, which is dedicated to protecting Siberian tigers in China (Bhambri and Singh, 2005; Bhambri and Gupta, 2005). The purpose of this partnership is to deploy AI-driven analysis and large datasets from camera traps to protect wild tigers and their habitats (Nadella, 2023). Intel's technology is capable of determining whether a tiger or any other wild animal is actually present in the photo before image capturing. By harnessing cloud computing technology, the images are transferred to a monitoring center allowing us to observe the situation in almost real time. What is significant is the technology helps to identify individual tigers on the basis of a unique pattern of stripes and the information is tagged to every image that accelerates data training of the model. WWF is also deploying infrared cameras, unmanned aerial vehicles, blockchain, and satellite image processing (Medium, 2018; Rana and Bhambri, 2024).

Another positive application of AI is applying machine learning to detect and count animals, used on the basis of millions of pictures from the Serengeti National Park in Tanzania (Ditria et al., 2022; Chisom et al., 2024). The Deep Mind project facilitates and accelerates the process of animal identification and is capable of recognizing most of the species with a high degree of accuracy (Xu, 2019). A. G. Shaji et al. present a phenomena of water consumption by ChatGPT and pinpoint a few steps that can be taken to reduce water consumption such as optimization of data center location to a cooler climate so less water will be used for cooling purposes; improvement of energy efficiency of data centers; usage of alternative cooling methods; minimization of unnecessary computations by optimization of algorithms and data processing techniques; monitoring and analyzing water consumption; encouraging environmental awareness; and implementation of water reuse systems. According to the authors the water footprints of AI models is a crucial issue requiring attention and action from the AI industry (Shaji et al., 2023).

13.4 THE ANALYSIS OF RESEARCH RESULTS

As research methodology is concerned, the study was conducted in January of 2024 on a research sample of 50 respondents (n = 50), international students of the Faculty of Management, Czestochowa University of Technology. The surveyed respondents indicated the advantages resulting from deployment of artificial intelligence solutions in the environment monitoring and conservation such as the ability to rapid and efficient processing of huge datasets related to environment protection (82%), together with pattern identification (80%); automation of environment monitoring processes delivered in real time and concerning quality of air, water, and soil (80%); monitoring population of endangered species (76%); increased accuracy in data processing and modeling (74%); development and optimization of more energy-efficient algorithms (70%); creation of new jobs based on human creativity (74%); the possibility

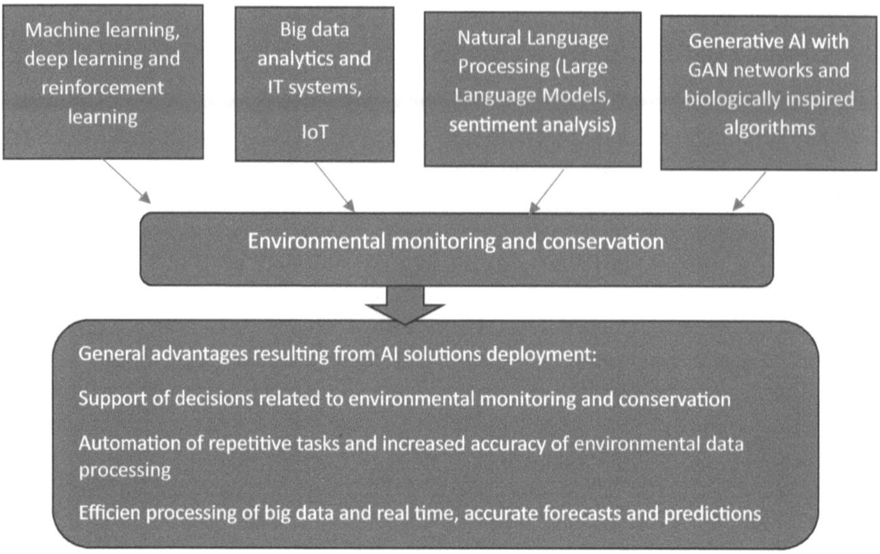

FIGURE 13.1 General framework of AI support in environmental monitoring and conservation. Source: Author's contribution.

of renewable energy source optimization (64%) (photovoltaic panel arrays, wind turbines, water-power-plants); better monitoring of wildlife and deforestation (with, e.g., satellite imagery) (64%); improved quality and speed of decision-making processes related to environment conservation (68%); and anomaly detection in a real time (66%). The other responses involved development of predictive modeling concerning, e.g., climate changes (60%); adaptability to various conditions; reduction of the cost of manual labor; scalability of AI solutions (58%); and smart agriculture (Figure 13.1).

The disadvantages resulting from AI application in environmental monitoring and conservation, according to respondents, include rising costs of AI usage together with high energy consumption (86%), which is connected with increased carbon footprint; ethics-related issues (78%); bias in data and in AI environmental models (80%); ethical concerns (78%); concerns related to data privacy and security of sensitive environmental data issues (breachers of systems and leakage of data) (72%); insufficient data quality for training models (72%); an increase of e-waste accumulation with hazardous chemicals contaminating soil and water (66%); the threat of losing jobs due to automation of multiple processes (64%); possible data misinterpretation (62%); proper expertise requirement due to solution's complexity (54%); and less human supervision on different processes' performances (40%) (Figure 13.2).

The surveyed respondents indicated the following future trends: development of AI solutions resulting in more accurate predictions and forecasts (82%); better deployment of various smart sensors within Internet of Things infrastructure with different types of sensors (e.g., acoustic) and 5G network usage (78%); better

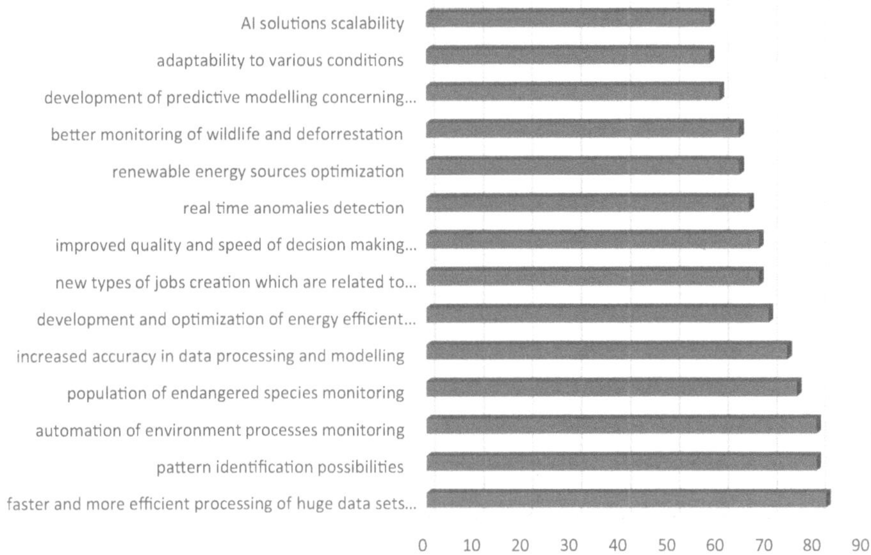

FIGURE 13.2 Advantages resulting from implementation of AI in environmental monitoring and conservation. Source: Author's study.

integration with other technologies (76%), e.g., enabling hybrid solution development; better and novelty data security solutions (68%); enforcement of law related to the improvement in conservation of environment; the use of blockchain to ensure transparency and integrity of information related to environment protection (50%); edge computing with decentralized processing and data analysis latency reduction (46%); and possibility of a quicker and more concise prediction of natural disasters such as wildfires or earthquakes (46%) (Figure 13.3).

CONCLUSION

AI solutions increase the speed, quality, and efficiency of decisions related to environment protection and conservation. Ensuring data accuracy leads to the increase of decision quality as well (Bhambri et al., 2005). AI revolutionizes many areas of environmental conservation including wildlife protection and monitoring deforestation, and species identification. These solutions allow for various trend identifications, threat detection, and also have multiple benefits which were mentioned in the research section of this chapter: cost optimization, automation of repetitive tasks, and increased accuracy of monitoring. It is noteworthy that some disadvantages may lead to potential human job losses but from the other side, the creation of new jobs. The chapter presented the current state of AI deployment in the environment monitoring and conservation field on the basis of literature review, analysis of case studies, and conducted research. On the basis of research, it discussed advantages and disadvantages resulting from application of AI solutions in the surveyed area.

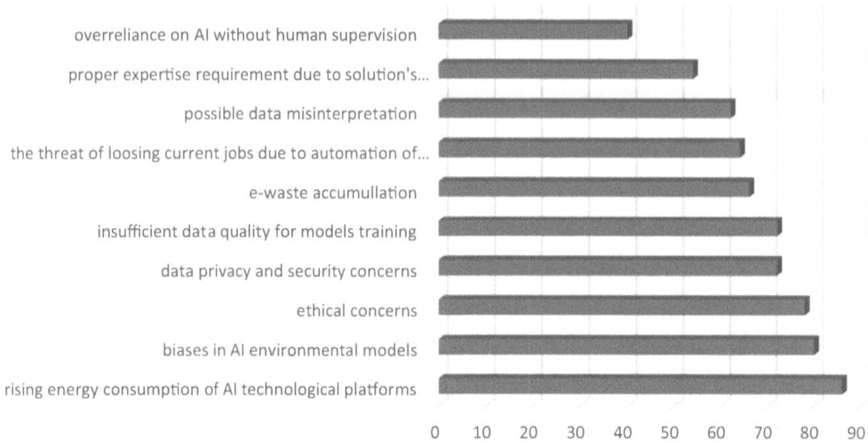

FIGURE 13.3 Disadvantages resulting from implementation of AI in environmental monitoring and conservation. Source: Author's study.

REFERENCES

Bhambri, P., & Gupta, S. (2005). A survey & comparison of permutation passibility of fault-tolerant multistage interconnection networks. In National Conference (Abstract Book) (p. 13). Presented at the "Application of Mathematics in Engineering & Technology" conference at MIMIT, Malout held during March 26-28, 2005.

Bhambri, P., Gupta, S., & Bhandari, A. (2005). Soft computing techniques. In UGC Sponsored National Conference (Abstract Book) (p. 100). Presented at the "Emerging Computing Technologies" conference at HRMMV, Jalandhar held during September 29-30, 2005.

Bhambri, P., & Singh, I. (2005). Electrical actuation systems. In National Conference (Abstract Book) (p. 8). Presented at the "Application of Mathematics in Engineering & Technology" conference at MIMIT, Malout held during March 26-28, 2005.

Chisom, O. N., Biu, P. W., Umoh, A. A., Obaedo, B. O., Adegbite, A. O., &Abatan, A. (2024, January 30). Reviewing the role of AI in environmental monitoring and conservation: A data-drivenrevolution for our planet. World Journal of Advanced Research and Reviews. https://doi.org/10.30574/wjarr.2024.21.1.2720

Ditria, E. M., Buelow, C., González-Rivero, M., &Connolly, R. M. (2022, July 28). Artificialintelligence and automated monitoring for assistingconservation of marineecosystems: A perspective. Frontiers in Marine Science. https://doi.org/10.3389/fmars.2022.918104

Dunay A., BálintIllés C., Ziora L. (2023): The applicability of generativeadversarial networks in the management of contemporary business organizations. 12-th International Conference of Management Sustainability – Security – QualityBook of Proceedings.

Dziembek, D., Ziora, L. (2023). Cloud-Based Business Intelligence Solutions in the Management of PolishCompanies. In: Silaghi, G.C., et al. Advances in Information Systems Development. ISD 2022. Lecture Notes in Information Systems and Organisation, vol 63. Springer, Cham. https://doi.org/10.1007/978-3-031-32418-5_3

Fan Z, Yan Z, Wen S. (2023) Deep Learning and ArtificialIntelligence in Sustainability: A Review of SDGs, Renewable Energy, and EnvironmentalHealth. Sustainability. 2023; 15(18):13493. https://doi.org/10.3390/su151813493

Filipsson, F. (2024, March 11). The Intersection of AI and Data Science: AnOverview. RedressCompliance, https://redresscompliance.com/the-intersection-of-ai-and-data-science-a-overview/

Florkin, J. (2024, January 25). Harnessing AI for a GreenerFuture in Environmental Monitoring. https://www.linkedin.com/pulse/harnessing-ai-greener-future-environmental-monitoring-julien-florkin-nz5ge

Harnessing the Power of AI for BiodiversityPreservation. (2023, February 15). https://www.linkedin.com/pulse/harnessing-power-ai-biodiversity-preservation-

How artificial intelligence is helping tackle environmental challenges (2022). UNEP. https://www.unep.org/news-and-stories/story/how-artificial-intelligence-helping-tackle-environmental-challenges

Jelonek, D.; Mesjasz-Lech, A.; Stępniak, C.; Turek, T.; Ziora, L. (2020) The ArtificialIntelligenceapplication in the management of contemporaryorganization: TheoreticalAssumptions, CurrentPractices and ResearchReview. In Proceedings of the Future of Information and Communication Conference. Arai, A.G.K., Bhatia, R., Eds.; Springer Nature: Cham, Switzerland, pp. 319–327

Jelonek, D.; Stępniak, C.; Ziora, L. The Meaning of Big Data in the Support of ManagerialDecisions in ContemporaryOrganizations: Review of SelectedResearch. In: Advances in Information and Communication Networks. FICC 2018. Advances in Intelligent Systems and Computing; Arai, K., Kapoor, S., Bhatia, R., Eds.; Springer: Cham, Switzerland, 2018; volume 886

Jones, J. P. G., Asner, G. P., Butchart, S. H. M., &Karanth K. U. (2013). *The "why", "what" and "how" of monitoring for conservation. KeyTopics in ConservationBiology 2, 327–343.* doi:10.1002/9781118520178.ch18

Konya A., Nematzadeh P., (2024) Recentapplications of AI to environmentaldisciplines: A review,Science of The Total Environment,Volume 906,167705,ISSN 0048-9697,https://doi.org/10.1016/j.scitotenv.2023.167705.(https://www.sciencedirect.com/science/article/pii/S0048969723063325

Lamba A., Cassey P., Segaran R.R., Koh L.P.: Deep learning for environmentalconservation, https://www.cell.com/current-biology/pdf/S0960-9822(19)31032-2.pdf

Medium (2018), Tech for tigers: WWF and Intel test AI technology for monitoring wildtigers in China. https://medium.com/wwftogetherpossible/tech-for-tigers-wwf-and-intel-test-ai-technology-for-monitoring-wild-tigers-in-china-2a5c93ece73c

Nadella, M. S. (2023, August 25). Exploring the Impact of AI in Wildlife with Examples. Analytics Vidhya. https://www.analyticsvidhya.com/blog/2023/03/wildlife-conservation-through-ai/

Pentapalli, S. (2023, October 5). BiodiversityConservation: EnhancingEffortsthrough AI. https://www.linkedin.com/pulse/biodiversity-conservation-enhancing-efforts-through-ai-sharmila-p

Perucica, N., & Andjelković, K. (2022, June 20). Is the future of AI sustainable? A casestudy of the European Union. TransformingGovernment. https://doi.org/10.1108/tg-06-2021-0106

Popescu S. M., Mansoor S., Wani O. A., Kumar S. S., Sharma V., Sharma A., Arya V. M., Kirkham M. B., Hou D., Bolan N., Chung Y. S. (2024): Artificialintelligence and IoTdriventechnologies for environmentalpollution monitoring and management. Frontiers in Environmental Science, Vol. 12, https://www.frontiersin.org/articles/10.3389/fenvs.2024.1336088,ISSN=2296-665X, DOI=10.3389/fenvs.2024.133608

Rana, R., & Bhambri, P. (2024). Artificial Intelligence and Internet of Things (IoT) Facilitated Digital Twin for Industry 4.0 Application Domains. In S. Rani, P. Bhambri, S. Kumar, P.K. Pareek, & A.A. Elngar (Eds.), AI-Driven Digital Twin and Industry 4.0: A Conceptual Framework with Applications (1st ed., pp. 168–181). CRC Press. https://doi.org/10.1201/9781003395416

Rattan, M., Bhambri, P., & Shaifali. (2005). Information retrieval using soft computing techniques. In National Conference (Pages 58–60) on Bio-informatics Computing. Presented at TIET, Patiala held during March 18-19, 2005.

Russel, S.; Norvig, P. (2021) ArtificialIntelligence. A modern Approach; 4th ed.; Pearson

Shaji A. G., Hovan A.S. G., Gabrio A.S.M. (2023, April 20). The EnvironmentalImpact of AI: A Case Study of Water Consumption by Chat GPT. Partners Universal International Innovation Journal (PUIIJ), Volume: 01 Issue: 02, March-April 2023 www.puiij.com, https://doi.org/10.5281/zenodo.7855594

Singh, M.P., & Bhambri, P. (2024). AI-Driven Digital Twin Conceptual Framework and Applications. In S. Rani, P. Bhambri, S. Kumar, P.K. Pareek, & A.A. Elngar (Eds.), AI-Driven Digital Twin and Industry 4.0: A Conceptual Framework with Applications (1st ed., pp. 127–141). CRC Press. https://doi.org/10.1201/9781003395416

Stem, C., Margoluis, R., Salafsky, N., & Brown, M. (2005). Monitoring and Evaluation in Conservation: a Review of Trends and Approaches. Conservation Biology, 19(2), 295–309. doi:10.1111/j.1523-1739.2005.00594.x

Xu J. (2019) Synced: AI In Wildlife Conservation. (n.d.). Welcome to Teamcore. https://teamcore.seas.harvard.edu/news/ai-wildlife-conservation

What is AI in Biodiversity Conservation? (2023, November 29). https://www.linkedin.com/pulse/what-ai-biodiversity-conservation-ai-news-lqpde

Ziora, L. (2019)The Role of the Internet of Things in the Management of Contemporary Business Organizations. Informatyka i zarządzanie na przełomie wieków. Metody, narzędzia, systemy, zastosowania (red); Parys, T., Ed.; Wydawnictwo Naukowe Wydziału Zarządzania Uniwersytetu Warszawskiego: Warsaw 2019; pp. 219–230.

Ziora L. (2022a), Natural Language Processing in the Support of Business Organization Management [in:] Intelligent Systems and Applications 2021 (red.) Arai K. Lecture Notes in Networks and Systems nr 294, Springer, Cham, pp. 76–83 (Intellisys 2021 Amsterdam).

Ziora L. (2022b), Machine-Learning Solutions in the Management of a Contemporary Business Organisation: a Case StudyApproach in a Banking Sector. [in:] Rational Decisions in Organisations. Theoretical and PracticalAspects (red.) Adam Frederic, Kuchta Dorota, Stanek Stanisław. Taylor and Francis Group. New York.

14 AI-Driven Sustainability in Agriculture and Farming

Julian Szymański, Karolina Nurzyńska, and Paweł Weichbroth

14.1 INTRODUCTION

The world's growing population, projected to reach 10 billion by 2050 (Gu et al., 2021), is placing significant pressure on the agricultural sector to increase agricultural production and maximize yields. Two options have emerged to address the impending food shortage: expanding land use (Kobayashi et al., 2020) and introducing large-scale agriculture (Huss et al., 2021), along with innovative practices and technological advances to increase productivity on existing farmland. The modern agricultural landscape is evolving and branching out in several innovative directions, including artificial intelligence, in the face of many obstacles to achieving desired agricultural productivity: climate change, environmental issues, limited land, labor shortages, and regulatory requirements, to name a few (Bhat & Huang, 2021).

Agriculture today is at a crossroads, facing a complex interplay of productivity demands and environmental challenges (Khan et al., 2021), witnessing significant advances in technology. However, the intensification of farming practices, reliance on chemical inputs, and large-scale monocultures have also raised environmental concerns. Soil degradation, water scarcity, and biodiversity loss are pressing challenges facing the agricultural sector (Kholaiq et al., 2022). In addition, climate change is exacerbating these issues, with unpredictable weather patterns affecting crop yields and threatening food security. Sustainable agricultural practices such as agro-ecology and precision agriculture are gaining traction as potential solutions to balancing global food demand with environmental stewardship. As the world population is still growing (Becker & Fanzo, 2023), the agricultural sector is called upon to innovate and adopt practices that prioritize ecological health alongside productivity.

Until fairly recent times, the words agriculture and artificial intelligence (AI) in the same sentence might have seemed like an odd pairing. But now they don't, as an ongoing debate involving international organizations (Dharmaraj & Vijayanand, 2018), local governments (Streich et al., 2020), businesses (Cavazza et al., 2023), and

DOI: 10.1201/9781003484226-14

a global research community discusses the applications of artificial intelligence in agriculture, farming, and other related fields (Eli-Chukwu, 2019).

The concept of AI-driven sustainability in agriculture (Alloghani, 2023) represents a transformative approach to addressing environmental challenges and improving the efficiency of food production. In general, AI technologies, including machine learning and advanced robotics (Balaska et al., 2023), have the potential to revolutionize several aspects of agriculture. By analyzing vast amounts of data and replacing human labor, AI can optimize resource allocation (Tyagi et al., 2020), improve crop management (Shankar et al., 2020), and reduce waste (Onyeaka et al., 2023). On the other hand, agricultural robots are used to automate slow, repetitive, and tedious tasks such as mowing, pruning, seeding, spraying and thinning, phenotyping, harvesting, and picking, sorting, and packaging (Loukatos et al., 2022).

14.2 THE NEED FOR SUSTAINABLE AGRICULTURE

In general terms, sustainable agriculture aims to meet the food and fiber needs of society today without compromising the ability of future generations to meet their own needs (Bhambri & Gupta, 2005). As the world's population grows and the need for more food increases, we must do business in an environmentally responsible way. Sustainable agriculture promotes social equity and economic viability while conserving natural resources. The importance of sustainable agriculture cannot be underestimated.

It results in higher yields and healthier products for consumers while reducing the environmental impact of traditional agriculture (Singh et al., 2005). Sustainable farming practices aim to conserve essential resources that are critical to agricultural success. These new production systems can benefit areas with limited or scarce resources.

Natural resources and the environment are often under significant pressure from agriculture. Sustainable agriculture practices aim to protect the environment, enhance the earth's natural resource base, and maintain and improve soil fertility. Based on a multipronged goal, sustainable agriculture seeks to (National Institute of Food and Agriculture, 2024):

- Enhance the quality of life for farming communities and families.
- Increase production to meet human demand for food and fiber.
- Increase profitable farm incomes.
- Promote environmental stewardship.

To make agriculture more sustainable, there are many sustainable farming techniques we can put into practice. Reducing our global reliance on pesticides and fertilizers and focusing on natural processes that can be used to limit pest populations is one of the big ways to achieve agricultural sustainability. Planting hedges along the edges of fields is one example of sustainable agriculture. This attracts pest predators (e.g., birds and other animals) to the area and keeps pest numbers under control, reducing the need for pesticides.

14.3 ROLE OF AI IN SUSTAINABLE AGRICULTURE

The role of AI in agriculture has been steadily growing in importance. The usage of intelligent algorithms brings innovative solutions to solve challenges in the industry. In this section we focus on the impact of Artificial Intelligence on modern farming practices, giving special attention to its contribution to sustainability. This section explores how AI is not just a tool for efficiency and productivity but a pivotal element in creating a more sustainable and environmentally conscious agricultural sector. We will examine the various ways algorithms aid in resource conservation, enhance crop yield with minimal environmental impact, and support the vital balance between meeting the growing food demands and preserving natural resources. Through this exploration, the section aims to highlight how modern technology is contributing to the agriculture industry.

14.3.1 Precision Farming for Sustainable Agriculture

One of the key aspects of sustainable agriculture is precision farming. This concept is also known as precision agriculture, which involves cutting-edge technologies, data acquisition and analysis to optimize crop yields, and waste reduction (Finger et al., 2019). This trend contrasts with traditional farming practices, which often apply resources uniformly across an entire field. Precision farming utilizes various data sources, such as GPS, cameras, sensors, and drones, to gather detailed information about the conditions in each part of the field. With this data, farmers can apply the exact amount of resources: water, fertilizers, and pesticides required for each specific area, rather than uniformly use on a whole field. This dedicated approach helps in conserving resources and also reduces the impact of farming on the environment (Schieffer & Dillon, 2015). For instance, applying the right amount of fertilizer in a specific area prevents excess nutrients from seeping into water, thus reducing pollution. Precision farming also leads to more efficient use of water, an increasingly critical resource in many parts of the world (Neupane & Guo, 2019). By using sensors to monitor soil moisture levels, farmers can irrigate only when necessary, thereby conserving water.

By increasing the efficiency of resource use, precision farming supports more sustainable food production (Abhilash Joseph et al., 2020). It helps in maintaining the health of the soil and minimizing greenhouse gas emissions. These factors allow for building a more sustainable and economical agricultural system, which is vital for meeting the growing global food demand while minimizing the environmental impact.

14.3.1.1 Agricultural Data and Acquisition

The forefront of the precision farming revolution are sensors, drones, and satellite imagery that can monitor crop health, and environmental conditions, in real time delivering data for the AI-powered algorithms (Dharmaraj & Vijayanand, 2018). Using data from IoT devices (Fastellini & Schillaci, 2020), farmers can remotely monitor various parameters like temperature, humidity, soil pH, and light intensity, and adjust farming settings through their smart devices. Unlike traditional methods, where observations are often general and manual, the sensors provide precise, real-time data that are vital for making decisions that strongly influence crop growth (Bhambri et al., 2005). This remote access saves farmers time, reduces costs, and also allows for rapid responses to any changes in farm conditions. For instance, if sensors detect low soil moisture levels, a farmer can activate irrigation systems with a simple command from their smartphone or tablet. This level of control and precision was unimaginable in traditional farming practices.

Drones and satellite imagery augment the capabilities of precision farming (Murugan et al., 2017). Sensors distributed on the field, cameras that can be moved, e.g., using drones, can quickly and efficiently monitor large farming areas. They can monitor crop health, identify pest infestations, and even assess crop maturity. Satellite imagery offers a broader perspective, allowing farmers to understand large-scale environmental changes and patterns. This aerial view is particularly useful for monitoring weather impacts and planning irrigation schedules (Inoue, 2020).

With the integration of AI, drones, satellite technology, and IoT devices in precision farming by closely monitoring crop health and soil conditions, farmers can anticipate problems before they escalate, leading to healthier crops and potentially higher yields.

14.3.1.2 AI Algorithms in Farming

The backbone of precision agriculture are algorithms that enable farmers to make decisions that are crucial for optimizing crop production and maintaining sustainable practices. These data-driven, sophisticated machine learning algorithms process and analyze vast amounts of information to provide insights and recommendations for farmers. The analysis based on intelligent algorithms brings farmers a detailed understanding of the farmland, leading to better crop management, reduced environmental impact, and improved agricultural productivity. As technology advances, the role of algorithms in agriculture is likely to become even more prominent, driving innovation and efficiency in this vital sector (Bhambri & Singh, 2005). Various algorithms employed in certain agricultural applications can be utilised to identify it by several types of algorithms used in specific applications in agriculture.

Predictive analytics use historical data, in modern farming it can be information on past crop yields, the way farmland was weathered, and the conditions of the soil. This information allows building of predictive models for approximating future outcomes (Kolipaka, 2020). They help in forecasting weather conditions, estimating crop yields, and predicting pest infestations. By doing so, they enable farmers to make proactive decisions, such as selecting the right time for planting and harvesting or implementing pest control measures at the optimal moment.

Image processing with the advent of drones and satellite imagery in agriculture (Naaz & Sucharitha, 2022) allows one to analyze images and identify issues like crop stress, disease, or nutrient deficiencies. The algorithms can differentiate between healthy and unhealthy plants, allowing for targeted intervention. For example, if a section of a field shows signs of disease (Ale et al., 2019), farmers can apply particular treatments specifically to that area, rather than treating the entire field.

Optimization algorithms are used for resource allocation and management (Dhaya & Kanthavel, 2022). They help determine the most efficient use of resources like water, fertilizers, and pesticides. By analyzing soil conditions, crop requirements, and weather forecasts, farmers can optimize irrigation schedules and fertilizer applications. It ensures that crops receive the exact amount of the required nutrients for optimal growth, as well as it allows to optimize supply chains and selling strategies.

Machine learning-based classification in agriculture is used in a wide range of tasks, eg., for identifying plant species, detecting weed types (Grace, 2021), and classifying soil types (Sirsat et al., 2017). This information is used for making decisions about crop rotation, weed control strategies, and soil management.

Regression methods are used, eg., for predicting the quantity of a harvest. They usually employ factors such as soil quality, weather conditions, and crop management practices (Kumar et al., 2020) that can support planning market demands and storage needs.

Agriculture heavily depends on seasonal and temporal patterns. Time series analyses like temperature, rainfall, and crop growth stages (Suryo et al., 2019) help in understanding seasonal patterns and trends, which are crucial for planning agricultural activities.

14.3.1.3 Modern Farming Equipment

Modern agriculture also relates to autonomous farming which represents a significant shift in agricultural practices, driven by advancements in robotics (Sparrow & Howard, 2021). This aspect of modern farming encompasses a range of activities from field cultivation to livestock management, all automated for greater efficiency, and precision.

Autonomous tractors and automatic farming equipment are at the forefront of this change (Rondelli et al., 2022). These machines can independently perform essential farming tasks like planting, harvesting, and weeding. Equipped with advanced sensors, GPS, and algorithms, they can navigate fields, recognize crop rows, and carry out tasks with minimal human intervention. This automation reduces the reliance on manual work, which is particularly beneficial given the labor shortages in many agricultural regions. Furthermore, these machines can operate around the clock, significantly increasing the efficiency of farming. Robots have revolutionized crop maintenance (Monteiro et al., 2021) with capabilities to distinguish between crops and weeds and can remove weeds precisely without harming the crops. Additionally, they can perform other maintenance tasks like pruning, thinning, and applying pesticides or fertilizers with high precision. This targeted approach to pest control and fertilization reduces the usage of chemicals and also minimizes environmental impact. The robots have been also designed to handle delicate tasks like picking fruits and vegetables, which significantly transform the harvest process (Tang et al., 2020). Usage of cameras and sensors to identify ripe produce and robotic arms or grippers to pick them without causing damage reduces labor costs but also enhances efficiency as these robots can work continuously, regardless of time or weather conditions.

In modern farming, the integration of algorithms with sensors and cameras has also revolutionized livestock management. These technologies enable continuous and precise monitoring of animal health and behavior, which results in benefits in productivity and animal welfare (Sparrow & Howard, 2021). The predictive analytics allows for detection of early signs of diseases or stress in animals, which helps with timely interventions, reducing the spread of illnesses, and ensuring the herd's health. Additionally, automatization plays a crucial role in routine tasks such as milking and feeding, which enhances consistency and efficiency in farm operations. Automated systems ensure precision in tasks like milking while continuously checking the health and well-being of the livestock. This not only improves farm productivity but also elevates animal welfare standards. Moreover, the analyzed data can provide clues to optimize overall farm management, leading to more effective resource utilization and better farming practices.

Smart irrigation systems represent another important area of autonomous farming (Bwambale et al., 2022) wherein IoT sensors placed in the field monitor soil moisture levels and relay this information to irrigation systems that are managed automatically. These systems take care of keeping optimal watering schedules and thus fulfill the exact needs of the plants, which leads to the optimization of crop yield and water usage.

14.3.1.4 Applications

A significant impact of the use of algorithms in smart farming is agricultural monitoring. One potential application is the detection of particular pests and weeds that have a substantial impact on their growth. In the past, herbicides and pesticides were applied broadly, affecting a wide range of organisms. Now, algorithms can distinguish between crops, weeds, and various pests, allowing for the selective application of chemicals. This specificity not only improves crop health but also contributes to an overall reduction of herbicide and pesticide use leading to healthier crops (Terence & Purushothaman, 2020) and also minimizes the environmental impact of farming by reducing chemical runoff and preserving beneficial organisms in the soil.

The analysis of images captured by drones or cameras is a cornerstone of modern crop monitoring (Maimaitijiang et al., 2020). These images provide a detailed view of the crops, enabling the AI to detect signs of diseases, nutrient deficiencies, and pest activities that might not be visible to the naked eye (Esgario et al., 2022). By analyzing color patterns, leaf shapes, and other visual indicators, the algorithms can identify problems at an early stage, which is crucial in agriculture as it allows for timely interventions, which prevents the spread of diseases and pest invasions on the farm.

Another facet of contemporary agriculture is significantly improved by the utilisation of machine learning models in farming monitoring, enabling accurate predictions of crop production (Agarwal & Tarar, 2021). These models can predict crop yields with a level of accuracy that was previously unattainable, fundamentally transforming how farmers plan and manage their agricultural operations. Models used in crop yield prediction take into account a wide array of factors, including historical data, current and forecasted weather conditions, and specific soil characteristics.

Historical data provides a foundational understanding of how different crops have performed under various conditions in the past. This data can include previous yield amounts, the types of crops planted, and the methods used in their cultivation, thus providing, more than ever, precise predictions.

The ability to predict crop yields accurately provides numerous benefits for farmers. It helps in planning agricultural operations more effectively. Knowing the likely yield outcomes enables farmers to manage their resources better, including labor, equipment, and inputs like seeds and fertilizers. This optimization leads to farming cost savings and more efficient use of resources. Accurate yield predictions are crucial for managing inventory and optimizing supply chain logistics. Farmers can plan for storage needs, align with market demands, and make informed decisions about selling their produce (Ishak et al., 2021). This is particularly important in today's globalized market, where timing and pricing can significantly impact profitability. Yield prediction models can assist in making strategic decisions. By understanding which crop varieties have performed well under similar conditions in the past, farmers can choose the most suitable crops for the upcoming season. This informed decision-making can lead to better production outcomes, maximizing yield, and potentially increasing revenue.

Supply chain optimization in modern agriculture has become increasingly sophisticated with the advent of technologies that have revolutionized the way agricultural products are transported and managed, significantly reducing wastage and improving the quality of produce reaching consumers (Khandelwal et al., 2021). Here, IoT technologies play a crucial role in tracking the movement and condition of agricultural products during their journey from farm to table. By embedding sensors in transportation vehicles and storage facilities, farmers and distributors can monitor the temperature, humidity, position, and other factors in real time. This real-time data is crucial in ensuring that perishable items like fruits, vegetables, and dairy products are stored in optimal conditions, thus preserving their freshness and extending their shelf life. For instance, if a shipment of strawberries is exposed to high temperatures during transit, IoT sensors can alert supply chain managers, who can take immediate action to rectify the situation (Onwude et al., 2020).

Analyzing data collected from various stages of the supply chain can predict potential bottlenecks and suggest the most efficient routes for transportation. This optimization not only minimizes waste by reducing the time products spend in transit but also cuts down on transportation costs. Algorithms can analyze historical traffic data, weather reports, and vehicle performance to determine the best routes and modes of transportation. This means that agricultural products can be delivered to the market in the shortest possible time, reducing the likelihood of spoilage and ensuring that consumers receive fresh produce.

Algorithms also significantly enhance inventory management. By predicting market demand based on factors like seasonality, consumer trends, and economic indicators, they support farmers and retailers in determining the optimal amount of produce to harvest and stock. This helps in avoiding overproduction or understocking, both of which can lead to financial losses and wastage.

In an industry as variable and hardly predictable as agriculture, market analysis and price prediction become important aspects of making profits by giving the farmers information to make decisions about selling their products. This kind of market analysis is based on vast amounts of data acquired from various sources, which includes historical market trends, weather patterns, consumer behavior, and global economic indicators. By processing and analyzing this data, algorithms can identify patterns and trends that are not visible to those who do not use advanced analytics. For example, it can predict the demand for certain crops in different markets, considering factors like seasonal changes, consumer preferences, and even geopolitical events (Gunda et al., 2017). Price prediction is particularly useful for farmers in planning their production and distribution. For instance, if the algorithm predicts a rise in the price of a particular crop in the coming months, farmers can adjust their planting schedules accordingly to maximize their profits (Reddy et al., 2022). Smart algorithms can also help farmers understand the best moments and locations to sell their products (Wolfert et al., 2017). By analyzing transportation costs, market demand, and local consumer preferences, it can find the most profitable markets for farmers' produce. This not only helps in reducing wastage due to overproduction or unsold stock but also maximizes the income potential for farmers.

14.3.2 Environment Monitoring for Self-Sustaining Farming

Environmental monitoring involves the systematic collection of data about the environment, followed by the analysis, interpretation, and implementation of strategies to manage and protect the environment effectively. It is crucial in agriculture as it provides information about the natural factors that affect crop production. One of these factors is the weather, which can be analyzed locally or globally to give farmers short-term forecasts of essential data like temperature, wind, and precipitation rate. This information helps farmers make aware decisions about the optimal application of fertilizers and pesticides as well as irrigation. Long-term weather trends can also be observed when historical data is available, which may indicate the need for crop changes due to climate change (Rana & Bhambri, 2024; Linaza et al., 2021). In addition to weather analysis, environment monitoring also encompasses air humidity, which affects water management and air pollution, which impacts the development of livestock and plants. Soil analysis, including its state, moisture content, texture, nutrient content, erosion, and enzyme activity, is also an essential aspect of environmental monitoring. This information allows farmers to make the most effective use of resources, including land division, for optimal crop rotation. Various artificial intelligence methods are applied in agriculture monitoring, starting from standard machine learning approaches such as artificial neural networks or support vector machines, and going through fuzzy logic solutions or genetic algorithms, finishing with optimization methods (Wakchaure et al., 2023). However, when analyzing the results, one can see that the outcomes achieved in one field may not be easily repeated elsewhere. It is also difficult to ensure regular data gathering in all possible destinations, as each field is different (Lianza et al., 2021). Thus, there is still a big challenge in Artificial Intelligence implementations to overcome those problems.

Data regarding the environment's state can be acquired either domestically or internationally. Sensors can gather information on local air and ground temperature, humidity, wind speed, and precipitation rate, and then send it to cloud servers for storage and analysis. The construction of the Internet of Things is a prevalent solution for such purposes (Shaikh et al., 2022). Drones can perform visual inspections to analyze the soil condition in a specific area, and the comparison of images can also be used to assess the air quality. Furthermore, satellite imagery, particularly hyperspectral data, allows for collecting intricate and multidimensional data that, when processed, can provide information about crop diseases, nutrient deficiencies, or soil properties (Subudhi et al., 2023). As we can see, technological advances allow for weather data collection with a large variety of sources with high temporal resolution (Hachimi et al., 2023). When these data are collected over large regions for extended periods, they create a massive dataset that requires algorithms that support Big Data analysis.

Artificial Intelligence methods that incorporate cloud computing use historical weather data to construct weather models for prediction (Abri et al., 2023). These models rely on time series analysis (Hachimi et al., 2023) rather than on a single point in time. With a long data collection period, short-term predictions become more accurate, resulting in better decisions concerning planning, harvesting, and irrigation. However, predicting the overall state of the global weather is crucial (Malone et al., 2022). There are numerous approaches for dealing with local characteristics, especially in severe weather situations. For instance, AI-based hardware solutions are designed for frost prediction. In contrast, others use fuzzy inference systems based on support vector machines and particle swarm optimization artificial neural networks for daily rainfall prediction. Some use multimodel ensembles for annual monsoon and winter weather forecasting (Shaikh et al., 2022), while others develop systems for low-temperature zone forecasting (Wakchaure et al., 2023). Satellite images and meteorological data are analyzed to examine the effects of extreme weather conditions like drought or heatwaves on crop growth (Kim et al., 2020). These applications demonstrate that smart weather data management is indispensable to precision agriculture and implementing sustainability (Hachimi et al., 2023).

Water management is very important in agriculture. It becomes necessary in regions with water shortages to ensure optimal irrigation and crop growth (Singh & Bhambri, 2024). One possible solution to predict the needed water quantity for plants is an estimation of reference evapotranspiration (a sum of soil surface evaporation and transpiration from plants) by models designed using artificial intelligence. Accurate estimates allow for amending the water quantity evaporated to the atmosphere (Hachimi et al., 2023). Other solutions use hyperspectral imaging to analyze spectral characteristics related to vegetation water content, allowing for a farmer to draw conclusions concerning the crops' water demands. All methods supporting water management ensure that the water is supplied in the right amount and when it is most necessary. This assures water-efficient utilization and reduces water waste (Subudhi et al., 2023).

Effective soil management is a crucial element of agriculture. Integrating the Internet of Things (IoT) with visual inspection techniques, such as drone or satellite imagery, provides valuable data for analysis (Eli-Chukwu, 2019). Artificial Intelligence can enhance soil irrigation and fertilization by accurately predicting soil moisture levels, texture, nutrient content, erosion effects, enzyme activity, and soil temperature (Zhou et al., 2022). This can result in optimal crop yield predictions, conservation of soil resources, and efficient pesticide management. The health and composition of the soil, including factors like pH levels, nutrient content, and moisture levels, are crucial for plant growth. Automatic soil quality forecasting is critical for proper land management and crop rotation, while decision support systems are essential for making informed decisions in dry and semi-dry regions (Javaid et al., 2023). The residual networks, along with bi-directional long short-term memory, were applied to predict the water content in the soil, which helps in selecting appropriate crops to increase the yield. Next, field spectroscopy allows offline and online soil parameter predictions (Shaikh et al., 2022). Hyperspectral imaging is used for soil characterization and analysis, giving the farmers detailed information about nutrient distribution, organic matter content, and soil composition. Those factors may influence the land distribution between various crops to increase the yield (Subudhi et al., 2023).

14.4 CONCLUSIONS AND FUTURE TRENDS

Artificial intelligence (AI) is emerging as a beacon of innovation in the field of technological evolution. Once the stuff of science fiction and fantasy, AI is now intertwined with various industries, transforming the way we live and work. As we have argued, agriculture, a sector as old as civilization itself, is no exception to the paradigm shift that AI is bringing everywhere. AI has redefined agriculture over the years, and technological advances have impacted the agriculture industry in more ways than one. Agriculture is the mainstay of many countries around the world, and as the world's population grows, from 7.5 to 9.7 billion by 2050, according to UN projections (United Nations, 2024), there will be more pressure on land, with only 4% more land available for cultivation by 2050. So farmers are having to do more with less.

Considering farm management, one can point to at least five emerging fields of innovation:

First, farm automation technologies have been a game-changer in agriculture for the past several years. These advances, such as self-driving tractors and equipment that detect weeds and pests, save farmers time and money.

The second is indoor vertical farming, an innovative approach to growing crops indoors without touching the soil. The goal is to reduce water use and save space. Farm managers can also control the environment and create a year-round growing season; this technology is highly valued in places with limited space or harsh weather conditions.

Third, precision agriculture is another major innovation, using advanced technologies such as drone imagery, satellite mapping, and soil sensors to optimize resource use and chemical application. This technology allows farmers to focus on a specific area of a field to apply only the necessary amount of pesticides and herbicides.

Fourth, areas of innovation that address the problem of high levels of organic matter and toxic substances that can harm crops and soils include wastewater treatment. Farm managers are exploring the use of closed-loop wastewater treatment systems. This technology can treat wastewater on-site and return it to crops, reducing reliance on freshwater resources and supporting sustainable agriculture.

Finally, innovation involves the application of machine learning models. Every day, farms generate a plethora of data points about soil, temperature, water use, and weather conditions. Machine learning models use this vast amount of data in real time to generate useful insights, such as choosing the right time to plant, crop selection, hybrid seeds for higher yields, and the like. By feeding these models, data-intensive approaches enable better decision-making, increased efficiency, and reduced waste while minimizing negative environmental impacts.

REFERENCES

Abhilash Joseph, E., Abdul Hakkim, V., & Sajeena, S. (2020). Precision farming for sustainable agriculture. *International Journal of Agriculture Innovations and Research*, *8*(16), 543–553.

Abiri, R., Rizan, N., Balasundram, S. K., Shahbazi, A. B., & Abdul-Hamid, H. (2023). Application of digital technologies for ensuring agricultural productivity. *Heliyon*, *9*(12), e22601.

Agarwal, S., & Tarar, S. (2021). A hybrid approach for crop yield prediction using machine learning and deep learning algorithms. *Journal of Physics Conference Series*, *1714*(1), 012012.

Ale, L., Sheta, A., Li, L., Wang, Y., & Zhang, N. (2019). Deep learning based plant disease detection for smart agriculture. In 2019 IEEE Globecom Workshops (GC Wkshps) (pp. 1–6). IEEE. https://doi.org/10.1109/GCWkshps45667.2019.902443

Alloghani, M. A. (2023). AI for sustainable agriculture: A systematic review. *Artificial Intelligence and Sustainability* (pp. 53–64), Signals and Communication Technology. Springer. https://doi.org/10.1007/978-3-031-45214-7_3.

Balaska, V., Adamidou, Z., Vryzas, Z., & Gasteratos, A. (2023). Sustainable crop protection via robotics and artificial intelligence solutions. *Machines*, *11*(8), 774.

Becker, S., & Fanzo, J. (2023). Population and food systems: What does the future hold? *Population and Environment*, *45*(3), 20.

Bhambri, P., & Gupta, S. (2005). A survey & comparison of permutation passibility of fault-tolerant multistage interconnection networks. In National Conference (Abstract Book) (p. 13). Presented at the "Application of Mathematics in Engineering & Technology" conference at MIMIT, Malout held during March 26–28, 2005.

Bhambri, P., Gupta, S., & Bhandari, A. (2005). Soft computing techniques. In UGC Sponsored National Conference (Abstract Book) (p. 100). Presented at the "Emerging Computing Technologies" conference at HRMMV, Jalandhar held during September 29–30, 2005.

Bhambri, P., & Singh, I. (2005). Electrical actuation systems. In National Conference (Abstract Book) (p. 8). Presented at the "Application of Mathematics in Engineering & Technology" conference at MIMIT, Malout held during March 26–28, 2005.

Bhat, S. A., & Huang, N.-F. (2021). Big data and ai revolution in precision agriculture: Survey and challenges. *IEEE Access*, *9*, 110209–110222.

Bwambale, E., Abagale, F. K., & Anornu, G. K. (2022). Smart irrigation monitoring and control strategies for improving water use efficiency in precision agriculture: A review. *Agricultural Water Management*, *260*, 107324.

Cavazza, A., Dal Mas, F., Paoloni, P., & Manzo, M. (2023). Artificial intelligence and new business models in agriculture: A structured literature review and future research agenda. *British Food Journal, 125*(13), 436–461.
Dharmaraj, V., & Vijayanand, C. (2018). Artificial intelligence (AI) in agriculture. *International Journal of Current Microbiology and Applied Sciences, 7*(12), 2122–2128.
Dhaya, R., & Kanthavel, R. (2022). Energy efficient resource allocation algorithm for agriculture IoT. *Wireless Personal Communications, 125*(2), 1361–1383.
Eli-Chukwu, N. C. (2019). Applications of artificial intelligence in agriculture: A review. *Engineering, Technology & Applied Science Research, 9*(4), 54–67.
Esgario, J. G., de Castro, P. B., Tassis, L. M., & Krohling, R. A. (2022). An app to assist farmers in the identification of diseases and pests of coffee leaves using deep learning. *Information Processing in Agriculture, 9*(1), 38–47.
Fastellini, G., & Schillaci, C. (2020). Precision farming and IoT case studies across the world. In *Agricultural Internet of Things and Decision Support for Precision Smart Farming* (pp. 331–415). Elsevier.
Finger, R., Swinton, S. M., El Benni, N., & Walter, A. (2019). Precision farming at the nexus of agricultural production and the environment. *Annual Review of Resource Economics, 11*, 313–335.
Grace, R. K. (2021). Crop and weed classification using deep learning. *Turkish Journal of Computer and Mathematics Education (TURCOMAT), 12*(7), 935–938.
Gu, D., Andreev, K., & Dupre, M. E. (2021). Major trends in population growth around the world. *China CDC Weekly, 3*(28), 604.
Gunda, T., Bazuin, J., Nay, J., & Yeung, K. (2017). Impact of seasonal forecast use on agricultural income in a system with varying crop costs and returns: An empirically-grounded simulation. *Environmental Research Letters, 12*(3), 034001.
Hachimi, Ch. El., Belaqziz, S., Khabba, S., Sebbar, B., Dhiba, D., & Chehbouni, A. (2023). Smart weather data management based on artificial intelligence and big data analytics for precision agriculture. *Agriculture, 13*(1), 95.
Huss, M., Brander, M., Kassie, M., Ehlert, U., & Bernauer, T. (2021). Improved storage mitigates vulnerability to food-supply shocks in smallholder agriculture during the COVID-19 pandemic. *Global Food Security, 28*, 100468.
Inoue, Y. (2020). Satellite-and drone-based remote sensing of crops and soils for smart farming–a review. *Soil Science and Plant Nutrition, 66*(6), 798–810.
Ishak, M., Rahaman, M. S., & Mahmud, T. (2021). FarmEasy: An intelligent platform to empower crops prediction and crops marketing, In 2021 13th International Conference on Information & Communication Technology and System (ICTS) (pp. 224–229). IEEE. https://doi.org/10.1109/ICTS52701.2021.9608436
Javaid, M., Haleem, A., Khan, I. H., & Suman, R. (2023). Understanding the potential applications of Artificial Intelligence in Agriculture Sector. *Advanced Agrochem, 2*(1), 15–30.
Khan, N., Ray, R. L., Sargani, G. R., Ihtisham, M., Khayyam, M., & Ismail, S. (2021). Current progress and future prospects of agriculture technology: Gateway to sustainable agriculture. *Sustainability, 13*(9), 4883.
Khandelwal, C., Singhal, M., Gaurav, G., Dangayach, G., & Meena, M. (2021). Agriculture supply chain management: A review (2010–2020). *Materials Today: Proceedings, 47*, 3144–3153.
Kholaiq, M., Benmessaoud, S., Kara, M., Assouguem, A., Abbasi, A. M., Al-Ghamdi, A. A., Elshikh, M. S., Rahimi, A., & Saber, N. (2022). Sustainability of coastal agriculture in the face of soil degradation: The influence of water salinization as an example. *Sustainability, 14*(20), 13641.
Kim, N., Na, S.-I., Park, Ch.-W., Huh, M., Oh, J., Ha, K.-J., Cho, J., & Lee, Y.-W. (2020). An artificial intelligence approach to prediction of corn yields under extreme weather conditions using satellite and meteorological data. *Applied Sciences, 10*(11), 3785.

Kobayashi, Y., Higa, M., Higashiyama, K., & Nakamura, F. (2020). Drivers of land-use changes in societies with decreasing populations: A comparison of the factors affecting farmland abandonment in a food production area in Japan. *PLoS One*, *15*(7), e0235846.

Kolipaka, V. R. R. (2020). Predictive analytics using cross media features in precision farming. *International Journal of Speech Technology*, *23*(1), 57–69.

Kumar, Y. J. N., Spandana, V., Vaishnavi, V., Neha, K., & Devi, V. (2020). Supervised machine learning approach for crop yield prediction in agriculture sector, In 2020 5th International Conference on Communication and Electronics Systems (ICCES) (pp. 736–741). IEEE. https://doi.org/10.1109/ICCES48766.2020.9137868

Linaza, M. T., Posada, J., Bund, J., Eisert, P., Quartulli, M., Döllner, J., Pagani, A., Olaizola, I. G., Barriguinha, A., Moysiadis, Th., & Lucat, L. (2021). Data-driven artificial intelligence applications for sustainable precision agriculture. *Agronomy, 11*(6), 1227.

Loukatos, D., Kondoyanni, M., Kyrtopoulos, I.-V., & Arvanitis, K. G. (2022). Enhanced robots as tools for assisting agricultural engineering students' development. *Electronics*, *11*(5), 755.

Maimaitijiang, M., Sagan, V., Sidike, P., Daloye, A. M., Erkbol, H., & Fritschi, F. B. (2020). Crop monitoring using satellite/UAV data fusion and machine learning. *Remote Sensing*, *12*(9), 1357.

Malone, B., Stockmann, U., Glover, M., McLachlan, G., Engelhardt, S., & Tuomi, S. (2022). Digital soil survey and mapping underpinning inherent and dynamic soil attribute condition assessments. *Soi Secuirity, 6,* 100048.

Monteiro, A., Santos, S., & Gonçalves, P. (2021). Precision agriculture for crop and livestock farming—Brief review. *Animals*, *11*(8), 2345.

Murugan, D., Garg, A., & Singh, D. (2017). Development of an adaptive approach for precision agriculture monitoring with drone and satellite data. *IEEE Journal of Selected Topics in Applied Earth Observations and Remote Sensing*, *10*(12), 5322–5328.

Naaz, A., & Sucharitha, B. (2022). Precision farming using image processing and machine learning. In *Artificial Intelligence Applications in Agriculture and Food Quality Improvement* (pp. 55–73). IGI Global.

National Institute of Food and Agriculture. (2024). Sustainable Agriculture. https://www.nifa.usda.gov/topics/sustainable-agriculture (Accessed on: 2024-02-09).

Neupane, J., & Guo, W. (2019). Agronomic basis and strategies for precision water management: A review. *Agronomy*, *9*(2), 87.

Onwude, D. I., Chen, G., Eke-Emezie, N., Kabutey, A., Khaled, A. Y., & Sturm, B. (2020). Recent advances in reducing food losses in the supply chain of fresh agricultural produce. *Processes*, *8*(11), 1431.

Onyeaka, H., Tamasiga, P., Nwauzoma, U. M., Miri, T., Juliet, U. C., Nwaiwu, O., & Akinsemolu, A. A. (2023). Using artificial intelligence to tackle food waste and enhance the circular economy: Maximising resource efficiency and Minimising environmental impact: A review. *Sustainability*, *15*(13), 10482.

Rana, R., & Bhambri, P. (2024). Artificial Intelligence and Internet of Things (IoT) facilitated digital twin for industry 4.0 application domains. In S. Rani, P. Bhambri, S. Kumar, P. K. Pareek, & A. A. Elngar (Eds.), *AI-Driven Digital Twin and Industry 4.0: A Conceptual Framework with Applications* (1st ed., pp. 168–181). CRC Press. https://doi.org/10.1201/9781003395416

Reddy, P. C. S., Suryanarayana, G., & Yadala, S. (2022). Data analytics in farming: Rice price prediction in Andhra Pradesh. In 2022 5th International Conference on Multimedia, Signal Processing and Communication Technologies (IMPACT) (pp. 1–5). IEEE. https://doi.org/10.1109/IMPACT55510.2022.10029009

Rondelli, V., Franceschetti, B., & Mengoli, D. (2022). A review of current and historical research contributions to the development of ground autonomous vehicles for agriculture. *Sustainability*, *14*(15), 9221.

Schieffer, J., & Dillon, C. (2015). The economic and environmental impacts of precision agriculture and interactions with agro-environmental policy. *Precision Agriculture, 16*, 46–61.

Shaikh, F., Memon, M., Mahoto, N., Zeadally, S., & Nebhen, J. (2022). Artificial intelligence best practices in smart agriculture. *IEEE Micro, 42*(1), 17–24.

Singh, M.P., & Bhambri, P. (2024). AI-Driven Digital Twin Conceptual Framework and Applications. In S. Rani, P. Bhambri, S. Kumar, P.K. Pareek, & A.A. Elngar (Eds.), AI-Driven Digital Twin and Industry 4.0: A Conceptual Framework with Applications (1st ed., pp. 127-141). CRC Press.

Singh, M., Singh, P., Kaur, K., & Bhambri, P. (2005). Database security. In Proceedings of National Conference (pp. 35–41). Presented at the "Future Trends in Information Technology" conference at SJPMLIET, Radaur, Yamuna Nagar held during March 6–7, 2005.

Singh, M.P., & Bhambri, P. (2024). AI-Driven Digital Twin Conceptual Framework and Applications. In S. Rani, P. Bhambri, S. Kumar, P.K. Pareek, & A.A. Elngar (Eds.), AI-Driven Digital Twin and Industry 4.0: A Conceptual Framework with Applications (1st ed., pp. 127-141). CRC Press. https://doi.org/10.1201/9781003395416

Sirsat, M. S., Cernadas, E., Fernández-Delgado, M., & Khan, R. (2017). Classification of agricultural soil parameters in India. *Computers and Electronics in Agriculture, 135*, 269–279.

Sparrow, R., & Howard, M. (2021). Robots in agriculture: Prospects, impacts, ethics, and policy. *Precision Agriculture, 22*, 818–833.

Streich, J., Romero, J., Gazolla, J. G. F. M., Kainer, D., Cliff, A., Prates, E. T., Brown, J. B., Khoury, S., Tuskan, G. A., & Garvin, M. (2020). Can exascale computing and explainable artificial intelligence applied to plant biology deliver on the United Nations sustainable development goals? *Current Opinion in Biotechnology, 61*, 217–225.

Subudhi, Sh., Dabhade, R. G., Shastri, R., Gundu, V., Vignesh, G. D., & Chaturvedi, A. (2023). Empowering sustainable farming practices with AI-enabled interactive visualization of hyperspectral imaging data. *Measurements: Sensors, 30,* 100935

Suryo, P. S. B. C., Mustika, I. W., Wahyunggoro, O., & Wasisto, H. S. (2019). Improved time series prediction using LSTM neural network for smart agriculture application. In 2019 5th International Conference on Science and Technology (ICST) (pp. 1–4). IEEE. https://doi.org/10.1109/ICST47872.2019.9166401

Tang, Y., Chen, M., Wang, C., Luo, L., Li, J., Lian, G., & Zou, X. (2020). Recognition and localization methods for vision-based fruit picking robots: A review. *Frontiers in Plant Science, 11*, 510.

Terence, S., & Purushothaman, G. (2020). Systematic review of Internet of Things in smart farming. *Transactions on Emerging Telecommunications Technologies, 31*(6), e3958.

Tyagi, S. K. S., Mukherjee, A., Pokhrel, S. R., & Hiran, K. K. (2020). An intelligent and optimal resource allocation approach in sensor networks for smart agri-IoT. *IEEE Sensors Journal, 21*(16), 17439–17446.

United Nations. (2024). World population projected to reach 9.8 billion in 2050, and 11.2 billion in 2100. https://www.un.org/en/desa/world-population-projected-reach-98-billion-2050-and-112-billion-2100 (Accessed on: 2024-02-10).

Wakchaure, M., Patle, B. K., & Mahindrakar, A. K. (2023). Application of AI techniques and robotics in agriculture: A review. *Artificial Intelligence in the Life Sciences, 3*, 100057.

Wolfert, S., Ge, L., Verdouw, C., & Bogaardt, M.-J. (2017). Big data in smart farming–a review. *Agricultural Systems, 153*, 69–80.

Zhou, Y., Xia, Q., Zhang, Z., Quan, M., & Li, H. (2022). Artificial intelligence and machine learning for the green development of agriculture in the emerging manufacturing industry in the IoT platform. *Acta Agriculturae Scandinavica, Section B - Soil & Plant Science, 72*(1), 284–299.

15 The Impact of ICT in Agriculture and Rural Development

Neelam Sharma and Pankaj Bhambri

15.1 INTRODUCTION

Information and communications technology (ICT) has become a transformative force in many fields, including agriculture and rural development. In recent years, the integration of ICT tools and applications has transformed traditional agriculture, providing new opportunities to improve productivity, sustainability, and survival in rural communities (Sharma et al., 2020). This chapter explores the important role of ICT in solving the unique challenges faced by farmers and rural areas and ultimately stimulating growth in agriculture, fields and cities.

Agriculture is the main economic activity in many countries, especially in developing countries where part of the population's livelihood depends on farming (Singh & Bhambri, 2023). However, agricultural production is often affected by many factors, including limited access to information, resources and markets, as well as climate change and environmental damage (Singh et al., 2005). To solve this problem, ICT technologies provide fresh approaches to resolving these issues and enabling agriculture to fulfill its promise as an engine of economic expansion and a means of reducing poverty.

Information and communication technology plays a major role in supporting agriculture by giving people access to pertinent information and services. Via a farm website, mobile device, or postal service, farmers may obtain weather forecasts, market pricing, best agricultural practices, and pest control techniques. With the availability of knowledge, farmers may reduce risk and maximize yield by choosing crops, planting dates, watering schedules, and managing pests.

15.2 TECHNOLOGY

1. **Precision farming:** Also referred to as smart farming , is a kind of agricultural management that increases crop diversity by utilizing technology and data-driven solutions (Bhambri et al., 2019). It entails monitoring, analyzing, and reacting to changes in hilly terrain by utilizing technology like Global Positioning Systems (GPS), Geographic Information Systems (GIS),

DOI: 10.1201/9781003484226-15

TABLE 15.1
Comparative Analysis of Literature Review

S.No.	Parameter	Remarks	References
1.	**Pesticides**	Precision farming to be used accurately.	(Ravikumar et al., 2021)
2.	**Nutrient deficiencies and stress detection**	Remote sensing to be used accurately.	(Wavrek et al., 2023)
3.	**Uneven distribution of resources (water, fertilizers, pesticides)**	IoT sensors collect data on soil moisture, temperature, humidity, and crop health. This information enables informed decisions regarding irrigation, fertilization, and pest control.	(Surendera Bhosle et al., 2019)
4.	**Transparency and authenticity in the supply chain**	Blockchain records every transaction, making it possible to trace the origin of agricultural products.	(Chandra Sekhar Bhusal, 2021)
5.	**Managing farms efficiently**	GIS integrates satellite and drone data to assess crop health.	(Chakkra Prakasam et al., 2020)
6.	**Reducing reliance on chemical pesticides**	ML models identify weed species and pests, allowing targeted treatments.	(S.S.,)
7.	**Minimizing soil erosion, water pollution, and habitat destruction**	Biotech crops with reduced tillage needs (e.g., herbicide-tolerant soybeans) promote soil conservation.	(Nahirñak et al., 2022)
8.	**Utilizing vertical space, multiple layers of crops can be grown indoors**	Controlled indoor environments minimize water wastage through evaporation and runoff.	(Wei et al., 2024)

sensors, drones, and data analysis. Among the advantages of precision farming are:

- **Enhanced productivity:** Precision farming allows farmers to apply pesticides, fertilizer, and water more effectively, saving waste and raising output.
- **Increasing output and quality:** By adjusting to specific circumstances, precision agriculture may raise crop quality.
- **Savings:** By employing more sophisticated machinery and fewer inputs, precision agriculture will result in cost savings.
- **Environmental sustainability:** Precision agriculture supports sustainability and protection of natural resources by using fewer resources and reducing environmental impact.
- **Data-driven decisions:** Information gathered from sensors, drones, and other Internet of Things devices is essential to precision agriculture.

Other technologies help farmers make judgments, react fast to changes, and improve the effectiveness of their operations. Drones and satellites are examples of field-sensing technology that may be used to monitor crops, measure land usage, and look into environmental changes. Farmers may use this information to make well-informed decisions regarding crop planning, pest control, and irrigation (Kumari et al., 2024).

2. **Remote sensing:** This term describes tools or systems, including satellites, airplanes, drones, or ground sensors, that gather data remotely. Electromagnetic radiation emitted or reflected from the earth's surface is detected using remote sensing instruments. Various radiation wavelengths are measured by various types of sensors, producing varying features and outcomes.
   - **File processing:** Once the data from the remote control has been obtained, eliminate any interference, optimise the quality, and transform the raw data into its native form as an image or document.
   - **Interpretation and analysis:** To learn more about the Earth's surface or atmosphere, remote sensing data is interpreted and analyzed. This might involve classifying and charting different types of land cover, the health of the plants, urban areas, groundwater, and geological formations.
3. **Internet of Things (IoT):** Instantaneous monitoring and management of agricultural activities are made possible by IoT devices including soil moisture sensors, weather stations, and autonomous irrigation systems. IoT technology enhances agricultural decision-making, lowers resource waste, and boosts output.

15.3 BENEFITS

1. **Supply chain visibility:** By monitoring conditions and activities layer by layer across the agricultural chain, the Internet of Things promotes uniformity and transparency in the agricultural supply chain (Tondon & Bhambri, 2017). It raises customer confidence and trust in food items while simultaneously improving food safety, quality control, and management.
2. **Access to markets and finance:** IoT-based applications and platforms connect farmers to markets, buyers and financial services, even in rural areas. Mobile applications, e-commerce platforms, and digital payment systems allow farmers to access information, sell products, and access credit and insurance services, causing their power to weaken and the economy to shrink.
3. **Protection and monitoring of the environment:** IoT sensors may assist farmers in implementing sustainable land management by monitoring the environment, including the cloud, clean air, clean water, and healthy soil. Through encouraging soil and water conservation, biodiversity conservation, and water conservation, IoT supports long-term environmental sustainability and agriculture (Singh et al., 2024).

4. **Blockchain:** Blockchain technology ensures transparency, traceability, and security of the agricultural supply chain. Blockchain allows farmers to track their products from farm to table, verify their authenticity, and get a fair price in the market.
   - **Fair trade:** Blockchain promotes direct collaboration between farmers and consumers or buyers, as well as middle installers such as wholesalers or retailers. This direct access to the market ensures that farmers receive a fair price for their products and allows consumers to directly support small-scale agriculture and permaculture.
   - **Smart contracts and payments:** Blockchain-based smart contracts can operate and execute agreements between parties without the need for an intermediary. Smart contracts can be used to pay farmers based on pre-determined criteria such as delivery or quality standards. This reduces transaction costs, delays, and risks associated with traditional payments.
   - **Access to financial services:** Blockchain technology can help small farmers in rural areas get financial assistance from loans, insurance, and other services. By providing transparent and immutable information about transactions and assets, blockchains can help farmers build trust and access financial products that were previously unavailable to them.
   - **Land tenure and property rights:** Blockchain technology may be utilized to provide safe documentation and proof of property rights and land ownership, particularly in places where land ownership is ambiguous or prohibited. Governments may enhance land management, lessen conflict, and increase land security for rural populations by digitizing land information on blockchain.
5. **Geographic information systems (GIS):** Farmers may examine geographical data on climate, land usage, and soil types with the use of GIS technology. Farmers can lower environmental hazards, plan crop rotations, and manage soil better with the use of GIS technologies.
   - **Land-use planning:** Using GIS, farmers and decision-makers may examine land-use trends and pinpoint locations that are ideal for forestry, farming, and other activities. By taking into account variables like land type, terrain, climate, and distance from employment, GIS may enhance landuse planning, making it less intrusive and more effective.
   - **Data science:** GIS allows farmers to evaluate the suitability of different crops and select suitable crops for specific areas based on climate, soil, and water. By overlaying layers of data and performing spatial analysis, farmers can determine which crops are best for their land, thereby increasing crop yields and profitability (Koliouska and Andreopoulou, 2013).
   - **Natural resource management:** The management of local resources, including water, forests, and biodiversity, is aided by GIS. Policymakers

may create sustainable management plans that strike a balance between economic development and conservation by utilizing models and ecosystems, as well as by assessing geographical data on resources that are accessible.
- **Disaster management and risk assessment:** GIS assists farmers and policymakers in assessing and mitigating risks related to natural catastrophes including floods, rain, and fire. GIS allows for proactive planning and aids in limiting disasters and builds stronger communities by its ability to map sensitive areas, key infrastructure, and sensitive regions.

6. **Artificial intelligence (AI) and machine learning:** These two fields combine to use algorithms to generate information on crop health, yield prediction, and pest testing through the analysis of large amounts of data by the use of algorithms. With the use of technology, farmers may enhance farm management and make well-informed decisions.
   - **Crop management and monitoring:** Drone, sensor, and satellite data may all be analyzed by Artificial Intelligence and machine learning algorithms to monitor crop health, reduce disease, and enhance fertilization and watering (Lakota and Stajnko, 2013). AI can assist farmers in making well-informed crop management decisions, leading to higher yields and lower entry costs, by seeing patterns and anomalies in large data.
   - **Yield forecasting:** To precisely forecast future yields, Artificial Intelligence and machine learning algorithms can 2021examine past data on crop performance, soil composition, and weather.
   - **Pest and disease control:** AI-powered image recognition algorithms can detect pests, diseases, and weeds in crops based on the visual perception of needle photography or drones. By detecting threats early and accurately, AI can reduce crop loss and pesticide use by helping farmers implement targeted interventions such as pesticide testing or crop rotation strategies (Supriya et al., 2024).
7. **Biotechnology and genetic engineering:** Innovations in biotechnology, such as genetically modified crops and gene editing technology, can increase product durability, insect resistance, and food content. These advances help increase agricultural productivity and food security.
   - **Crop improvement:** Scientists can confer advantageous traits on crops such as resistance to pests, diseases, and environmental stress, via the use of biotechnology and genetic engineering techniques, such as genetic modification (GM). Genetically engineered crops that boost yields, lower losses, and enhance food quality have resulted from these techniques.
   - **Disease resistance:** Crops resistant to specific diseases and pests can be produced through biotechnology. For instance, by transferring genes from disease-resistant plants to vulnerable plants, genetic engineering may defend against pests and illnesses and eliminate insects without the need for insecticides.

- **Drought and salt tolerance:** Genetic engineering can help crops become more resilient to drought and salt, enabling them to flourish in arid climates with high soil salinity or no water. By inserting genes that control water efficiency and salt tolerance, scientists can create crops that are resistant to both climate change and water shortage.
- **Increasing nutritional value:** By adding more vitamins, minerals, and other nutrients to crops, biotechnology can raise crop nutritional value. For instance, genetic engineering can improve public health by addressing nutritional inadequacies by increasing the iron and zinc content of basic crops like wheat, rice, and maize.
- **Environmental impact:** Biotechnology offers effective solutions to agricultural problems by reducing the use of chemical products such as pesticides and fertilizers. Crops replaced with disease-free species require less spraying, reduce environmental pollution, reduce health risks, and increase biodiversity.
- **Biofortification:** Crops that have been biofortified with high concentrations of nutrients like vitamin A, iron, and zinc are the result of biotechnology. Malnutrition and micronutrient deficiencies are addressed by these items, particularly for disadvantaged populations in nations with limited access to a variety of dietary sources.

8. **Vertical farming and hydroponics:** Vertical farming and hydroponics use technology to grow crops in a controlled environment. This system can optimize space, save water, and reduce environmental impacts, and is suitable for urban agriculture and areas with limited land.

The most effective use of available space is through **vertical farming**, which enables crops to be grown in layers to maximize vertical space. Farmers can make the most use of their land, thanks to this vertical structure, particularly in metropolitan locations where space is limited. Vertical farming allows for the cultivation of crops in densely populated places while reducing the requirement for a certain land size since it grows crops vertically rather than outwards.

- **Annual production method:** Hydroponic and vertical farming systems are often managed inside, in vertical towers or greenhouses. With the help of these control centers, farmers can modify temperature, humidity, and lighting to create the ideal growth environment for crops all year round.
- **Water efficiency:** Compared to soil farming, vertical farming often uses less water, thanks to its use of soilless systems. Since it is a closed water cycle, hydroponic farming is an effective method of conserving water since it minimizes water loss due to evaporation or runoff. Furthermore, air mist or water flow are frequently used in vertical farming techniques to lower total water use and improve water efficiency.
- **Reduced environmental effect:** When compared to traditional farming, hydroponics and vertical farming have a smaller environmental effect. These systems use a closed environment to manage growth, which lowers

the need for pesticides and herbicides and lowers the danger of contamination and soil deterioration (Galionis et al., 2013). To further lessen their negative environmental effects, hydroponics and vertical farming may be used with eco-friendly techniques like organic farming, renewable energy, and water recycling.

- **Local food production:** Localized food production is made possible by hydroponics and vertical gardening, which brings food closer to urban customers. By lowering transit costs and the carbon footprint associated with food delivery, vertical farming and hydroponics assist local food production, promote food security, and aid in urban conservation. Furthermore, local production guarantees food independence and lessens reliance on long-term food supplies for the urban population.

15.4 GOVERNMENT INITIATIVES

E-NAM Portal: The National Agricultural Market Portal's electronic goods contribute to provide a one-stop shop for every window. Businesses engaged in buying and selling respond to the market based on the availability and cost of items (Botsiou and Dagdilelis, 2013). It gives farmers in all of India's states and regions a single platform to market their agricultural goods. By disclosing data on agricultural product prices, supplies, and demand, the agricultural market may become more transparent. By ensuring that they obtain business information and select the most suitable company to market their goods via the platform, farmers may directly gain and intermediaries are involved less. Additionally, it makes digital payments easier, enabling farmers to get paid swiftly and safely for their goods. Farmers can access a wider market through the portal and expand their business beyond the local market.

National Policy on Crop Residue Management (NPMCR): Multicrop harvesting causes a lot of residue on and off the field. The Ministry of New and Renewable Energy estimates that approximately 500 metric tons (million tons) of grain crops are produced annually. Grain crops (beef, rice, corn, millet) account for 70% of the remaining crops, while single crops account for 34% of the remaining crops. Sugarcane residue, including tops and leaves, produces 12 metric tonnes, or 2% of India's crop. Crops are mostly used as animal bedding, animal feed, mulch, soil, biogas, roofs of rural houses, etc. However, the crops are usually burned in the ground to be planted in the next season. The problem of crop "burnt area" has become a growing problem in recent years due to labor shortages and the high cost of resolving crops and using machines to harvest crops.

15.5 TECHNICAL INTERVENTIONS

- Utilize conservation agriculture to incorporate agricultural wastes into the soil and reduce soil erosion brought on by wind and water.
- Utilize crops to create pesticide and bioaccumulate compost for use in farming.

- Cultivated mushrooms, specifically Volvariella Volva Cea (straw mushroom) and Bosporus (white mushroom), are produced on crops. Power plants require fibre for their operations. Non-vegetarian ethanol manufacturing products promote the use of straw and straw crops in the production of packaging, paper, board, and panels. It facilitates the process of gathering crops and provides financial assistance for transporting them to areas lacking food access.Products such as non-vegetarian ethanol manufacturing encourage the use of straw and straw crops in packaging, paper, board, and panels. It helps with harvesting and offers financial aid for delivering crops to places without access to food.

15.6 DIVERSIFIED USE OF CROP RESIDUE

- According to the ongoing project, various crops will be used as fuel in power plants, cellulosic ethanol production, etc.
- Encouragement and promotion of the use of rice straw and agricultural waste in paper, board, panels, and packaging materials.
- Encourage hay gathering by feeding, building bricks, and other means, and expand subsidies for the transportation of agricultural leftovers to areas lacking in fodder.

PMFBY: The government of India introduced the ground-breaking Pradhan Mantri Fasal Bima Yojana (PMFBY) to safeguard farmers' interests against crop losses resulting from a variety of natural catastrophes and extraordinary occurrences. This crop insurance offers producers protection from finance and covers risks from pre-planting to post-harvest such as inclement weather, pests, and illnesses. The PMFBY program makes crop insurance inexpensive and available to all farmers by having them pay a small cost that is subsidized by the federal and state governments. The program uses technology to address issues, concentrate, assure efficiency, and enhance the insurance process. Examples of these technologies include mobile applications and satellite imaging. Here are a few standouts:

- **Insurance:** PMFBY provides insurance to farmers for all food crops, oilseeds, and cash crops grown in the region. This project covers risks from pre-planting to post-harvest, including crop loss due to bad weather conditions and natural disasters such as floods, storms, wind, hail, insects, and disease.
- **Premium subsidy:** Under PMFBY, farmers pay a nominal premium for crop insurance, which is supported by the central and state governments, and is valuable and accessible to all farmers. The premium price is determined by the type of crop, the area of cultivation, and the size of the problem.
- **Insurance companies:** PMFBY is used by insurance companies selected through competitive bidding. These insurance companies provide crop insurance to farmers and resolve claims quickly and transparently.

- **Recommendation:** When crops are lost or damaged, farmers can submit the insurance form to the insurance company in time. Applications are evaluated and determined based on harvest data received from government agencies through the Crop Harvest Experiment (CCE).
- **Technology integration:** PMFBY uses technologies such as satellite imagery, remote sensing and mobile applications to streamline the insurance process, accurately detect crop damage, and make information available.

SHC: The Indian government's Soil Health Card (SHC) program aims to support sustainable development by giving farmers access to soil health information. In order to combat the issue of declining soil fertility and support farmers in making knowledgeable choices about soil management and agriculture, the Soil Health Card was created. Farmers are required to provide soil samples for analysis under the Soil Health Card Act in order to test a variety of factors, including organic content, pH balance, and nutrient levels (Bournaris et al., 2013). Farmers will receive a soil health report with advice for controlling soil nutrients and particular crop techniques based on the findings of the soil test. With the help of the Soil Health Report, farmers may lower input costs and their environmental effect while increasing production and optimizing fertilizer use. The Soil Health Card program aims to promote permaculture practices, protect natural resources, and provide safe food for future generations by encouraging the balanced and rational use of fertilizers and other agricultural chemicals.

Per Drop More Crop (PDMC) is an initiative of the Government of India that aims to promote good water management in agriculture to increase crop production. Grow more while protecting water resources. The PDMC project was launched with the aim of addressing the growing challenges of water shortage, as well as the inefficient utilisation of water in agriculture. Here are some highlights:

- **Water conservation:** PDMC encourages the application of water-conserving technologies such as micro-irrigation, rainwater collecting, laser leveling, spray irrigation, and drip irrigation.
- **Training and support:** Farmers receive assistance, training, and financial support to implement water-saving technologies and practices in their fields (Javeed, 2020).
- **Maximizing crop yield:** The program aims to maximize the productivity of each basin, minimizing water waste while maximizing crop yield.
- **Supporting farmers' income:** PDMC helps increase farmers' income and livelihood by improving irrigation and crop production.
- **Water saving:** PDMC contributes to water conservation by promoting the sustainable use of water in agriculture.

Indian Farmers Fertilizer Cooperative Limited (IFFCO) created IFFCO Kisan, a digital platform and initiative, to offer farmers in India a range of agricultural services and support. The key characteristics of IFFCO Kisan are:

- **Digital platform:** Through its website and mobile application, IFFCO Kisan functions as a digital platform that offers farmers a range of services and information.
- **Information services:** This platform offers farmers best practices and professional guidance on planting, using pesticides and herbicides, maintaining healthy soil, and other agricultural-related topics (Boopathi, 2024).
- **Weather updates:** To assist farmers in making well-informed decisions on planting and other field-related agricultural operations, IFFCO Kisan offers real-time weather forecasts and updates.
- **Training and seminars:** The platform arranges training sessions, seminars, and workshops to enhance farmers' understanding and proficiency in modern agricultural practices.
- **Community participation:** IFFCO Kisan encourages farmers' community participation by providing seminars, group discussions, and community centers where farmers can meet, discuss, and exchange ideas.
- **Problem solutions:** The platform offers personalized advice and solutions based on farmers' specific needs and problems, their location, their favorite crops, and agriculture.
- **Partnerships:** IFFCO Kisan collaborates with government agencies, agricultural experts, research institutes, and other partners to provide support and services to farmers.
- **Empowerment:** By providing information, resources, and tools, IFFCO Kisan empowers farmers to make informed decisions, adopt permaculture practices, and improve agricultural products and livelihoods.

e-Chaupal: They connect directly with buyers through digital channels. It provides new experiencessuch as weather conditions, soil fertility, and best agricultural practices.

Main supply chain which should be implemented:
Agricultural Produce Market Committee Act (APMC Act): State governments in India formed the Agricultural Produce Market Committee (APMC) as a regulating agency. APMC provides a controlled marketplace for the purchase and sale of agricultural commodities by acting as a middleman between farmers and purchasers.

Here are some important APMC facts:

- **Market regulation:** Within its assigned market region, the APMC is in charge of regulating the purchase and sale of agricultural goods. To safeguard the interests of both farmers and consumers, this law establishes rules for fair trading practices, quality requirements, and price.
- **Market infrastructure:** Market yards, warehouses, cold storage facilities, and auction platforms are just a few examples of the structures that fall under the purview of APMC's development and upkeep. These establishments offer crucial services for the grading, and storage.

- **Market fees:** For transactions that take place under its purview, APMC imposes fees and charges on the market. These fees, which are normally collected from both buyers and sellers, go toward paying for administrative costs as well as the upkeep and expansion of the market's infrastructure.
- **Market committees:** Farmers, merchants, commission agents, and government officials are among the stakeholders represented on the Market Committees that oversee each APMC (Rana et al., 2024). When it comes to making decisions about how to execute policies and run the market, these committees are extremely important.

Although information and communication technology (ICT) has great potential to transform agriculture and urban remote development, there are still some **limitations** and **challenges** in its use:

- **Digital divide:** People in rural and urban regions are different because many rural communities still have limited access to ICT infrastructure, including computers, cellphones, and the internet. Rural communities and farmers are unable to use ICT solutions due to a lack of technology.
- **Infrastructure challenges:** Reliable network connections, storage, and technical support are just a few of the things that rural regions frequently need in order to facilitate the use of ICT. Inadequate infrastructure impedes ICT leaders' efficacy and efficiency in rural development and agriculture.
- **Technology literacy:** It is possible that farmers and rural communities lack the abilities and know-how needed to make efficient use of ICT technologies. Inadequate technological expertise may impede the adoption and utilization of ICT solutions, thereby diminishing their influence on rural development and agricultural output.
- **Barriers related to language and culture:** It is possible that ICT solutions aren't available in the local tongue or haven't been tailored to the cultural norms of rural populations (Bhambri, 2024). ICT tool adoption and uptake by farmers and rural communities might be impeded by language and cultural barriers that restrict access to and use of these technologies.
- **Cost and affordability:** ICT solutions can be prohibitively expensive for small company owners and rural communities with limited resources since they frequently involve large expenditures in hardware, software, and connection. Adoption and sustainability of ICT projects in agricultural and rural development will be hampered by high prices.
- **Data security and privacy:** As ICT solutions are used in agriculture, worries about data security, ownership, and privacy will grow. It is possible that rural stakeholders and farmers won't share crucial information.
- **Sustainability and scalability:** Because of funding limitations, capacity issues, and reliance on outside assistance, ICT projects in rural and agricultural development may find it difficult to achieve long-term, sustainable development. Strong planning, cooperation with organizations, and

cooperative effort at home are necessary to achieve sustainability and increase the capability of ICT interventions.

- **Collaboration and integration:** ICT solutions can function independently without affecting current agricultural methods. The effect of any ICT initiatives might be limited by poor coordination and performance, which can result in rivalry in efforts, inefficiencies, and issues with sharing and exchanging information.

15.7 RESULTS

The improvement in information and service accessibility is one of the most significant outcomes. They may make better farming judgments and make informed decisions, thanks to updates, market pricing, agricultural guidance, and professional expertise. Drones, sensors, and GPS are examples of precision farming instruments that assist farmers in managing resources, cutting input costs, and boosting profitability, all while conserving water and lessening their environmental effect. Furthermore, ICT makes direct market connections easier by giving farmers the ability to communicate with customers, bargain for reasonable pricing, and access marketplaces throughout the nation, which boosts productivity and lessens reliance on middlemen. Additionally, ICT has improved crop management processes, reduced post-harvest losses, and ensured timely delivery to markets. Digital financial services continue to increase financial inclusion and resilience by providing farmers with access to formal financial services, allowing them to save, borrow, and hedge risk. Overall, the adoption of ICT in agriculture promotes permaculture practices, encourages innovation and entrepreneurship, and provides farmers with the necessary tools and knowledge to improve their livelihoods and build rural communities.

REFERENCES

Bhambri, P. (2024). NFTs: Transforming digital ownership in the Web 3 era. In D. Darwish (Ed.), *Decentralizing the Online Experience With Web3 Technologies* (pp. 151–167). IGI Global. https://doi.org/10.4018/979-8-3693-1532-3.ch007

Bhambri, P., Sinha, V. K., Dhanoa, I. S., & Kaur, J. (2019). Genome DNA sequence matching using HBM algorithm. *International Journal of Control and Automation*, 12(5), 531–539. http://sersc.org/journals/index.php/IJCA/article/view/2686/1887.

Bhusal, C. S. (2021). Systematic review on social engineering: Hacking by manipulating humans. Journal of Information Security, 12, 104–114. Available at SSRN: https://ssrn.com/abstract=3821594

Boopathi, S. (2024). *Sustainable Development Using IoT and AI Techniques for Water Utilization in Agriculture*. IGI Global.

Botsiou, M., & Dagdilelis, V. (2013). Aspects of incorporation of ICT in the Greek agricultural enterprises: The case of a prefecture. In Proceedings of the 6th International Conference on Information and Communication Technology in Agriculture, Food and Environment.

Bournaris, T., Basila, M., Moulogianni, C., Kiomourtzi, F., & Tandini, M. (2013). Measuring users satisfaction of an e-Government portal. HAICTA.

Deore, B., & Bhosale, S. (2022). Intrusion detection system based on RNN classifier for feature reduction. SN Computer Science, 3, 114. https://doi.org/10.1007/s42979-021-00991-0

Galionis, D., Charou, E., Stefouli, M., Kabassi, K., & Martinis, A. (2013). Mapping burnt forest areas and monitoring their regeneration using MERIS images. HAICTA.

Javeed, I., Narayan, S., Malik, A. A., Kumar, A., Rahman, R., Nisar, S., Akhter, A., Indrabi, S. A., & Sultan, A. (2020). Role of information and communication technology in agriculture.

Koliouska, C., & Andreopoulou, Z. (2013). Assessment of ICT adoption stage for promoting the Greek National Parks. In Proceedings of the International Conference on Information and Communication Technologies in Agriculture, Food, and Environment (HAICTA).

Kumari, S., Pandey, G. K., & Tiwari, S. (2024). *Data Visualization Techniques in Smart Agriculture Implementation*. IGI Global.

Lakota, M., & Stajnko, D. (2013). Using of GIS tools for analysis of organic waste management in Slovenia region Pomurje. HAICTA.

Nahirñak, V., Almasia, N. I., González, M. N., Massa, G. A., Décima Oneto, C. A., Feingold, S. E., Hopp, H. E., & Vazquez Rovere, C. (2022). State of the art of genetic engineering in potato: From the first report to its future potential. *Frontiers in Plant Science*, 12, 768233. https://doi.org/10.3389/fpls.2021.768233

Prakasam, C., Aravinth, R., Kanwar, V. S., & Nagarajan, B. (2020). Comparative study between weighted overlay and fuzzy logic models for landslide vulnerability mapping—A case study of Rampur Tehsil, Himachal Pradesh. In V. Kanwar & S. Shukla (Eds.), Sustainable civil engineering practices (Vol. 72, pp. 103–120). Lecture Notes in Civil Engineering. Springer. https://doi.org/10.1007/978-981-15-3677-9_16

Rana, R., Bhambri, P., & Chhabra, Y. (2024). Evolution and the future of industrial engineering with the IoT and AI. In *Integration of AI-Based Manufacturing and Industrial Engineering Systems with the Internet of Things* (pp. 19–37). CRC Press.

Ravikumar, D., Zhang, D., Keoleian, G., Miller, S., Sick, V., & Li, V. (2021). Carbon dioxide utilization in concrete curing or mixing might not produce a net climate benefit. *Nature Communications*, 12, 855. https://doi.org/10.1038/s41467-021-21148-w

Sakthivel, S., Akash Ram, R. K., Sidarth Sai, B., Vijaya Kesavan, K. M., & Ganesh Kumar, C. (2022). AIoT-based smart cart system. In 2022 International Conference on Smart Technologies and Systems for Next Generation Computing (ICSTSN) (pp. 1–6). IEEE. https://doi.org/10.1109/ICSTSN53084.2022.9761346

Sharma, R., Bhambri, P., & Sohal, A. K. (2020). Energy bio-inspired for MANET. *International Journal of Recent Technology and Engineering*, 8(6), 5580–5585. https://www.ijrte.org/wp-content/uploads/papers/v8i6/F8522038620.pdf. https://doi.org/10.35940/IJRTE.F8522.038620.

Singh, G., & Bhambri, P. (2023). Simulation analysis of AODV and DSDV routing protocols for secure and reliable service in Mobile Adhoc Networks (MANETs). In *Integration of AI-Based Manufacturing and Industrial Engineering Systems with the Internet of Things* (pp. 205–216). CRC Press.

Singh, M., Sachchan, T. K., Sabharwal, P. K., & Singh, R. (2024). *Intelligent Robots and Drones for Precision Agriculture*. Springer Science and Business Media LLC.

Singh, M., Singh, P., Bhambri, P., & Sachdeva, R. (2005). A comparative study: Security algorithms. In U.G.C. Sponsored National Seminar (Abstract Book) (p. 26). Presented at the "Network Security and Its Implementations" seminar at Doaba College, Jalandhar held on March 1, 2005.

Supriya, M., Tyagi, A. K., Tiwari, S., & Richa. (2024). *Sensor-Based Intelligent Recommender Systems for Agricultural Activities*. IGI Global.

Tondon, N., & Bhambri, P. (2017). Novel approach for drug discovery. *International Journal of Research in Engineering and Applied Sciences*, 7(6), 28–46. www.euroasiapub.org.

Wavrek, M. T., Carr, E., Jean-Philippe, S., & McKinney, M. L. (2023). Drone remote sensing in urban forest management: A case study. *Urban Forestry & Urban Greening*, 86, 127978. https://doi.org/10.1016/j.ufug.2023.127978

Wei, B., Zhao, C., Luo, M., Jia, Z., & Xia, H. (2024). Beyond the scenery: Understanding the impact of rural tourism development on household consumption in China. Review of Development Economics, 28(3), 1152–1191. https://doi.org/10.1111/rode.13096

16 IoT and Sustainable Transportation

Paula Bajdor

16.1 INTRODUCTION

This chapter is devoted to the Internet of Things and Sustainable Transportation issues. Internet of Things (IoT), artificial intelligence (AI), machine learning (ML), and fifth-generation networks (5G) are now integral parts of the broad field of information and communication technologies. At the same time, sustainable transportation is a permanent element of the area related to transport.

The Internet of Things is the result of continuous progress in the field of computerization and the development of IT systems. Initially, there was the internet, the popularization and unlimited availability of which contributed to a radical change in people's lifestyles. Thanks to the internet, interpersonal relationships have moved from the level of real contacts to virtual contacts, including both professional and private ones. This ubiquity of the internet has contributed to the creation and development of many concepts, among which the Internet of Things plays one of the most important roles.

Transportation is a comprehensive domain that encompasses the movement of products, individuals, and commodities. Transport is one of the most essential branches of the national economy in many countries, contributing to its development. Transport is a field that has accompanied people from the very beginning, enabling their needs related to moving or transporting loads from one place to another. Due to the growing awareness of the negative impact of transport on the natural and social environment, sustainability is becoming increasingly important in transportation–perceiving transport from the perspective of long-term ecological and economic stability. It is associated with promoting effective transport that meets society's expectations and is economically efficient, while also minimizing the poor or even harmful impact of means of transport on public health, the natural environment, the economy, and urban planning.

However, combining these two concepts, the Internet of Things and sustainable transportation, contributes to implementing IoT solutions in the transport field, transforming it into sustainable transportation. Therefore, this chapter covers the characteristics of both concepts and presents IoT solutions that can be implemented in the transport industry – currently in the planning phase and those solutions that have already been implemented and are being used.

DOI: 10.1201/9781003484226-16

16.2 THE FUNDAMENTALS OF THE INTERNET OF THINGS

The term Internet of Things was first used by Kevin Ashton in 1999. Kevin Ashton, a British entrepreneur and one of the founders of the Auto-ID Center at the Massachusetts Institute of Technology (MIT), used the term during a presentation to describe the concept of connecting the physical world to the internet by identifying and monitoring objects using Radio Frequency Identification (RFID) (Ashton, 2010). IoT has since evolved and become a widely used concept, covering various fields such as smart homes, health, industry, transportation, etc.

It is a concept that refers to the connection of the physical world to the internet, enabling everyday objects to communicate and exchange data. The main goal of IoT is to allow objects to collect, process, and share information in real time without direct human participation. IoT is one of the most significant concepts arising from the development and spread of the internet. It assumes the connection of almost all types of devices into a network. The essence of IoT is to create a network of interconnected devices to carry out communication processes (Rot & Blaicke, 2016). The Internet of Things concept can be presented as a network enabling communication in three dimensions: anytime, anyplace, and anything (Figure 16.1).

Gartner views the Internet of Things as an emerging concept that covers a wide range of applications, from smart homes and cities to industry and health services. Gartner's forecast of 26 billion devices connected to the internet by 2020 has come to fruition, leading to a vast increase in the amount of data that must be safely stored and processed (Middleton et al., 2013). Research by McKinsey & Company

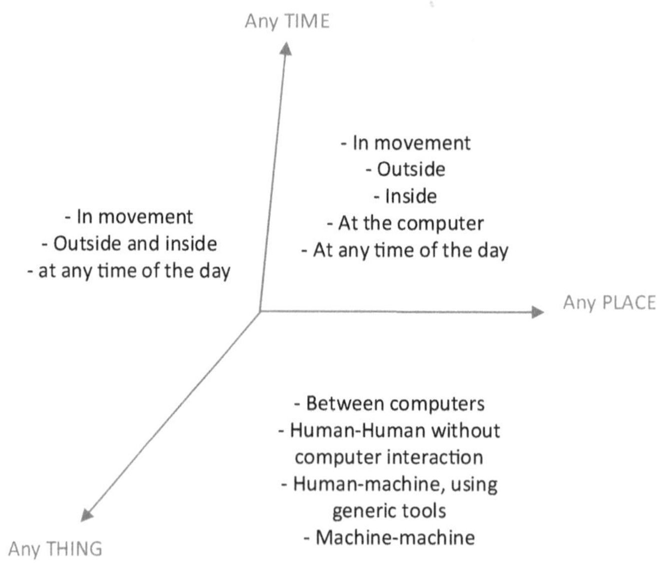

FIGURE 16.1 The Internet of Things concept is based on three criteria. Source: Brachman A. (2013). Internet of Things. Report of the ICT Observatory, Science and Technology Park "Technopark Gliwice", Gliwice.

estimates that IoT may be a source of significant economic benefits for the entire global economy; in 2025, they may amount to around 2.7-6.2 trillion US dollars (McKinsey, 2015). The International Telecommunications Union (ITU) defines IoT as a global infrastructure for the information society, enabling access to advanced services through the connection (physical or virtual) of objects based on existing and developed interoperable information and communication technologies. European Research Cluster on the Internet of Things, International Electronics Recycling Congress (IERC), defined IoT as a dynamic global network infrastructure with self-configuring capabilities based on standard and interoperable communication protocols in which existing physical and virtual "things" have an identity, physical characteristics, and virtual personality, use intelligent interfaces, and are seamlessly integrated into the information network (Duartem, 2023). Interactive Advertising Bureau defines IoT as an ecosystem in which objects communicate with computers using sensors, which can communicate with or without human intervention (Lipski, 2015). Miłasiewicz & Ostapowicz, (2011) define IoT as a phenomenon that has arisen to reflect the growing number of intelligent objects connected to each other. In turn, Cisco Internet Business Solutions Group pointed out that IoT is when the number of objects connected to the internet exceeds the population (Evans, 2013). In 2017, over 8 billion objects were connected via IoT, which increased to over 17 billion in 2023 (van der Maulen, 2017). By 2030, these devices are expected to increase to almost 30 billion. Considering that the population in 2023 was around 8 billion, it can be seen that there are more than twice as many devices connected to IoT. Moreover, IoT also presents a vision of a future reality in which digital and physical devices or everyday objects are bound by appropriate infrastructure to provide many new applications and services (Brachman, 2013).

16.2.1 Architecture of IoT

The architecture of the tools that make up IoT, in the simplest terms, is based on (Choroś, 2015):

- Physical devices equipped with sensors, actuators, microcontrollers, and other components enabling data collection and interaction with the environment.
- Connectivity in the form of Bluetooth, Zigbee, RFID, LTE, LoRa, NB-IoT, and others that enable IoT devices to communicate with each other and with other systems.
- Intermediary devices between IoT devices and, e.g., a computer cloud, enabling data transfer.
- Cloud computing, a central platform where data collected by IoT devices is stored, processed, analyzed, and made available to applications and users (Bhambri et al., 2005).
- Data analysis mechanisms, such as artificial intelligence algorithms, machine learning, and analytical tools, which allow obtaining valuable information from data collected by IoT devices.

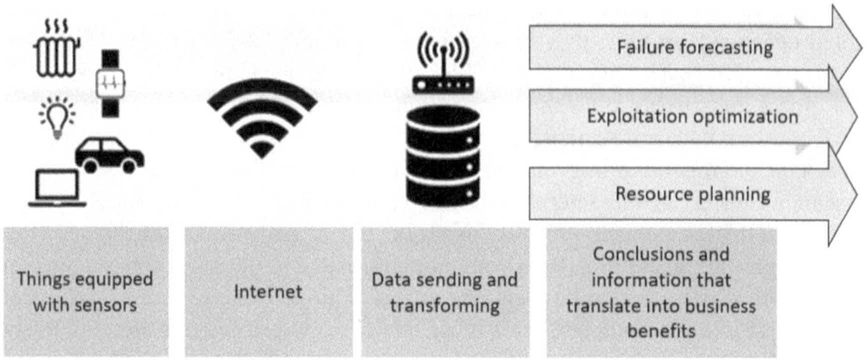

FIGURE 16.2 IoT – main idea.

- Applications and interfaces that enable users to monitor, manage, and interact with IoT devices and data.
- Security mechanisms such as encryption, authentication, access management, and security protocols designed to protect data and prevent unauthorized access.
- Communication standards such as Message Queuing Telemetry Transport (MQTT), Constrained Application Protocol (CoAP), and Hypertext Transfer Protocol (HTTP), which define the communication rules and protocols used by IoT devices.
- Management systems that enable IoT device registration, configuration, software updates, and status monitoring.
- Mechanisms for integrating IoT with existing IT systems to enable seamless collaboration between different platforms.

Figure 16.2 shows the general idea of how IoT works.

16.2.2 Selected Areas of Application of the Internet of Things Concept

IoT's versatility and possibilities mean it is used in many areas of life and economic sectors (Singh et al., 2005). Based on analyses and reports, it is possible to create a comprehensive list of IoT applications, which not only highlight the raised usefulness of this concept but also the strategic dimension in the area of Information and Communication Technologies and the entire internet. The most important areas include (Satish & Varma, 2017):

1. Smart life means innovative and cutting-edge technology to simplify life with a greater sense of security. Smart life covers:
    - Healthcare – a business approach focused and tailored to a given patient.
    - Banking – new models of banking and personal finance.
    - Public services – improving efficiency and convenience for administration, managers, and citizens.

2. Smart mobility enables route management in real time and offers solutions to make travel more pleasant and increase the reliability of transport services. Connecting vehicles with IoT provides many new possibilities and applications that not only facilitate movement but also increase the level of safety. Smart mobility covers:
   - Electric and autonomous vehicles.
   - Car-sharing and ride-sharing.
   - Smart transportation systems.
   - Solutions for pedestrians and bike riders.
   - Mobile applications.
   - Integrated payment systems.
3. Smart city means innovations aimed at improving the quality of life in cities. Smart city covers:
   - Intelligent management of urban infrastructure using analysis tools data specific to big data.
   - Cooperation of various administration bodies using the cloud technologies.
   - Real-time data collection and processing using mobile technology, enabling immediate response.
   - Increasing the sense of security by improving enforcement rights and a more effective response to crises.
   - Sustainable city planning–improved schemes and project management spatial planning.
   - "Connected media" – intelligent metering devices and network management, gas, water, etc.
   - Development of construction–greater automation, better management and safety.
4. Smart manufacturing consists of production and logistics solutions designed to optimize processes, control, and quality. Smart manufacturing covers:
   - Process automatization.
   - Digital factory.
   - Industry 4.0 technologies.
   - M2M (machine-to-machine) communication.
   - Big data and data analysis.
   - Supply chain management.
   - Cybernetic security.
   - 3D printing.
5. Smart energy/smart grid – these technologies significantly change how energy is produced and transmitted to the end user. They are characterized by a high level of security and are used in both concentrated and distributed networks. Smart energy/smart grid covers:
   - Energy savings through the use of more reliable and intelligent sensors and actuators.
   - Scalability of security functions.
6. Smart home, smart buildings, and infrastructure are solutions and technologies that use wireless internet access, enabling automation of the

functioning of buildings (Sharma et al., 2020). These solutions are implemented into consumer electronics/home appliances, making them part of the home internet network. Smart home, smart buildings, and infrastructure covers:

- Integration of intelligent devices and equipment in buildings or apartments with, e.g., entertainment systems.
- Health care monitoring.
- Wireless monitoring of energy consumption in the context of a house or building.
- Intelligent building management systems.

IoT is also widely used in many economic sectors (Table 16.1).

Table 16.1 shows that IoT is used in many industries and sectors. Therefore, it can be assumed that if IoT is used in transport, it can also be used in sustainable transportation, the importance of which is increasing.

16.2.3 Internet of Things – Threats

On the one hand, the practical application of IoT brings many benefits and improvements in implementing many different processes, whether from business or human activity. However, in addition to this concept's visible benefits and advantages, new threats pose a challenge to IoT users. Indeed, the fact that IoT is based on the use of the internet makes the issue of digital security and protection against hacker attacks significant, who can use the gap between individual devices to carry out cyberattacks (Rot & Sobińska, 2013). Therefore, the most frequently mentioned threats resulting from the use of IoT include problems related to data privacy, resulting from, for example, insufficiently strong points in authorization or authentication systems, unsecured web interfaces, software vulnerabilities, and errors (Lipski, 2015). I. Maj (2015) also points out many threats posed by this technology:

- Threats resulting from insufficient testing of a specific system. The dynamics of changes in the system's environment can create a threat that was not previously known (not to mention simply not covering all cases of potential system behavior).
- Danger resulting from the failure of the communication medium itself, which is the internet.
- Threats resulting from technical defects of the interfaces used and other peripheral devices.
- Threats resulting from error algorithms, including decision-making algorithms.
- Danger resulting from the lack of standardization of software that is publicly available, including update options. We are currently dealing with the copyright phenomenon solutions whose protection against copying and theft is, among other things, the lack of openness and access to the software (also securing firmware).

TABLE 16.1
Application Areas of the Internet of Things

Branch	The Use of IoT
Construction	- heating, ventilation, and air conditioning control, - access control, - lighting, - security systems in buildings
Energy	- extraction of raw materials, - searching for alternative energy sources, - design and production of electricity supply devices
Consumption	- home security systems (alarms, monitoring), - control of devices responsible for energy use or home lighting, - entertainment
Healthcare and Environmental Science	- telemedicine, - devices monitoring the elderly, sick people, or those with an implanted pacemaker, - research and development of new drugs and medical equipment
Industry	- monitoring and tracking of devices, machines, and products, - location analysis for a wide range of production processes
Transport	- vehicle fleet management (navigation system, telematics, search for missing vehicles) - management of transport and distribution systems, - information systems for passengers and drivers, - payment systems for the use of transport infrastructure (highways) and urban infrastructure (parking lots)
Retail	- network systems and devices enabling efficient management of supply chains, - automatic vending machines (food, drinks, cigarettes), - parking meters, - automatic amusement machines (slot machines), - display devices (billboards, information points)
Public Safety	- monitoring the state of the natural environment, - obtaining meteorological and climatic information, - tracking people, animals, and shipments, - military security
IT	- photocopiers, printers, and scanners, - infrastructure for mobile transmission - data centers, - energy and air conditioning maintenance and management systems, - e-commerce

- Danger resulting from a lack of compatibility between the offered systems in terms of hardware, software, and properties of energy-efficient systems provided.
- Threats resulting from limited address space.
- Threats resulting from limited internet bandwidth.
- Threats arising from external interference by third parties in the system, with the intention of gaining specific advantages and causing unforeseen undesirable consequences, such as unauthorised disruptions in the supply of electricity, gas and water, or disruptions in the management of road traffic in the city or air traffic at the airport.

16.3 SUSTAINABLE TRANSPORTATION – DEFINITIONS AND BASIC ASSUMPTIONS

Transport is one of the sectors of the economy that is of fundamental importance for increasing competitiveness, social development, and world integration (Rattan et al., 2005). It provides a high level of mobility and constantly increases efficiency in speed, comfort, safety, and accessibility when traveling (Motowidlak, 2017). Transportation is the process of moving people, goods, or information from one place to another. It may refer to the movement of people, vehicles, and means of transport, but also to the transfer of data or energy between places (Rana & Bhambri, 2024). This includes various means and methods such as road, rail, air, sea transport, public, private, freight, passenger transport, and many others. Transport is a critical element in the functioning of societies and economies, enabling trade, mobility, cultural exchange, and other aspects of modern life. It covers various activities, from everyday travel to advanced logistics operations.

However, in addition to the benefits offered by transport (the possibility of transport of goods and people, the mobility of societies, and time-saving), transport is also responsible for many adverse effects, for example, its negative impact on the environment. Today, many societies and local authorities know that pursuing Sustainable Transportation will bring tangible results for the environment, the economy, and society. Sustainable Transportation is an approach to the movement of people and goods that minimizes environmental, social, and economic impacts while ensuring efficiency, accessibility, and safety. The essence of sustainable transport includes various factors that focus on balancing the needs of modern society with minimal negative effects on the environment (Bajdor & Grabara, 2013).

Key elements of sustainable transport include:

1. Minimizing the negative impact on the environment by reducing greenhouse gas emissions, air pollution, noise, and consumption of natural resources.
2. Promoting the efficient use of energy through the use of energy-saving technologies and minimizing energy losses in the transport process.
3. Encouraging use of public transport such as buses, trains, and trams to reduce the number of individual vehicles on the road and reduce traffic jams.

4. Investing in infrastructure that enables safe movement of pedestrians and cyclists, thus promoting physical activity and a healthy lifestyle.
5. Introducing electric vehicles (EVs) to reduce greenhouse gas emissions and dependence on fossil fuels.
6. Encouraging the use of rail and sea transport, which can be more efficient and less harmful to the environment than road or air transport.
7. Use of advanced technologies such as intelligent traffic lights, traffic management systems, and mobile applications for route planning to optimize road traffic and minimize traffic jams.
8. Coordination of different means of transport so that they are complementary and enable easy transfers, which increases travel efficiency.
9. Involving local communities in transport decision-making processes to take into account their needs and opinions.
10. Road traffic safety, a safe environment for pedestrians and cyclists, as well as the safety of public transport users.

Sustainable transport strives to balance efficiency, social benefits, and minimizing negative environmental impacts, contributing to building more sustainable societies.

The dimensions of sustainable transport include various social, economic, and environmental aspects that create a holistic approach to the movement of people and goods (Table 16.2).

The Organization for Economic Cooperation and Development (OECD) approaches sustainable development issues, including transport, in the context of balance between three main dimensions: economic, social, and environmental. In

TABLE 16.2
Dimensions of Sustainable Transportation

Social Dimension	Economic Dimension	Ecological Dimension
- mobility - accessibility - flexibility - safety - social cohesion - transport system integrity	- improving cost and operational efficiency - route optimization - delivery planning - innovative technological solutions - development of transport infrastructure - integration of different means of transport	• environmental friendliness of transport (minimization of environmental impact) • counteracting and eliminating the transport effects of environmental threats

practice, sustainable transport in the OECD approach could include strategies and actions that aim to minimize the negative impact of transport on the environment while considering social and economic aspects (Guidelines, 2002).

The Renewed Sustainable Development Strategy of the European Union defines sustainable transport as a transport system that minimizes negative environmental impacts while ensuring efficiency and accessibility (COM, 2006). The elements of sustainable transport in an EU context include emission reduction, electrification of transport, development of public transport, technological innovation, integrated spatial planning, and road safety (Bhambri & Khang, 2024).

The most representative definition of Sustainable Transportation is the definition adopted in 2004 by the European Ministers of Transport and by the Center for Sustainable Transportation in 2005 (CST, 2005):

A Sustainable Transportation system is defined as one that is:

- Ensuring the safe meeting of essential access and development requirements for individuals, businesses, and societies, aligning with the well-being of both humans and ecosystems and fostering fairness across present and future generations.
- Being cost-effective, operating reasonably and efficiently, providing transportation options, and contributing to a competitive economy and well-balanced regional growth.
- Restraining emissions and waste within the Earth's capacity to absorb them, utilizing renewable resources at or below their regeneration rates, and managing non-renewable resources at or below the development rates for renewable alternatives, all while minimizing the impact on land use and noise generation.

The following features must characterize Sustainable Transportation:

- Its functioning must contribute to improving the health of society and raising the standard of living.
- Public transport must be preferred.
- The needs of pedestrian road users must be taken into account, as well as cyclists.
- Local communities should play an essential role in the stage of creating strategies, plans, and transport policies.
- Energy prices in transport must take into account all costs (including external costs), which will cause rational investment decisions.
- When designing transport networks, it is essential to take into account the presence of ecologically significant places.

Analyzing Sustainable Transportation in detail, it is worth keeping in mind that Sustainable Transportation is available for transport services consistent with health and ecological safety requirements while considering the principle of justice, using land space, or the need to reduce noise. Sustainable Transportation aims to improve the standard of living and safety, give preference to public transport, consider pedestrians' needs, and promote civil society's role (OECD, 2004). According to this assumption, transport should not threaten human health or the ecosystem (Brzustewicz, 2012). It is for achieving economic and social goals for using and protecting the natural environment. In contrast, unsustainable transportation consumes huge amounts of energy, causes environmental pollution, increases expenditure, and does not improve the financial situation of specific social and economic groups (Miłasiewicz & Ostapowicz, 2011).

16.3.1 The Future Trends in Sustainable Transportation

For over 30 years, questions have been asked about transport's responsibility for exploiting non-renewable resources and the state of the environment. Transport plays a significant role in the consumption of some non-renewable resources, mainly oil, as well as in the increase in air pollution, which is why transport is one of the key aspects of the concept of sustainable development. Transport of people and goods is the most significant contributor to the UK's carbon emissions, accounting for 28%. Also, in the United States, transportation is the second largest source of greenhouse gas emissions (after energy production) and accounts for 70% of all domestic oil consumption (Russ, 2023).

The search for environmentally friendly forms of transport has become necessary, which involves increasing rail transport or developing new types of engines powering road transport. Since the spread of commercial flights, transportation itself has not changed radically. Currently, it is assumed that within the next decade, there will be another revolution in transport, which will be characterized by radical inventions in the form of engines powered by renewable energy or the introduction of solutions that significantly reduce carbon dioxide emissions, which is one of the main adverse effects of transport processes.

This revolution is slowly taking place through the popularization of car-sharing systems, Bolt and Uber, electric vehicles, electric scooters, drones, and autonomous buses. However, while these revolutionary systems concern passenger transport, there are also several revolutionary solutions in the case of freight transport. Drone deliveries have been carried out for over ten years, aiming to reduce the number of delivery vehicles used and, consequently, reduce the pollution they generate while maintaining the convenience of home delivery. Amazon was the first to introduce drones and has been operating the Amazon Prime service for 17 years as of now. At that time, delivering parcels to people wherever they wanted – within a few days – was revolutionary and new. Amazon intended to use drones to increase delivery speeds and the overall safety and efficiency of the entire transportation network.

Another change, although perhaps not so revolutionary, will be the switch to electric vehicles. The factors determining this change include upcoming bans on the sale of vehicles with combustion engines, changing consumer behavior, and continuous improvements in battery and charging technology. By 2035, the world's major automotive markets – the United States, the European Union, and China – are expected to sell only electric vehicles (EVs). By 2050, 80% of global vehicle sales will be electric. Electric vehicles are a vital element in achieving climate neutrality (in Europe, for example, emissions over the lifecycle of an electric vehicle are approximately 65 to 85% lower than for a vehicle with an internal combustion engine) and improving the quality of life in cities by reducing air emissions and noise pollution. The desire to decarbonize transport and eliminate air pollution in urban areas means that electric vehicles are widely perceived as the optimal solution for moving away from fossil fuels. Today, the rapid development of electric vehicles (EVs) is part of a fundamental shift in transportation that promises significant benefits for people, businesses, communities, and entire nations. Electric cars, equipped with more efficient electric batteries and an increasingly dense network of charging stations, are already becoming an increasingly popular means of moving people. In freight transport, this change is a bit slower, but the regulations recently adopted by the EU introducing the goal of reducing CO_2 emissions of new passenger cars and delivery vans by 100% from 2035 result in a ban on the sale of new vehicles with combustion engines. The regulations adopted in March 2023 specify the following CO_2 emission reduction goals:

- 55% for new passenger cars and 50% for new commercial vehicles in 2030-2034 compared to 2021 levels.
- 100% reduction of CO_2 emissions for both new passenger cars and delivery vans from 2035.

In some countries, the sale of cars with combustion engines will be banned in 2030. In return, the governments of these countries are allocating very high amounts (up to $1.5 billion) to expand the infrastructure for intelligent charging of electric vehicles. Some transport companies, e.g., DHL, are already modernizing their fleets to replace traditional vehicles with electric ones, mainly electric scooters and bicycles (Duarte, 2023). However, there are already several forms of electric transport:

- Logistics vehicles, from large trucks to cargo handling – several companies are creating electric logistics vehicles. However, at the moment, the too-high cost of electric batteries for these vehicles means they constitute a percentage of the company's entire transport fleet. However, it was indicated that developing charging networks could solve this problem.
- Electric buses are used for public transport. Electric buses are already very popular in China, while Europe is accelerating the pace of their implementation. Around 38% of the UK's rail network is already electrified. Electric trams are also gaining popularity.
- The technology of electric planes is still in its early stages. According to experts, existing passenger planes would require batteries that weigh 30 times more than their present fuel usage in order to fly. This makes it unlikely for them to take off. Currently struck by the coronavirus pandemic, the global aviation industry aims to reduce aircraft carbon emissions by replacing traditional fuels with synthetic jet and hydrogen and developing new futuristic engines and aircraft designs. However, even in this field, some progress has

been made and electric planes exist. The first battery-powered aircraft built and capable of flying is the Alice aircraft, built by Eviation. It is a passenger plane that made its maiden flight in September 2022. So far, it is the only passenger aircraft developed from scratch and capable of flying on battery power.
- Boats – battery power is also a problem, especially for larger boats covering longer distances. However, prototypes are starting to emerge that prove that electric vehicles can exist in many different formats.

Autonomous vehicles (AV) are also the future of transport. AV combined with artificial intelligence (AI), e.g., autonomous public buses operating based on advanced algorithms, will pick up people going in the same direction. Central artificial intelligence software can control city traffic systems, directing traffic flow so that traffic jams and accidents are a thing of the past. In the UK, pilot projects involving the use of autonomous vehicles in urban transport have already been carried out several times: DRIVEN–a trial fleet of fully autonomous vehicles used in urban and intercity areas, which will enable independent travel in Oxford and London; RoboPilot–autonomous driving of an electric delivery vehicle that could be introduced into city buses and larger trucks StreetWise – a technology offering independent personal mobility services that may, in the future, replace the city car used to commute to work. Similar work has been underway in the US: Ford, Argo AI, and Walmart have collaborated on a delivery service based in Austin, Miami, and Washington that will use Ford's autonomous vehicles combined with Argo AI's self-driving system technology. However, this project was interrupted at the end of 2022 due to high financial losses.. It is assumed that as time passes and work progresses, this technology will be used more often in freight transport.

It is also expected that autonomous ships sailing on the seas will be used in the future. Advanced artificial intelligence will allow for calculating optimal routes and maximizing speed using weather data and sea currents. The use of maritime data generated by intelligent ship systems is seen as the most significant advance in commercial shipping since the introduction of containers.

In turn, considering the negative impact of road transport on the natural environment, two concepts have been developed: eco-driving and zen-driving, which are perceived as indirect methods for reducing pollutants emitted by transport.

Eco-driving:The assumptions of this concept were first formulated in the 1990s, mainly in Switzerland, Finland, Germany, and the Netherlands Campaigns implemented by the European Union played an essential role in its dissemination: the Eco-driving Europe program, EcoDriven (European Campaign On Improving Driving Behaviour, Energyefficiency and Traffic Safety), and EcoWill. However, until today, no single valid definition of eco-driving has been developed, most likely due to the discrepancies between the recommendations proposed by individual countries (Słowiński et al., 2016). Most often, this concept takes the form of specific recommendations. Still, eco-driving can also be defined as a driving style that reduces fuel consumption and, consequently, reduces the level of pollutants emitted and the number of accidents. The following definition treats eco-driving as a strategy for changing the behavior of drivers acting following specific rules.

The effective implementation of eco-driving should be preceded by an analysis covering three levels: strategic, tactical, and operational. At the strategic level, basic decisions and actions are taken that have a long-term impact, the effects of which will also translate into effects at other levels. These decisions concern the selection of a specific vehicle that will be used for transport, operational activities, and ongoing maintenance. All decisions regarding route planning are made at the tactical level, including selecting specific roads, considering potential traffic restrictions or traffic intensity, and the proper arrangement and securing of cargo. Decisions made at the operational level cover a specific route: without unnecessary acceleration, sudden braking, or sudden maneuvers. According to eco-driving, the vehicle's movement should be smooth, acceleration should occur with the gas pedal almost pressed, the vehicle's movement should start gently, gears should be changed following the requirements of a given engine, a constant speed should be maintained, and the car should be left idle during stops. The engine is on or off, depending on the recommendation for a given engine type.

In turn, zen-driving means not only the driving technique but also the culture, behavior, and way of thinking while driving. From a systemic perspective, it leads to increased fluidity of the vehicle flow and harmonized behavior of all road traffic participants. The basic principle of this concept is constant observation of the surroundings and smooth, anticipatory adaptation to occurring changes, as well as proper planning of the route so that the time used for the journey is optimally used. Also important is the so-called attentive driving, which is understood as the ability to concentrate on stimuli occurring in the environment that may potentially influence the course of driving. Zen-driving is also a traffic navigation system, including road infrastructure and traffic participants.

While both of these concepts constitute guidelines or recommendations for transport companies at the moment, in the future, their provisions may become mandatory, as in the case of Euro standards. Indeed, based on the above content, it can be predicted that soon, transport will undergo another revolution, the main effect of which will be to reduce the negative impact of transport on the natural environment, and the transport itself will fully meet the Sustainable Transportation standards.

16.4 THE USE OF IOT IN SUSTAINABLE TRANSPORTATION

The possibilities offered by IoT solutions meant that it was only a matter of time before this concept would be used in sustainable transportation. Currently, many IoT solutions are being implemented to transform the transport sector into a sustainable one. IoT is increasingly important in improving efficiency, safety, and environmental responsibility in the transport sector. These solutions take the following form:

1. Intelligent traffic management systems are solutions that obtain information, skillfully process it, and then use it in such a way as to achieve the expected goal. Using IoT in the framework, these systems manifest as follows:

a) Installation of visual cameras, traffic, light, and sound sensors at intersections, highways, and main urban arteries, which are responsible for monitoring the movement of vehicles, bicycles, and pedestrians; and collecting the necessary data on traffic intensity, vehicle speed, travel times, and other variables related to road traffic. These data are then sent to the management center by movement.
b) Using appropriate sensors at intersections enables data acquisition to detect potentially dangerous situations that may contribute to accidents. In addition, these sensors can be equipped with modules that, in the event of accidents, notify the appropriate services, which results in their faster arrival at the scene of the incident.
c) The use of appropriate software whose task is to analyze data obtained from cameras and sensors. This software can enable the generation of highly accurate road traffic maps while forecasting the possibility of congestion or identifying areas requiring attention.
d) Adapting traffic lights to current road conditions, optimizing traffic flow by increasing its fluidity while minimizing congestion;
e) Vehicle drivers use mobile applications to select the best or alternative route in real-time.

It follows from that IoT in traffic management systems enables effective response to changing road conditions, reduces congestion, improves the safety of all road users, and contributes to the overall improvement of transport.

2. Monitoring and optimising fuel use entails tracking two variables: the fuel level in the tank and the fuel consumed by the engine. The characteristics of IoT applications, including case IoT, are as follows:
   a) Installing appropriate vehicle sensors that collect data on engine operation, fuel consumption, speed, driving style, road conditions, temperature, and emissions. This data is then sent to the company's central fleet management system.
   b) Using a central fleet management system in the enterprise that receives transmitted data and then processes and stores it.
   c) Monitoring drivers' driving styles enables the identification of aggressive driving styles, excessive speeding, and excessive braking or acceleration. This data is then used to assess the driver's driving efficiency and identify areas for improvement.
   d) Real-time analysis of engine efficiency: IoT sensors enable the identification of problems, e.g., related to excessive fuel consumption, and analyze emission data, helping to meet environmental standards.
   e) Optimization of routes and speeds, generated based on collected data, which can significantly affect fuel efficiency.
   f) Remote vehicle monitoring enables, among others, the ability to identify potential mechanical problems or areas requiring maintenance. Also, in the event of a failure, remote monitoring can automatically generate service orders.

g) Generating reports and trend analyses based on data obtained by IoT sensors. The reports and trends created are useful for implementing a more economical driving style.

All this means that, for example, transport companies can save operational costs and reduce their ecological footprint by reducing vehicle emissions, transforming transport activities into Sustainable Transportation.

3. Car-sharing systems: solutions enabling the sharing of vehicles by several drivers. The use of IoT in these systems involves, among others:
   a) Equipping vehicles with GPS sensors and location technologies so that users can easily locate available vehicles using a mobile application or website.
   b) Reservations and payments that are fully automated and IoT sensors track the time of vehicle use, charging fees per established rates.
   c) Remote locking and unlocking of vehicles, in which users who have booked a vehicle receive a digital key that allows them to access the vehicle at a specific time.
   d) IoT sensors monitor the technical condition, fuel level, tire pressure, and battery condition (in the case of electric vehicles). When a problem is detected, the sensors notify the service center or deactivate the vehicle, preventing it from starting;
   e) Collecting ratings and opinions of users who not only rate vehicles but also other users, which helps maintain high standards of behavior and care for vehicles;
   f) Smart parking: by using IoT sensors that locate available parking spaces, users can leave their vehicle in any parking space in a specific area;
   g) Collect and analyze data from IoT sensors to optimize the deployment of the entire vehicle fleet and adapt vehicle availability to user demand. By analysing the acquired data, it is possible to foresee trends in vehicle use. This enables the adaptation of the offer to meet the evolving demands of customers.
   h) The gradual introduction of electric vehicles into the fleet, whose energy consumption and emissions are monitored by IoT sensors, will ultimately lead to further sustainability of the transport fleet.

Car-sharing systems become more efficient, flexible, and tailored to the needs of users and also support sustainable development goals by reducing the number of private vehicles on the road.

4. E-mobility (Electromobility) is a term defining the development of electric motors. The aim is to avoid using fossil fuels and carbon emissions in vehicle design. Also, in this area, IoT solutions are widely used:
   a) Electric chargers are equipped with IoT sensors that enable remote monitoring and management of the charging process. Charger users can check the availability of chargers, plan to charge depending on energy prices, and receive notifications about the completed charging process.

b) Intelligent charging management systems are e-mobility platforms that use IoT technology to intelligently manage the charging process depending on the energy demand in a given location. They allow you to adjust the charging power to optimize energy consumption in the network, avoiding its overload.
c) Remote monitoring of electric vehicles to check battery status and energy consumption. The data obtained by IoT sensors is then sent for further analysis and optimization of the vehicle management process.
d) Planning routes with charging using mobile applications equipped with IoT sensors, supporting route planning by taking into account the availability of charging stations.
e) Integration with payment systems.Thanks to built-in IoT sensors, it is possible to integrate chargers with online payment systems that enable cashless payments, which significantly facilitates and speeds up the payment process.
f) Energy efficiency monitoring enabled by IoT sensors makes it possible to track the energy efficiency of electric vehicles. The reports obtained in this way inform users about energy consumption, operating costs, and environmental impact.
g) Integration with renewable energy systems, which allows vehicles to be charged with energy from renewable sources.
h) Smart grid networks enable flexible management of the supplied energy while supporting the energy infrastructure.

IoT solutions, such as e-Mobility, are becoming an increasingly common method of transport, which is becoming more effective and environmentally friendly. And the vehicle charging process is more tailored to the needs of users.

5. Fleet Management (FM) describes the practices and technologies companies use to manage commercial vehicles in their fleet. FM includes various activities such as vehicle maintenance, fuel management, driver safety, and logistics planning. Including aspect IoT solutions are there useful in:
   a) IoT sensors installed in vehicles to send data in real time to the central fleet management system, enabling precise tracking of each vehicle.
   b) Optimizing routes and managing drivers' working time is possible, thanks to sensors that collect information about current road conditions. Based on the collected data, it is possible to plan routes that minimize travel time and optimize fuel consumption.
   c) Monitoring the technical condition of vehicles is possible due to IoT sensors that track parameters such as tire pressure, fuel consumption, engine condition, and others, which allows for early detection of potential problems and maintenance planning, which translates into more effective management of operating costs and vehicle maintenance.
   d) Systems for appropriate vehicle use equipped with IoT sensors enable driving analysis and checking whether a given journey is carried out

economically. Data acquired by sensors can also be used for driver training and to assess and reward appropriate vehicle use.

e) Integration with payment systems allows you to make payments, e.g., related to journeys on toll sections. As a result, this makes it easier to manage operating costs and track transactions associated with the use of vehicles.

f) Analyzes and reports based on data obtained from IoT sensors making it possible to increase the efficiency of the vehicle fleet and identify areas requiring further improvement or ways to reduce the negative impact of the fleet on the natural environment.

IoT solutions enable transport companies to manage their vehicles effectively and sustainably while improving operational efficiency and environmental impact.

6. Smart parkingis an IoT-based system that transmits data about free (and occupied) parking spaces via a wired or wireless system using a web or mobile application, including:
   a) Parking space monitoring systems that inform drivers about the number of available spaces, enabling them to find parking quickly and efficiently.
   b) Mobile applications for parking management enable drivers to reserve a parking space and pay for the service online, and, additionally allow planning routes to a free space, which saves time.
   c) Parking price and zone management allow you to adjust parking prices depending on demand and availability of spaces.
   d) Integration with payment systems allows drivers to pay for parking via a mobile application, contactless card, or vehicle registration system.
   e) Parking time monitoring tracks parking time and automatically generates information, e.g., about the approaching end of the payment period or lack of funds in the account. They also help in avoidance of mandates behind exceeding time parking.
   f) Intelligent security systems, when an event is detected, such as attempted theft or attempt to damage a vehicle, activate alarms and notify services responsible for the safety of parked vehicles.
   g) Analyzes and reports based on data obtained from IoT sensors making it possible to determine usage trends and optimize the availability of parking spaces, which allows for adapting the parking infrastructure to actual demand.
   h) Sustainable parking is manifested by offering preferential parking conditions for electric vehicles or rewarding ecological forms of transport, such as bicycles or car-sharing.

All this means that intelligent parking based on IoT improves the efficiency of using parking spaces and positively impacts the sustainable development of cities, eliminating unnecessary circulation of vehicles in search of parking spaces and reducing exhaust emissions.

7. Emission trading systems are mechanisms where enterprises, like transport companies, receive limited greenhouse gas emission permits. Including the IoT area, ETS is used for:
   a) Emission monitoring, thanks to automatic data collection.
   b) Permit management, ETS can track emission permit transactions.
   c) Integration with systems for proper energy use, which enables process optimization and emissions minimization.
   d) Integrating with correct energy use systems enables enterprises to optimize energy use.
   e) Analyzing trends and forecasting emission levels to support actions aimed at achieving desired emission levels.
   f) Supporting ecological innovations, the implementation of which will contribute to increasing energy efficiency while reducing emission levels.
   g) Reporting that increases transparency and facilitates monitoring progress in achieving emission goals.

IoT technology is already widely used in the transport sector, significantly contributing to achieving an increasingly higher level of sustainability. Transport companies, city authorities, and individual users can successfully use IoT technology. Below are some examples of the practical use of IoT in public transport in selected cities:

1. In Amsterdam, the Netherlands, electric vehicle charging infrastructure has been developed and is an essential element of Sustainable Transportation in the city. The city of Amsterdam strongly supports electric mobility, which translates into numerous charging stations in various parts of the city and initiatives promoting the use of electric vehicles. Thanks to IoT technology in electric vehicle charging systems in Amsterdam, it is possible to manage the charging infrastructure effectively. Thanks to sensors and monitoring networks, you can track charging stations' availability and occupancy and effectively control charger access. Amsterdam's electric vehicle charging systems often offer the option to pay for charging via mobile apps, which makes using chargers convenient and accessible for residents and tourists. In addition, IoT systems allow you to monitor energy consumption during charging, optimizing electricity consumption, and increasing the efficiency of charging electric vehicles.
2. In Barcelona, Spain, intelligent traffic management systems play a crucial role in optimizing urban transport and promoting the city's sustainable development. Thanks to IoT technologies such as sensors, CCTV cameras, motion detection systems, and intelligent software, Barcelona focuses on innovative solutions in road traffic management. IoT systems in Barcelona allow you to collect data on traffic intensity, vehicle speed, weather conditions, and other factors affecting road traffic. This data is analyzed in real time, enabling quick decisions to optimize the road situation and ensure traffic flow. Intelligent traffic management (ITM) systems in Barcelona

use advanced algorithms and artificial intelligence to control traffic lights, monitor vehicle flows, and coordinate traffic situations. ITM makes it possible to reduce traffic jams, reduce exhaust emissions, and increase the safety of both drivers and pedestrians. With the support of IoT technology, traffic management systems in Barcelona are becoming increasingly intelligent and sustainable, supporting the development of ecological mobility in the city and improving the quality of life of residents and visitors.

3. Air quality monitoring is crucial to sustainable transportation and environmental protection in Copenhagen, Denmark. Thanks to advanced IoT technologies, the city can monitor air quality in real time, allowing it to track pollution levels and take action to ensure better air quality for residents. IoT systems in Copenhagen include networks of sensors located in various parts of the city that monitor the levels of substances such as nitrogen dioxide (NO_2), PM2.5, PM10, and other pollutants. The collected data is transmitted and analyzed in real time, which allows city authorities to take immediate action if air quality standards are exceeded. By monitoring air quality using IoT, Copenhagen can track the impact of traffic, industry, and other factors on air quality and make policy decisions to reduce pollution.
4. Additionally, residents have access to information on air quality through mobile applications or websites, which allows them to make informed decisions about their daily activities. Thanks to its involvement in air quality monitoring using IoT, Copenhagen aims to improve the quality of life of its inhabitants, reduce emissions of harmful substances, and promote the city's sustainable development. This initiative is crucial to protecting the environment and public health.
5. Cycling systems are essential to sustainable urban transport in Paris, France. Thanks to IoT technology and advanced management systems, Paris promotes using bicycles as an ecological means of transport, which helps reduce exhaust emissions and street congestion. IoT systems in Paris include extensive networks of bicycle stations that enable easy rental and return of bicycles. Using mobile applications or contactless cards, residents and tourists can use city bikes conveniently, which helps develop the culture of using low-emission means of transport. Thanks to IoT technology, bicycle systems in Paris allow monitoring of the availability of bicycles and the technical condition of the fleet, as well as analyzing bicycle use patterns. IoT enables optimal management of resources, arrangement of bicycle stations, and adaptation of the offer to user preferences. Integrated cycling systems in Paris allow for easy integration with other public transport options, enabling residents to switch seamlessly between different modes of transport. Thanks to this, city bikes become an attractive alternative to cars and contribute to improving air quality and reducing traffic jams in the city. Paris is actively investing in developing cycling systems using IoT, which helps promote sustainable urban transport and create a friendlier and greener environment for residents and visitors.

6. In Singapore, the intelligent public transport system (ITS) is an advanced public transport management system that uses IoT technology to optimize travel efficiency, safety, and comfort. Singapore is a world leader in using technology in public transport, which helps improve the traveler experience and reduce travel times. ITS in Singapore includes many advanced solutions such as traffic monitoring systems, intelligent traveler information systems, cashless payment systems, and optimization of vehicle schedules and routes. By integrating these IoT-based systems, Singapore can ensure more efficient and eco-friendly transportation of people and goods. IoT systems in Singapore allow for monitoring road traffic, managing traffic lights, informing passengers about schedules and delays, and analyzing transport data. IoT allows for quick response to changes in road conditions, route optimization, and increased traffic flow.

Additionally, cashless payment systems enable easy use of various means of public transport. Thanks to advanced ITS systems based on IoT technology, Singapore can maintain its position as a place of innovative transport solutions, which contributes to reducing street congestion and exhaust emissions and improving residents' quality of life. This is a perfect example of using modern technology to create an intelligent, sustainable, and effective transport system in a large metropolis.

Summarizing the above considerations, it is clearly visible that IoT is widely implemented in Sustainable Transportation. It can be assumed that with the emergence of new solutions in the field of IoT and Sustainable Transportation, both of these concepts will be inextricably linked over time.

REFERENCES

Ashton, K. (2010). That 'Internet of Things' Thing. In the Real World, Things Matter More Than Ideas. *RFID Journal*, 6, 1. http://www.rfidjournal.com/articles/pdf?4986, access on 20/01/2024.

Assessment & Decision Making for Sustainable Transport. European Conference of Ministers of Transport, OECD 2004, p. 17.

Auerbach, P., & Kukla, W. (2013). The Essence and Principles of Eco-driving, i.e., an Integral Part of Training in Enterprises Providing Transport Services. *Scientific Journals of the Poznań University of Technology. Organization and Management*, 60, 5–19.

Bajdor, P., & Grabara, J. (2013). The Transport Infrastructure Investment's Impact on the Level of Sustainable Transport Indicators in the Czestochowa City. *Applied Mechanics and Materials*, 309, 316–323.

Bhambri, P., Gupta, S., & Bhandari, A. (2005). Soft Computing Techniques. In UGC Sponsored National Conference (Abstract Book) (p. 100). Presented at the "Emerging Computing Technologies" conference at HRMMV, Jalandhar held during September 29–30, 2005.

Bhambri, P., & Khang, A. (2024). The Human-Machine Nexus With Art-Making Generative AIs. In S. Hai-Jew (Ed.), *Making Art With Generative AI Tools* (pp. 73–85). IGI Global. https://doi.org/10.4018/979-8-3693-1950-5.ch004.

Brachman, A. (2013). Internet of Things. Report of the ICT Observatory, Science and Technology Park "Technopark Gliwice", Gliwice.

Brzustewicz, P. (2012). Sustainable Solutions in Urban Transport - Directions of Development. *Acta Universitatis Nicolai Copernici*, 413, 86–87.

Chaari, H., & Ballot, E. (2012). *Fuel Consumption Assessment in Delivery Tours to Develop Eco-driving Behavior.* London: Association for European Transport.

Choroś, P. (2015). Using Business Analytics in the Internet of Things. In M. Grodner, W. Kokot, P. Kolenda, K. Krejtz, A. Legoń, P. Rytel, & R. Wierzbiński (Eds.), *Internet of Things in Poland* (pp. 4–7). Warsaw: IAB Polska.

Defining Sustainable Transportation, The Center for Sustainable Transportation. http://cst.uwinnipeg.ca/documents/Defining_Sustainable_2005.pdf, access on 22/02/2024.

Duarte, F. (2023). Number of IoT Devices. https://explodingtopics.com/blog/number-of-iot-devices, access on 2/02/2024.

Ecodriving. A Guide for Fleet Managers. ZenDriving Foundation, 2016.

Evans, D. (2013). Beyond Things: The Internet of Everything, Explained in Four Dimensions. *The Huffington Post*, September 24. http://www.huffingtonpost.com/dave-evans/cisco-beyond-things-the-interne_b_3976104.html, access on 10/02/2024.

Guidelines Towards Environmentally Sustainable Transport. Paris: OECD, 2002.

Korosec, K. (2022). Ford, VW-backed Argo AI is Shutting Down. https://techcrunch.com/2022/10/26/ford-vw-backed-argo-ai-is-shutting-down/?guccounter=1&guce_referrer= aHR0cHM6Ly93d3cuZ29vZ2xlLmNvbS8&guce_referrer_sig=AQAAAHqbRNs3NRflwSjhxzXxom7IWsiKP5xZPXsujwG2q5Hww6VDgmAI140QZAaYsNmpPR6cErADu1tTEl3eI_6hSUiusFFqskDCvl- 0i4LzffD7uQkIOe_nt-IPbBxa8Fyg8YGDggO-fDG131H2yli8JwHQwulYmAoKWWIg1ENgZ-CU, access on 15/01/2024.

Lipski, J. (2015). Internet of Things as Applied to Production Control. In R. Knosala (Ed.), *Innovations in Management and Production Engineering*, vol. 2. Opole: Polish Society of Production Management.

May, I. (2015). Internet of Things and the Threats Associated with it. *Security*, 3, 51–58.

McKinsey & Company: McKinsey Global Institute, 2015, The Internet of Things: Mapping The Value Beyond The Hype. http://www.mckinsey.com/business-functions/digital-mckinsey/our-insights/the-internet-of-things-the-value-of-digitizing-the-physical-world, access on 30/01/2024.

Middleton, P., Kjeldsen, P., & Tully, J. (2013). Forecast: The Internet of Things, Worldwide 2013. Gartner. http://www.gartner.com/doc/2625419/forecast-internet-things-worldwide, access on 19/02/2024.

Miłaszewicz, D., & Ostapowicz, B.. (2011). Conditions of Sustainable Transport in Light EU Documents. *Studies and Works of the Faculty of Economic Sciences and Management*, 4, 103–118.

Motowidlak, U. (2017). Transport Development and the Sustainable Development Paradigm, StudiaEkonomiczne. *Scientific Journals of the University of Economics in Katowice*, 337, 138–152.

Podolski, P., & Szafulska, M. (2017). Eco-driving and Zen-driving in Transport as an Element of Modern Logistics. In R. Rogaczewski, Sz. Zimniewicz, & A. Zimny (Eds.), *Transport and Logistics in the Enterprise, City and Region* (pp. 19–30). Selected issues, Sophia Scientific Publishing House.

Rana, R., & Bhambri, P. (2024). Artificial Intelligence and Internet of Things (IoT) Facilitated Digital Twin for Industry 4.0 Application Domains. In S. Rani, P. Bhambri, S. Kumar, P. K. Pareek, & A. A. Elngar (Eds.), *AI-Driven Digital Twin and Industry 4.0: A Conceptual Framework with Applications* (1st ed., pp. 168–181). CRC Press. https://doi.org/10.1201/9781003395416.

Rattan, M., Bhambri, P., & Shaifali. (2005). Institution for a Sustainable Civilization: Negotiating Change in a Technological Culture. In National Conference (Abstract Book) (p. 7). Presented at the "Technical Education in Globalized Environment-Knowledge, Technology & The Teacher" conference at SBBSIET, Padhiana, Jalandhar held during February 25–26, 2005.

Rot, A., & Blaicke, B. (2016). Threats Resulting from Implementing the Internet of Things Concept. Recommendations for Organizations and Solution Providers. Economic IT. *Scientific Works of the Wrocław University of Economics*, 3(41), 76–91. https://doi.org/10.15611/ie.2016.3.07.

Rot, A., & Sobińska, M. (2013). IT Security Threats in Cloud Computing Sourcing Model. In M. Ganzha, L. Maciaszek, & M. Paprzycki (Eds.), *Proceedings of the 2013 Federated Conference on Computer Science and Information*. Kraków: PTI.

Russ, A. (2023). UK Transport System and How It Compares With the USA. https://wanderwisdom.com/transportation/uk-public-transport, access on 9/02/2024.

Satish, G. N., & Varma, P. S. (2017). Internet of Things – Opportunities, Applications, and Challenges in the Prospective Smart World. *International Journal of Computer Science and Information Technologies*, 4(3), 8–16.

Sharma, R., Bhambri, P., & Sohal, A. K. (2020). Energy Bio-inspired for MANET. *International Journal of Recent Technology and Engineering*, 8(6), 5580–5585. https://www.ijrte.org/wp-content/uploads/papers/v8i6/F8522038620.pdf. https://doi.org/10.35940/IJRTE.F8522.038620.

Singh, M., Singh, P., Kaur, K., & Bhambri, P. (2005). Database Security. In Proceedings of National Conference (pp. 35–41). Presented at the "Future Trends in Information Technology" conference at SJPMLIET, Radaur, Yamuna Nagar held during March 6–7, 2005.

Sivak, M., & Schoettle, B. (2011). *Eco-driving: Strategic, Tactical, and Operational Decisions of the Driver that Improve Vehicle Fuel Economy*. Ann Arbor: The University of Michigan. Transportation Research Institute. Of the driver that improves vehicle fuel economy.

Słowiński, P., Brudzik, R., Folęga, P., & Domin, J. (2016). Eco-driving – A New Approach to Transport. https://www.artur-zimny.pl/nauka/monografie/transport_logistyka_w_przedsiebiorstwie_miescie_regionie.pdf

Van de Burgwal, H. C., & Gense, N. L. (2002). Interruption of Driving Style Tips. TNO Automotive Report 02.OR.VM.004.1/HVD.

van der Meule, R. (2017). Gartner Says 8.4 Billion Connected "Things" Will Be in Use in 2017, Up 31 Percent From 2016. https:/doi.org/10.1017/CBO9781107415324.004.

17 Call to Action
Driving Digital Sustainability

Ilona Pawełoszek and Pankaj Bhambri

17.1 INTRODUCTION

In the era of fast technological progress, digital transformation is a crucial strategic priority for businesses. Debates about the digital economy often revolve around two main issues: the potential of information and communication technologies (ICTs) and their impact on sustainability.

Digital sustainability means activities that aim to mitigate the impact of information technology advancements on the natural environment. Leveraging digital technology responsibly can tackle several sustainability challenges. Digital sustainability refers to using technology in an environmentally responsible, socially equitable, and economically efficient way. Sustainable digital transformation is an essential aspect of achieving the Sustainable Development Goals (SDGs) established by the United Nations. The set of goals is a universal call to action aiming at ending poverty, protecting the planet, and ensuring that everyone enjoys peace and prosperity.

Change management is a key task in achieving digital sustainability. It involves preparations, support, and helping individuals and organizations to adapt to changes. As it is a multifaceted task, often meeting resistance, misunderstanding, and uncertainty, a good action plan and convincing incentives for implementation are needed. With digital transformation, the scale of change is often large and fast-paced, making it even more problematic and essential.

This chapter aims to explore the potential of digital technology in advancing Sustainable Development and to inspire individuals, businesses, and governments to use this potential to create a sustainable and inclusive future. We will also delve into the environmental challenges posed by the digital age, such as the high energy consumption of ICT infrastructure and the increasing problem of electronic waste and its disposal. By addressing these issues, we aim to highlight the importance of sustainable practices in the digital sector and encourage the development and adoption of greener technologies.

17.2 SUSTAINABLE DEVELOPMENT GOALS

The Sustainable Development Goals (SDGs) are a plan of action developed by the United Nations (UN) to achieve broadly understood sustainability around the world. The 2030 Agenda, including the SDGs, was adopted by all 193 UN member states on

 DOI: 10.1201/9781003484226-17

September 25, 2015, in New York (United Nations, 2015). It includes 17 sustainable development goals and 169 related tasks that correspond with the three dimensions of sustainable development: economic, social, and environmental. These goals focus on achievements in five areas, the so-called 5xP: people, planet, prosperity, peace, partnership (Figure 17.1).

The SDGs vary in the number of targets and indicators assigned to each goal. Each goal consists of a set of specific targets that address different aspects of the central theme. These targets are further broken down into indicators used to measure progress toward achieving the respective goal (Sharma et al., 2020). The Sustainable Development Goals framework does more than just outline the goals; it also provides a set of measures for implementing sustainable management. The SDGs are undoubtedly a global call to action for all developed, developing, and least developed countries to promote cooperation and partnership for sustainable development. The implementation of SDGs has a huge impact on the future of our planet. It aims to

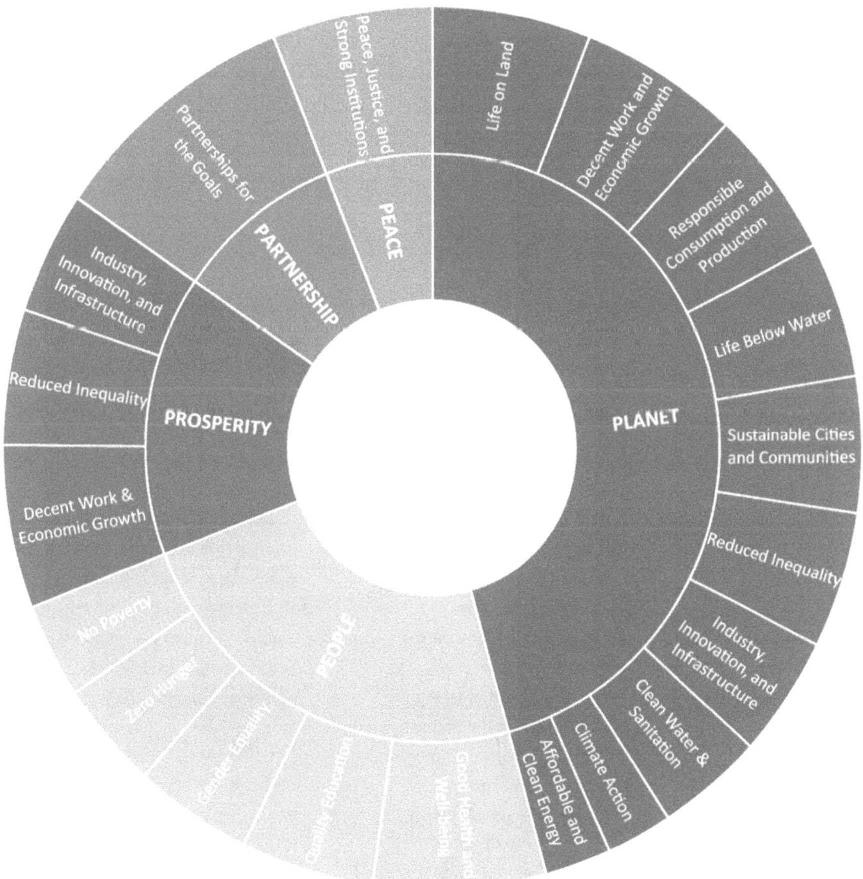

FIGURE 17.1 Sustainable development goals and areas.

eliminate poverty and hunger, ensure good quality education and health, achieve gender equality, and access to clean water and energy, as well as promote economic growth and decent work are necessary to create a better and more sustainable world for future generations.

17.3 DIGITAL TECHNOLOGIES AND SDGS

Information technology permeates all areas of the economy and individuals' lives. The importance of ICT is undeniable in every sector, where it is used to run daily operations, maintain competitive advantage, and innovate. The need for processing power is continuously growing with the proliferation of big data and artificial intelligence analytics.

According to the European Commission (2022), the ICT sector accounts for about 7% of global electricity consumption, and it is estimated to rise to 13% by 2030. This energy footprint currently represents 3–5% of global carbon emissions, which is comparable with the aviation sector. Sustainability has become a core aspect of global digital technology industry operations in recent years. ICT is related to advancing several SDGs outlined by the United Nations. Table 17.1 illustrates how digital tools can be harnessed to address specific challenges and contribute to achieving Sustainable Development Goals.

TABLE 17.1
Sustainable Development Goals and the Role of ICT

SDG	The Role of ICT	Example Digital Tools
SDG 4: Quality Education	Enhancing access to education, facilitating online learning, and providing educational resources to remote areas.	E-Learning platforms, Augmented and Virtual Reality, AI chatbots and search engines.
SDG 5: Gender Equality	Bridging the gender gap by providing women with access to information, education, and economic opportunities.	Mobile apps and web portals for women's empowerment. Providing information on health, education, rights, knowledge, and resources.
SDG 8: Decent Work and Economic Growth	Encouraging economic development by fostering innovation, improving productivity, and creating job opportunities in the technology sector.	Job matching online platforms that connect job seekers with suitable employment opportunities, optimizing the job market.
SDG 9: Industry, Innovation, and Infrastructure	Driving innovation and infrastructure development, contributing to the growth of industries, and enhancing overall economic progress.	Internet of Things (IoT) solutions that monitor and optimize infrastructure, improving efficiency, and reducing resource consumption.

(Continued)

TABLE 17.1 (CONTINUED)
Sustainable Development Goals and the Role of ICT

SDG	The Role of ICT	Example Digital Tools
SDG 10: Reduce Inequality within and among Countries	Enabling access to education, digital inclusion, and employment opportunities in ICT sector and e-commerce.	Digital literacy initiatives, mobile banking and payment services, telemedicine platforms, community Wi-Fi initiatives. E-government services.
SDG 11: Sustainable Cities and Communities	Enabling smart city initiatives, mastering urban planning, resource efficiency, and environmental protection in urban areas.	IoT sensors, data analytics, and AI to enhance urban planning, traffic management, and resource efficiency.
SDG 16: Peace, Justice, and Strong Institutions	Enhancing transparency, accountability, and access to justice through digital governance and information systems.	Digital governance platforms, e-services, Open Data portals, Blockchain technology.
SDG 17: Partnerships for the Goals	Facilitating global collaboration, information exchange, and partnerships, essential for achieving all SDGs.	Collaboration tools and platforms.

ICTs can potentially contribute to lasting, positive impacts on education, economic prosperity, social inclusion, and environmental protection. They can also assist nations in achieving the SDGs by 2030.

UNESCO's YouthMobile is an example of an initiative that aligns with the SDG framework by utilizing technology, specifically mobile phones. The initiative focuses on the African continent and acknowledges the unique challenges and opportunities in this region.

The aim is to empower young people with computer science competencies. The teaching program focuses on coding mobile applications to address local challenges and enhance learning skills. YouthMobile encourages students to develop, promote, and potentially sell mobile applications. This way, it contributes to fostering entrepreneurship and self-sustainability. The initiative is strongly related to Sustainable Development Goal 10, which focuses on reducing inequalities and promoting inclusive digital societies. YouthMobile also addresses gender equality concerns, ensuring equivalent access to education for girls and boys.

Continuous monitoring and evaluation are essential to enhancing the initiative's effectiveness (Rattan et al., 2005). This involves assessing the impact of the mobile applications developed by students, tracking the progress of gender inclusivity, and evaluating the socio-economic outcomes in the targeted regions. By incorporating feedback and lessons learned, YouthMobile can continuously evolve and adapt its strategies to maximize positive outcomes.

17.4 NEGATIVE IIMPACT OF ICT ON SUSTAINABILITY

In previous sections, we discussed how digital technologies tackle several environmental, social, and economic challenges. However, while digital technologies offer solutions to these challenges, they also pose certain risks and have a negative impact on sustainability. This is often referred to as the "dark side" of information and communication technology (ICT).

17.4.1 Production of Electronic Devices

The production of electronic devices fuels technological advancements in many industries and provides numerous conveniences The electronics manufacturing process involves using hazardous materials and consumes a significant amount of energy. The critical environmental concerns of electronic device production include resource extraction, energy consumption, and chemical pollution. The extraction of raw materials, such as copper, nickel, cobalt, or tin, contributes to habitat destruction, deforestation, and soil degradation. Mining activities associated with obtaining these materials often result in environmental degradation, disrupting ecosystems and biodiversity.

The energy-intensive nature of electronic device manufacturing further intensifies the negative environmental impact. From the raw materials extraction to the components assembly, the entire production process consumes substantial amounts of energy, predominantly derived from fossil fuels (Bhambri & Bhandari, 2006). This contributes to greenhouse gas emissions, causing climate change and its associated environmental challenges.

Chemical pollution is a significant issue in electronic device manufacturing. The production process involves the use of various toxic substances, including heavy metals, solvents, and flame retardants. Improper disposal of these chemicals poses risks to ecosystems and human health, as they can pollute air, water, and soil. Additionally, the release of the contaminants during manufacturing contributes to air pollution, affecting both the immediate vicinity and regions farther away.

Addressing these environmental challenges necessitates a comprehensive approach that incorporates sustainable practices throughout the electronic device production lifecycle. From a management perspective, adopting circular economy principles can promote the reuse and recycling of materials, reducing the demand for virgin resources. Sustainable design practices, focusing on product longevity, reparability, and recyclability, are crucial for minimizing the environmental footprint of electronic devices.

Business strategies should prioritize eco-friendly manufacturing processes, incorporating renewable energy sources, and minimizing waste generation. Implementing extended producer responsibility (EPR) policies can hold manufacturers responsible for the end-of-life disposal of their products, encouraging sustainable practices throughout the supply chain.

Information technology and artificial intelligence can be leveraged to optimize manufacturing processes, reduce energy consumption, and develop innovative solutions for waste reduction. The integration of smart technologies in production facilities can enhance efficiency and minimize environmental impact.

The production of electronic devices poses significant environmental challenges, from resource extraction to chemical pollution (Bhambri et al., 2005). Addressing these issues requires a collaborative effort of authorities, international organizations, and entrepreneurs. Incorporating sustainable practices in design, manufacturing, end-of-life disposal, and leveraging technology for eco-friendly innovations, we can work towards mitigating the environmental impact of electronic device production.

17.4.2 E-waste Consequences

Electronic waste (e-waste) has become a pressing concern from the sustainability perspective, drawing attention from researchers, policymakers, and environmentalists alike. With the proliferation of electronic devices and their rapid technological advancements, electronic product lifespans are getting shorter. Consumers increasingly replace their electronic devices with newer models. Consumer demand for the latest features and functionalities drives manufacturers to continually release new models, further accelerating the quick obsolescence of mobile devices. Discarded computers and mobile phones constitute a significant share to the growing e-waste problem, necessitating sustainable strategies for recycling and responsible disposal. This situation has led to an unprecedented surge in e-waste generation. The scope of the problem is underscored by the fact that e-waste is increasing three times faster than the world's population (Forti et al., 2020). Electronic equipment manufacturers face a difficult trade-off between increasing economic efficiency and sustainable operation. Equipment aging promotes the vendors' prosperity and ensures continuous demand for products. Nothing seems to stop consumers from purchasing new electronic devices even though the old ones still work. In this situation, strategies for disposing of unused devices and recovering raw materials for further production become particularly important.

The improper disposal and management of electronic products contribute significantly to environmental degradation and health risks (Singh & Bhambri, 2023). Addressing this issue requires a holistic approach that integrates insights from management, business strategies, and technological advancements.

From a management perspective, sustainable practices can be integrated into the product lifecycle. Companies should adopt circular economy principles, focused on product design for durability, repairability, and recyclability. In the business context, innovative models such as product-as-a-service and leasing arrangements can incentivize manufacturers to create long-lasting products. Additionally, establishing take-back programs and incentivizing consumers to return old electronics can contribute to the benefit of the ecosystem.

On the other hand, e-waste management processes can be significantly supported by Information technology and artificial intelligence (Bhambri, 2024). Intelligent systems can track the entire lifecycle of electronic products, facilitating efficient recycling and recovery of valuable materials. Blockchain technology, for instance, can enhance transparency and traceability in the disposal and recycling chain.

The World Health Organization (WHO) highlights the critical issue of electronic waste (e-waste) in a publication dated October 18, 2023. Key facts presented include the alarming growth of e-waste as the fastest-growing solid waste stream globally, with an estimated 53.6 million tons produced in 2019. However, only 17.4% of this was officially collected and recycled. Lead, a common substance in e-waste, poses an environmental hazard when recycled using inferior methods such as open burning.

Despite its hazardous nature, e-waste contains valuable resources, such as precious and semi-precious metals. Other useful components can be re-used, such as connectors, LCD screens, lenses, microphones, speakers, and screws. Recovering raw materials and components from electronics is an important income stream for individuals and communities, especially in low- and middle-income countries (LMICs). However, people face significant risks due to inadequate regulations, poor recycling infrastructure, and poor training. Exposure to e-waste occurs through various ill-considered practices such as scavenging, dumping, burning, and manual disassembly. These activities release toxic substances, impacting air, soil, and water, posing public health threats. Children are particularly vulnerable, engaging in hazardous activities like waste picking and burning. The potential health effects include neurodevelopmental issues and respiratory problems.

Prevention and management strategies advocated by WHO include adopting and enforcing high-level international agreements, implementing national e-waste management legislation, monitoring e-waste sites, and prohibiting child labor. The Basel Convention, controlling the transboundary movement of hazardous waste, plays a crucial role, supported by regional conventions like the Bamako and Waigani Conventions.

WHO's response involves the Initiative on E-waste and Child Health, contributing to international programs and pilot projects to protect children's health from e-waste exposures. The initiative aims to increase awareness of the health impacts of e-waste, enhance health sector readiness, and facilitate monitoring and interventions (ILO & UNICEF, 2021). Moreover, electronic waste is a growing concern due to the rapid obsolescence of devices and inadequate recycling infrastructures.

17.4.3 Power Consumption

Another critical issue is the energy consumption of data centers, which power the internet and cloud computing. These facilities, working 24/7, require vast amounts of electricity, not only for running servers but also for cooling systems to prevent overheating. This high energy demand contributes to greenhouse gas emissions unless powered by renewable energy sources.

Several studies in the scientific literature delve into the power consumption patterns of various ICT components (Gelenbe, 2023; Khomh, Abtahizadeh, 2018; Pothitou, Hanna, & Chalvatzis, 2017). It is widely recognized that data centers, which form the backbone of digital services, are substantial contributors to energy consumption in the ICT sector. The constant need for processing power, cooling systems, and data storage drives the energy requirements of these facilities. Figure 17.2 presents the average power consumption of the data center components.

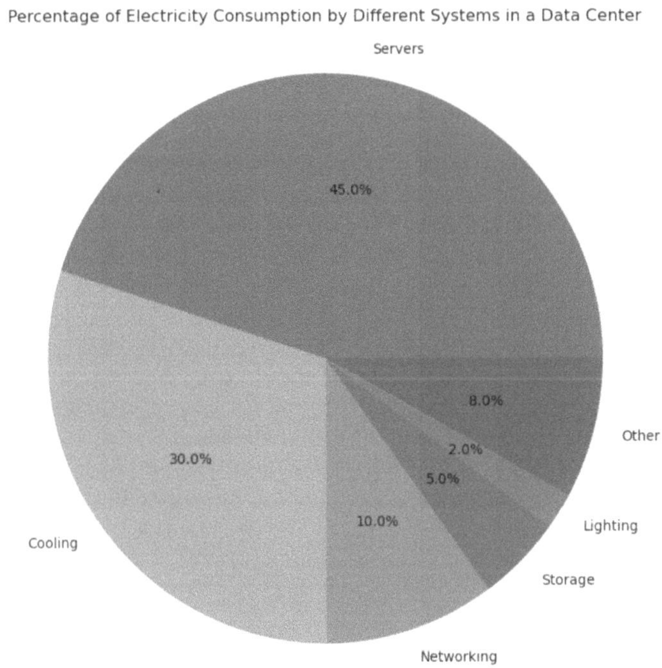

FIGURE 17.2 Indicative share of energy consumption by different subsystems in a data center.

Artificial intelligence plays a dual role in this scenario. On one hand, AI technologies can optimize energy usage in data centers through intelligent management and resource allocation. On the other hand, the training and running of complex AI models often demands substantial computational power, contributing to the overall energy footprint.

From a managerial and business perspective, organizations are under increasing pressure to reduce carbon dioxide emissions by adopting sustainable practices. Consequently, there is a growing focus on developing energy-efficient ICT solutions and implementing green IT strategies (Jayaprakash, Nagarajan, & Prado, 2021). The adoption of renewable energy sources for powering data centers, along with advancements in energy-efficient hardware design, represents a business opportunity for companies aiming to align with sustainability goals.

Mitigating power consumption of IT infrastructure is, first of all, a pressing challenge for ICT infrastructure providers and software companies. The tasks of these two crucial players in the ICT market in terms of improving the energy consumption of data centers are shown in Table 17.2.

From a managerial and business standpoint, adopting these practices aligns with the growing societal emphasis on corporate responsibility and environmental sustainability. Implementation of sustainable ICT practices also represents a strategic advantage in the competitive digital landscape.

TABLE 17.2
The Contribution of ICT Infrastructure Providers and Software Developing Companies to a More Sustainable and Energy-efficient Digital Ecosystem

ICT Infrastructure Providers	Software Developers
Data Center Optimization: Advanced cooling technologies, such as liquid cooling systems; virtualization and consolidation techniques.	**Optimized Code and Algorithms:** Energy-efficient code and algorithms, reducing the computational load, leading to lower power requirements for software applications.
Renewable Energy Integration Shift to renewable energy sources, such as solar, wind power, and hydrogen fuel cells.	**Cloud Resource Management:** Intelligent resource allocation and workload distribution in cloud environments. Powering down during periods of low demand.
Energy-Efficient Hardware: Energy-efficient hardware, including processors, storage devices, and networking equipment.	**Edge Computing:** Processing data closer to the source minimizes the need for data to travel long distances, reducing overall energy consumption.
AI for Energy Management: Artificial Intelligence (AI) algorithms dynamically adjusting resources based on demand patterns and workload fluctuations.	
Lifecycle Assessment: Assessment of ICT equipment to identify energy-intensive phases, enabling targeted improvements in manufacturing, usage, and disposal.	**Green Software Development Practices:** Optimizing software for energy efficiency during the coding phase and incorporating green metrics into the software development life cycle.
Collaboration: Aligning software development with energy-efficient hardware, ensuring that software is optimized for the underlying hardware architecture.	

17.4.4 Social Issues of Digital Sustainability

Digital sustainability could not be achieved without careful consideration of the social implications of advancing information technologies. In this section, we will examine the roots, implications, and mitigation strategies of the digital divide, privacy and ethical concerns, job displacement, and the cultural impact of digital technologies.

17.4.4.1 Digital Divide

The digital divide refers to the gap between those with access to information and communication technologies (ICT) and those without or limited access. It also relates to the difference in computer literacy hindering some groups from using the benefits of digital transformation. As the research shows, the economic barrier is one of the most significant hurdles in bridging the digital divide. An excellent comparative indicator is the Network Readiness Index (NRI), which is published annually by the Portland Institute and World Economic Forum (2023). The NRI is a framework used to assess the readiness of countries to exploit information and communication technologies for social and economic transformation. The NRI is a compound measure calculated using a combination of quantitative and qualitative data on infrastructure, regulatory environment, business and innovation capability, and social and individual usage. These pillars are further divided into subcomponents and each subcomponent is assigned a score. The overall NRI score is then derived from aggregating these subscores. Figure 17.3 presents the NRI ranking.

The digital gap manifests in various forms such as:

- Lack of affordable or reliable internet connectivity, particularly in rural or underprivileged areas.
- Difficulty in affording devices like computers or smartphones.
- Insufficient skills and knowledge to use technology effectively.

The poor availability of digital technology in certain countries can be attributed to many interconnected factors. Economic disparities play a crucial role. Lower-income nations such as African countries often encounter challenges in investing adequately in ICT infrastructure. Inadequate educational systems and a lack of digital literacy programs hinder people's ability to effectively engage with technology (Shanmuga & Bhambri, 2024). Additionally, regulatory and political environments can impact the development and accessibility of digital infrastructure. Geographical factors, such as remote or rural locations, may pose logistical challenges for establishing robust connectivity. Sociocultural factors and societal inequalities can further exacerbate the digital divide, as specific demographics may face barriers in adopting and utilizing digital tools. Bridging this gap requires a holistic approach that addresses economic, educational, regulatory, and social dimensions to ensure equitable access to digital technology on a global scale.

Foreign investments in ICT are now a powerful tool for development in poor countries. However, they present both potential benefits and drawbacks. The obvious benefits are:

- Possibility to build essential ICT infrastructure like broadband networks, mobile towers, and data centers, which are crucial for economic and social growth.
- Creating jobs in ICT sectors like software development, call centers, and IT support, reducing unemployment.
- Bringing expertise and advanced technologies and knowledge transfer.
- Facilitating access to education, healthcare, financial, and e-government services.
- Improving the efficiency and competitiveness of local businesses with possibility to gain new markets.

The dark sides of foreign investment should also be noted, which include:

- Unequal distribution of benefits widening the technological gap.
- Lack of strong legal regulations, leading to unfair contracts or unsustainable resource extraction by foreign companies.

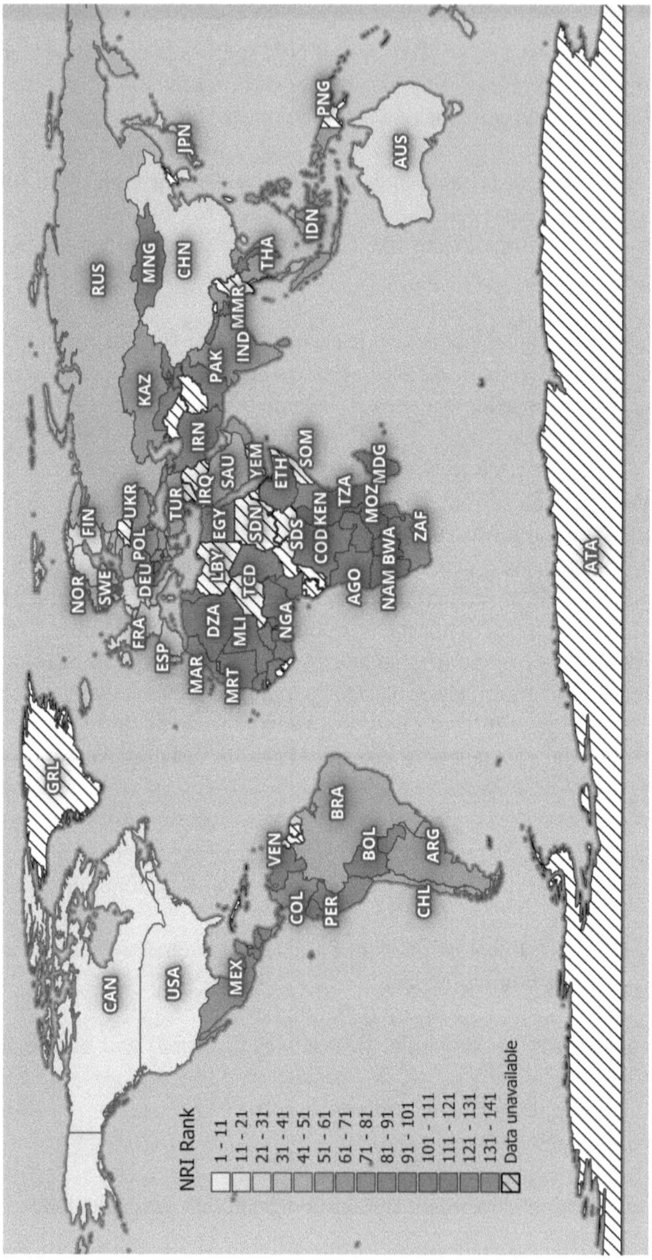

FIGURE 17.3 The Network Readiness Index 2023 ranking by country.

- Job losses in certain sectors like agriculture and manufacturing due to automation brought by new technologies. The need for further investment in workforce retraining and adaptation.
- Concerns about privacy protection, especially in countries with weak data protection laws.
- Dependence of crucial services and infrastructures on foreign entities, limiting local control of the ICT sector.

While ICT investment across Africa is rising, it's not evenly distributed among countries. The leading investment destinations are South Africa, Egypt, Nigeria, Kenya, and Morocco. As we can see on the map presented earlier in Figure 17.3, these countries are higher in the Network Readiness Index than smaller African economies such as the Central African Republic, Eritrea, Comoros, Somalia, or South Sudan.

Analyzing the factors encouraging foreign investors, it is easy to conclude that the favorable factors are an established economy, stable political environment, large population, active government attitude and support, and geographical proximity to Europe.

17.4.4.2 Privacy and Ethical Concerns

Privacy and ethics of data use are complex and evolving issues of sustainable digital transformation. The rapid advancement of technology outpaces the development of ethical standards (Dhirani et al., 2023) and legal regulations. This is particularly relevant in the context of Industry 5.0, where emerging technologies such as autonomous robots and cloud data analytics raise significant ethical dilemmas (Dhirani et al., 2023). Using these technologies in digital transformation also presents challenges in preserving human rights and ensuring development for all (Klein, 2022). Therefore, successfully adopting new technologies is contingent on maintaining privacy and security, which requires a combination of technical measures, organizational culture, and leadership commitment (Jonathan et al., 2020).

Take artificial intelligence, for instance. There are numerous examples where AI algorithms have demonstrated biased behavior, reflecting the biases present in the data they were trained on.

A good example is an issue of facial recognition algorithms that overrepresent white people and may create errors when recognizing people of color (Perkowitz, 2021).

Another disturbing phenomenon is bias in recruitment tools, which are more and more often supported by AI algorithms. On one hand, these tools are designed to reduce the human bias that has long plagued recruitment. On the other hand, if the training data is labeled inconsistently or reflects existing biases in society, the AI model can spread and even amplify those biases. Especially if historically marginalized groups are underrepresented in the training data, the AI model may not accurately assess the candidate's qualifications, potentially excluding qualified people from consideration.

The advancements in virtual and augmented reality have led to a recent proliferation of virtual realms. Virtual space is a new arena for business, work, entertainment, and social relations; it is also a new, uncontrolled area conducive to criminal activities. Research on crimes in virtual space has highlighted the diverse range of malicious activities that can occur in this environment. Sudzina and Pavlicek (2020) emphasize the presence of criminal behavior in virtual worlds such as abuse, stalking, and intimidation. Dremliuga, Prisekina, and Yakovenko (2020) explore the unique properties of crimes in virtual reality, including the potential for new types of offenses. The mentioned studies underscore the need for a comprehensive legal framework to address crimes in virtual space. The interdisciplinary collaboration between technology, law, and ethics is indispensable to ensure a harmonious development that benefits society.

In recent years, Artificial Intelligence (AI) has become increasingly popular in marketing (Pathak & Sharma, 2022). AI algorithms profile consumers based on their purchase history, demographic data, economic trends, and social media activity (Nalini et al., 2021). The incredible development of natural language processing (NLP) and generative artificial intelligence (GenAI) has made it possible to entrust customer service tasks to intelligent chatbots (Hildebrand & Bergner, 2019). This solution is beneficial for enterprises because it saves the labor costs of customer consultants and increases sales. However, several problems arise here. The first and most apparent is the advanced processing of customer data, which may lead to the disclosure of private preferences and sensitive data without informed consent.

Another problem is the so-called "sleazy salesman" phenomenon. There is the potential for AI to be misused in marketing for unethical sales tactics that manipulate or exploit customers for business gain. AI can generate fake reviews, endorsements, or personalized messages that mislead customers (Houde et al., 2020). GenAI algorithms have access to immeasurable resources of information and knowledge on manipulating the customer. NLP algorithms can also indistinguishably imitate human conversation. This, combined with the customer profile, gives great opportunities to tailor highly personalized messages to exploit individual vulnerabilities or biases for sales.

From a sustainability perspective, two problems arise here. The first is companies' unethical behavior to increase profits. The second is generating excessive purchases. Companies play on customers' emotions, so they often buy things that are too expensive or unnecessary. This leads to excessive consumption, the direct effect of which is the carbon footprint and, later in the product lifecycle, the accumulation of waste.

There are several strategies that can help mitigate the impact of the "sleazy salesman" problem. Ethical guidelines for implementing AI in marketing, as well as corresponding legal regulations, are essential. Given the novelty of this issue, information campaigns are necessary to enhance transparency and awareness among both business owners and consumers. Additionally, tools are required to empower users with greater control over their data and its utilization, ensuring accessibility and ease of implementation. Maintaining human oversight and accountability for AI training and decision-making processes is crucial.

17.4.4.3 Job Displacement and Employment Shifts

Automation and artificial intelligence, which are now regarded as crucial concepts of Industry 4.0 and 5.0, may lead to job displacement in specific industries, contributing to unemployment and economic instability (Karangutkar, 2023). The unavoidable shift in the labor market presents multifaceted challenges from a sustainability perspective.

The shift towards automation and AI-driven technologies may disproportionately affect certain segments of the population. Numerous jobs across various sectors are vulnerable to displacement. These are, first of all, routine and repetitive tasks, such as those in manufacturing, customer service, and data entry. For example, the advent of autonomous vehicles may threaten jobs in transportation, while advancements in machine learning could automate specific tasks in finance and accounting. However, the infiltration of generative artificial intelligence into creative domains such as graphic design, music composition, and acting raises significant questions about the future of these professions. Traditionally considered highly creative endeavors relied solely on human ingenuity and emotion. Advancements in AI have enabled the development of generative systems capable of producing artwork, music, and even performances that are as good as those created by humans. Displaced workers may face difficulties in transitioning to new roles, potentially leading to income inequality and social unrest.

The digital divide, described in Section 17.4.4.1, further exacerbates these challenges. The problem is not solely about access to technology but also encompasses disparities in digital literacy and IT skills. Those lacking access to ICT may face increased vulnerability to job displacement and limited opportunities for reskilling or upskilling. Addressing the challenges of job displacement caused by automation and AI needs collaborative effort from many stakeholders. Figure 17.4 presents the mitigation strategies for employment in the digital era involving many stakeholders such as government and local authorities, academia, and business.

The need for constant upskilling and reskilling to keep pace with technological changes seems to be the right strategy for sustainable future labor market. The benefits of this strategy are two-fold, the demand for lifelong learning and continuous professional development creates new jobs at each level of the education system. Mastering educational content that encourages a mindset of continuous professional development can empower individuals to navigate continuously evolving job markets. Collaboration between educational institutions, industries, and governments is vital to establish robust frameworks for upskilling initiatives. These collaborations can ensure that the skills taught align with industry needs and that individuals can seamlessly transition into new roles. In spite of the mentioned challenges, many researchers present an optimistic vision where potential new employment opportunities in the field of AI and machine learning outweigh the potential negative impacts (Tiwari, 2023; Badet, 2021).

Addressing issues related to job displacement involves not only education systems and technological solutions but also ethical frameworks that prioritize human well-being. Focusing on human welfare and the meaningfulness of work instead of solely maximizing profit facilitates transitioning towards a sustainable economy. Implementing these strategies requires significant financial and political resources. This is a main barrier for low-developed countries and those suffering from internal conflicts and unstable political situations.

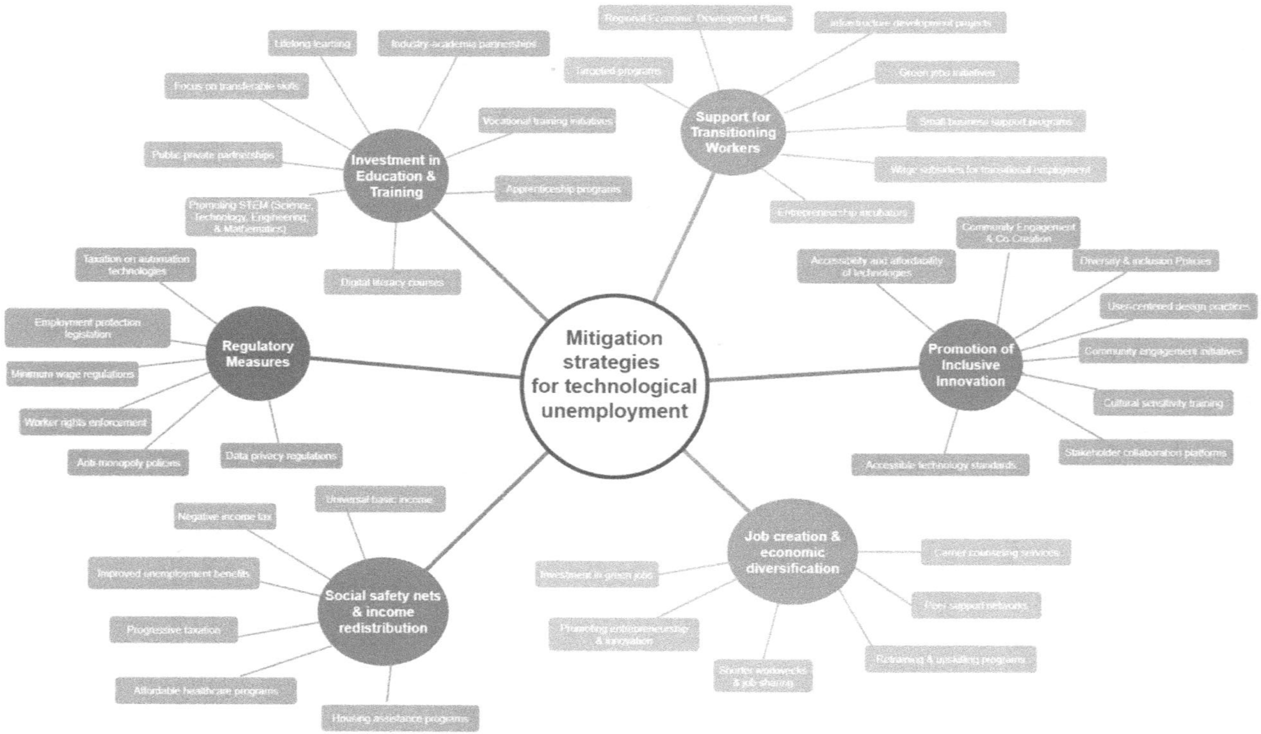

FIGURE 17.4 Mitigation strategies for technological unemployment.

17.5 CONCLUSIONS

In this chapter, we emphasized the urgent need for proactive strategies and measures to connect digital technologies' proliferation with sustainable development's imperatives. As we explained throughout the chapter, the Sustainable Development Goals serve as a vital outline for guiding efforts to harness digital innovations for the benefit of global sustainability objectives. Despite the huge potential of digital technologies in advancing the SDGs, it is imperative to acknowledge and address their negative impacts on sustainability. The production and disposal of electronic devices contribute to environmental degradation and exacerbate the e-waste problem, necessitating robust strategies for responsible manufacturing and recycling. Moreover, the significant energy consumption associated with digital infrastructure underscores the importance of prioritizing energy efficiency and renewable energy sources in ICT deployment.

Furthermore, we highlighted the social dimensions of digital sustainability, including concerns surrounding the digital divide, privacy infringements, ethical considerations, and the displacement of traditional employment structures. These issues underscore the need for inclusive and equitable approaches to digital development that prioritize human rights, cultural preservation, and the empowerment of marginalized communities.

In light of these discussed challenges, the chapter issues a compelling call to action for policymakers, industry stakeholders, civil society organizations, and individuals alike. Collaboration across sectors is essential to develop and implement policies, regulations, and initiatives that promote the responsible use of digital technologies while mitigating their adverse impacts on sustainability. Moreover, fostering digital literacy, promoting ethical design principles, and fostering dialogue on the societal implications of digitalization are essential steps toward achieving digital sustainability.

This chapter is just an introduction to developing future research directions on digital sustainability, which is constantly changing, thanks to the rapid progress of technologies such as big data and artificial intelligence. We are of the opinion that the pursuit of digital sustainability is not only a moral duty but also a strategic necessity for safeguarding the well-being of present and future generations. By embracing innovation while remaining alert to its potential pitfalls, society can harness the power of digital technologies as a force for positive change and inclusive development.

REFERENCES

Badet, J. (2021). AI, automation and new jobs. *Open Journal of Business and Management*, 9, 2452–2463. https://doi.org/10.4236/ojbm.2021.95132

Bhambri, P. (2024). NFTs: Transforming digital ownership in the web 3 era. In D. Darwish (Ed.), *Decentralizing the Online Experience With Web3 Technologies* (pp. 151–167). IGI Global. https://doi.org/10.4018/979-8-3693-1532-3.ch007

Bhambri, P., & Bhandari, A. (2006). Different protocols for wireless security. In National Conference on Advancements in Modeling and Simulation at LLRIET, Moga held during January 20–21, 2006.

Bhambri, P., Singh, I., & Gupta, S. (2005). Robotics systems. In UGC Sponsored National Conference (Abstract Book) (p. 107). Presented at the "Emerging Computing Technologies" conference at HRMMV, Jalandhar held during September 29–30, 2005.

Dhirani, L. L., Mukhtiar, N., Chowdhry, B. S., & Newe, T. (2023). Ethical dilemmas and privacy issues in emerging technologies: A review. *Sensors*, 23(3), 1151. https://doi.org/10.3390/s23031151

Dremliuga, R., Prisekina, N., & Yakovenko, A. (2020). New properties of crimes in virtual environments. *Law, Computer Science*, 5(6), 1727–1733. https://doi.org/10.25046/aj0506206

European Commission. (2022, October 18). Accompanying the document COMM (2022) 341 final Commission Staff Working Document. https://eur-lex.europa.eu/legal-content/EN/TXT/?uri=SWD:2022:341:FIN

Forti, V., Balde, C. P., Kuehr, R., & Bel, G. (2020). The global E-waste monitor 2020. United Nations University, International Telecommunication Union, International Solid Waste Association.

Gelenbe, E. (2023). Electricity consumption by ICT: Facts, trends, and measurements. In Proceedings of the 2023 International Conference on Computer Systems and Information Technology (CSIT) (pp. 321–328). ACM.

Hildebrand, C., & Bergner, A. (2019). AI-Driven sales automation: Using chatbots to boost sales. *NIM Marketing Intelligence Review*, 11(2), 36–41. https://doi.org/10.2478/nimmir-2019-0014

Houde, S., Liao, V., Martino, J., Muller, M., Piorkowski, D., Richards, J., Weisz, J., & Zhang, Y. (2020). Business (mis)Use cases of generative AI. In W. Geyer, Y. Khazaeni, & M. Shmueli-Scheuer (Eds.), *Joint Proceedings of the Workshops on Human-AI Co-Creation with Generative Models and User-Aware Conversational Agents co-located with 25th International Conference on Intelligent User Interfaces* (IUI 2020). CEUR. https://doi.org/10.48550/arXiv.2003.076

International Labour Office & UNICEF. (2021). Child labour: Global estimates 2020, trends and the road forward. International Labour Office and United Nations Children's Fund.

Jayaprakash, S., Nagarajan, M. D., Prado, R. P. D., Subramanian, S., & Divakarachari, P. B. (2021). A systematic review of energy management strategies for resource allocation in the cloud: Clustering, optimization and machine learning. *Energies*, 14, 5322. https://doi.org/10.3390/en14175322

Jonathan, G. M., Gebremeskel, B. K., & Yalew, S. D. (2020, November). Privacy and security in the digitalisation era. In 2020 11th IEEE Annual Information Technology, Electronics and Mobile Communication Conference (IEMCON) (pp. 0837–0844). IEEE.

Karanagutkar, A. A. (2023). The impact of artificial intelligence on job displacement and the future of work. *International Journal of Advanced Research in Science, Communication and Technology*, 3(1), 635–638. Publisher: Naksh Solutions. https://doi.org/10.48175/ijarsct-12096

Khomh, F., & Abtahizadeh, S. A. (2018). Understanding the impact of Cloud patterns on performance and energy consumption. *Journal of Systems and Software*, 141, 151–170.

Klein, A. Z. (2022). Ethical issues of digital transformation. *Organizações & Sociedade*, 29(102), 443–448. https://doi.org/10.1590/1984-92302022v29n0020en

Nalini, K., Radhakrishnan, P., Yogi, G., Santhiya, S., & Harivardhini, V. (2021). Impact of artificial intelligence (AI) on marketing. *International Journal of Aquatic Science*, 12(2), 3159–3167. https://doi.org/10.1109/IC3I56241.2022.10072425

Pathak, A., & Sharma, S. D. (2022). Applications of Artificial Intelligence (AI) in marketing management. 2022 5th International Conference on Contemporary Computing and Informatics (IC3I), 1738–1745. https://api.semanticscholar.org/CorpusID:257667405

Perkowitz, S. (2021). The bias in the machine: Facial recognition technology and racial disparities. *MIT Case Studies in Social and Ethical Responsibilities of Computing*. https://doi.org/10.21428/2c646de5.62272586

Pothitou, M., Hanna, R., & Chalvatzis, K. (2017). ICT entertainment appliances' impact on domestic electricity consumption. *Renewable and Sustainable Energy Reviews*, 69, 843–853.

Rattan, M., Bhambri, P., & Shaifali. (2005). Institution for a sustainable civilization: Negotiating change in a technological culture. In National Conference (Abstract Book) (p. 7). Presented at the "Technical Education in Globalized Environment- Knowledge, Technology & The Teacher" conference at SBBSIET, Padhiana, Jalandhar held during February 25–26, 2005.

Shanmuga, S. M., & Bhambri, P. (2024). Pulmonary and lungs nodule classification using deep learning. In P. Bhambri, S. Rani, & M. Fahim (Eds.), *Computational Intelligence and Blockchain in Biomedical and Health Informatics* (pp. 320–332). CRC Press, Taylor & Francis Group. https://doi.org/10.1201/9781003459347

Sharma, R., Bhambri, P., & Sohal, A. K. (2020). Energy bio-inspired for MANET. *International Journal of Recent Technology and Engineering*, 8(6), 5580–5585. https://www.ijrte.org/wp-content/uploads/papers/v8i6/F8522038620.pdf. https://doi.org/10.35940/IJRTE.F8522.038620

Singh, G., & Bhambri, P. (2023). Simulation analysis of AODV and DSDV routing protocols for secure and reliable service in Mobile Adhoc Networks (MANETs). In *Integration of AI-Based Manufacturing and Industrial Engineering Systems with the Internet of Things* (pp. 205–216). CRC Press.

Sudzina, F., & Pavlicek, A. (2020). Virtual offenses: Role of demographic factors and personality traits. *Information*, 11(4), 188. https://doi.org/10.3390/info11040188

Tiwari, R. (2023). The impact of AI and machine learning on job displacement and employment opportunities. *International Journal of Engineering Technologies and Management Research*, 7, 10.55041/IJSREM17506.

United Nations. (2015). Transforming our world: The 2030 agenda for sustainable development. https://sdgs.un.org/2030agenda

World Economic Forum. (2023). Network readiness index 2023 benchmarking the future of the network economy. https://networkreadinessindex.org/

18 Case Studies in Successful Digital Sustainability Ventures

K.S. Chakradhar, K. Neelima, Satyam, and Pankaj Bhambri

18.1 INTRODUCTION

The sustainability of any organization is essential as it affects community, as well as its economic, social, and entrepreneurial aspects. Entrepreneurs often encounter a moral conundrum. Notwithstanding their proclivity for rational decision-making and self-improvement, they prioritize private economic gains. Traditional entrepreneurship fails to address socio-ecological issues. Sustainable entrepreneurs capitalize on overlooked social and environmental concerns in order to generate value and capitalize on entrepreneurial prospects (Singh & Bhambri, 2023). They implement financially viable and novel business models in order to enhance and equalize the value proposition for consumers. They also employ indirect initiatives and youth education to promote the conservation of biodiversity in society (Sharma et al., 2020). However, due to the dual paradox of sustainable entrepreneurship, which involves balancing economic and socio-ecological benefits, new ventures tend to have a low preference for it. Nevertheless, a small number of researchers have discovered that a significant improvement in sustainability can be attained by fostering a high degree of creativity and ensuring strong institutional quality in entrepreneurial endeavours (Youssef, Boubaker, & Omri, 2018). Corporate entrepreneurship has become a paramount concern with the incorporation of digital transformation. Initiating digital sustainable entrepreneurship, small- and medium-sized enterprises (SMEs) have consequently altered the regulations governing sustainable entrepreneurship significantly.

A socio-ecological value proposition for consumers can be integrated into pre-existing technologies, products, and services (Ciriello, Richter, & Schwabe, 2018). Digital technology possesses a number of attributes, including connectivity, analyticity, perceptibility, and traceability, which facilitate the monitoring and evaluation of pollutant discharges, recycling, and so forth. As a result of pandemic situations that have enabled research into biometric-based authentication and secure transactions for even the smallest of values, the emphasis on digitization has shifted. This characteristic has overcome the constraints imposed by geographic space and exhibited remarkable scalability. Given the widespread adoption of the Sustainable

 DOI: 10.1201/9781003484226-18

Development Goals (SDGs) by numerous nations, digital sustainable entrepreneurship research has assumed a more prominent position. The integration of the environment, society, and economy has been bolstered by the development of digital technology, despite the introduction of a number of new sustainability concerns (Rattan et al., 2005). Entrepreneurs have benefited from the exponential growth of data size and the utilization and exploitation of rare metals, the recycling and treatment of wastes, and the substantial dissipation of electric power. Enterprises must maintain a focus on digital innovation in order to access and implement digital sustainability and ensure long-term viability. Undoubtedly, successful integration and transformation will generate newly identified opportunities that offer a valuable perspective for the organization's digital initiatives. Additionally, it contributes to the practice of digital innovation and minimizes the sustainability dilemma to the greatest extent possible. In addition, the cognition and behavior of the manager and organizer are interdependent and interact in the analysis of decision-making for sustainable corporate strategic initiatives.

According to a number of case studies and literature evaluations, it is evident that numerous facets of sustainability co-occur, as depicted in Figure 18.1. A considerable

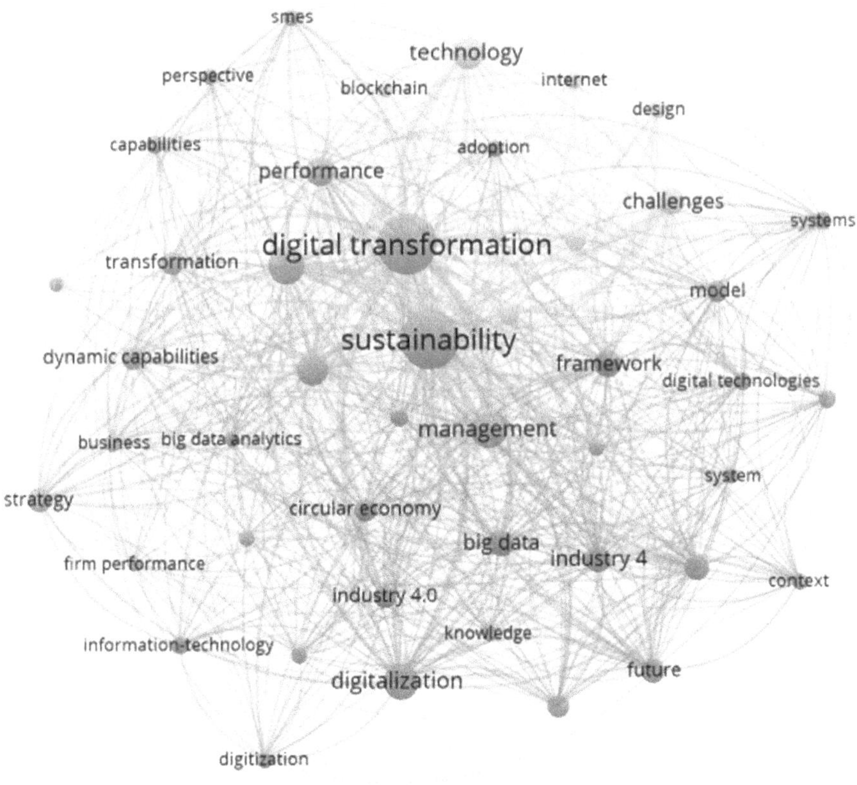

FIGURE 18.1 Sustainability co-occurrence (Martínez-Peláez, R., et al, 2023).

number of these contemporary technologies exert a profound influence on socio-economic and environmental conditions (Bhambri, 2024). The incorporation of sustainability principles into the operations of modern businesses, such as micro-, small-, and medium-sized enterprises (MSMEs), via digital transformation, is crucial for gaining a competitive edge and attracting investors and consumers who share these values. A number of studies indicate that in order to support decisions and strategies, owners or senior management must modify the organizational culture; stakeholders, meanwhile, must encourage the innovation process to enhance the sustainability-related competitiveness of local and global businesses.

18.2 ENTREPRENEURIAL ORGANIZATIONS INTEGRATED WITH DIGITAL TECHNOLOGIES

Entrepreneurs are encouraged to contribute to and safeguard the environment through the practice of sustainable entrepreneurship (Neumeyer & Santos, 2018). These business owners place a significant emphasis on the rights of stakeholders, including but not limited to employees, suppliers, and consumers. Sustainable entrepreneurs have a greater impact as a result of their value networks being intertwined with environmental innovations, which prioritize economic, environmental, cultural, and social values; in contrast to sustainability-oriented innovative firms that prioritize people in terms of economic, environmental, and social dimensions.

The case studies include:

- Desi Hangover – Hands That Make Footwear: as an engineer, Hitesh Kenjale fervently rethought the business concept to emphasize eco-friendly footwear while amassing vast social capital in the form of co-founders, suppliers, influential connections, and more. The Kolhapuris' artisanal footwear facilitated improvements in the cobbler community's standard of living and financial resources. In addition, the Radio Frequency Identification (RFID) chip that was integrated into the footwear furnished consumers with details pertaining to the artisans. The environment was protected by the utilization of sun-dried and hand-tanned leather upcycled from meat industry refuse (Desi Hangover, 2021). Additionally, organic substances such as turmeric and sal wood are used to color the leather.
- H.G. Group – Creating a Cultural Connect: Dushyant Parikh exerted a substantial influence on the textile manufacturing sector in India for a number of years. Once the designs have been approved, he proceeds with the production of one-of-a-kind fabrics for every brand (H.G. Group, 2021). Each artisan is notified of the confirmed orders, which include the necessary meters of fabric, design, and color. Garments crafted from silk, cotton and Tencel are embellished with shibori prints, dyed fibers, artisanal hand-prints, wax, and digital prints. The garments are therefore 99% recyclable.

A multitude of industries, including transportation, agriculture, and hospitality, have been profoundly impacted by digital innovations such as the Internet of Things (IoT), blockchain, and artificial intelligence (AI). Technological advancements have presented businesses with unprecedented benefits and novel opportunities to improve their operations.

As shown in Figure 18.2, the influence of digital technologies spans multiple sectors, with social and environmental value constituting the most critical aspect of sustainability, followed by stakeholder inclusion, venture viability, and entrepreneurial individuals, all of which are of equal importance.

As depicted in Figure 18.3, sustainable entrepreneurs utilize digital technology to construct an integrated value proposition within their business models (Shanmuga & Bhambri, 2024). In the absence of streamlined production systems and organizational structures, digital technologies can be employed to generate greater value efficiency through the mitigation of setup and maintenance expenses. Alternative sources of funding, such as crowdfunding, offer the opportunity to secure funding through online transactions that transcend geographical, temporal, informational, and financial limitations. These technologies facilitate dynamic pricing mechanisms by accelerating and simplifying financial transactions. Therefore, by democratizing financial markets, a profound understanding of digital solutions, such as (social) cryptocurrencies and their associated initial coin offerings, can represent a constructive shift in society.

18.3 NEW VENTURES

The early or new venturing phases of fledgling companies are influenced by the following characteristics (Trautwein & Fichter, 2018, p. 281):

- Informal and rapidly evolving working and management structures pose a challenge to the continuity and reproducibility of assessments (Picken, 2017), which result in variations in the categories of data utilized over time.
- Limitations in resources within sizable organizations (Lepoutre & Heene, 2006; Skala, 2019). The range of error in emission measurements for businesses is 30% to 40% (BCG, 2021).
- Business model and value chain volatility resulting from unpredictability and uncertainties such as acquisitions, mergers, alliances, and the like (Clarke-Sather et al., 2011; Hübel et al., 2022).
- When new ventures are in their early stages of evaluation and are new to the market, they frequently depend on data acquired from third parties. However, this reliance on such data may lead to inaccuracies or disappointments (Judl et al., 2015).
- Insufficient expertise (Lepoutre & Heene, 2006) as novel enterprises evaluate utilizing uncomplicated and user-friendly methodologies.

These enable the alignment of new endeavors with emergent benchmarks, such as future-oriented evaluation that places emphasis on potential results rather than current ones by employing predictive models (Bhambri & Bhandari, 2006).

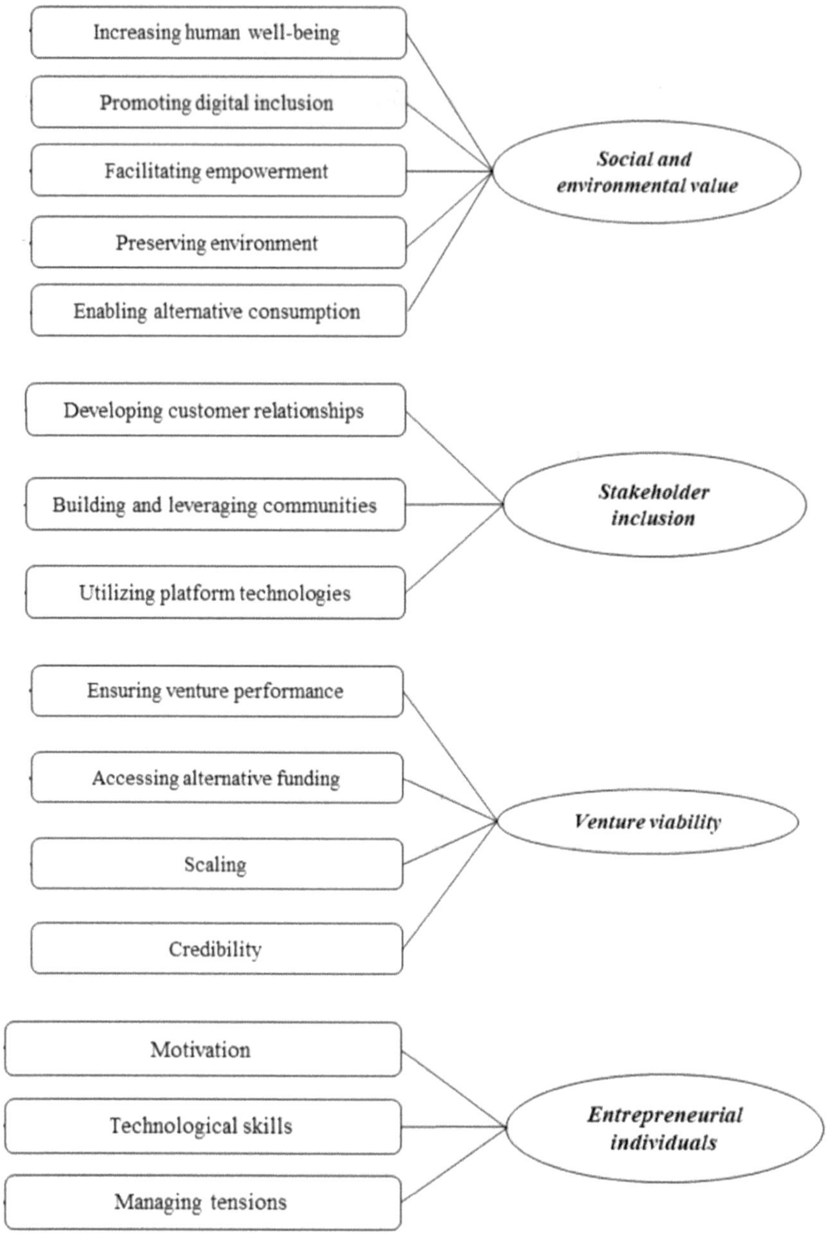

FIGURE 18.2 Impact of digital technologies (Patrick Holzmann, Patrick Gregori, 2022).

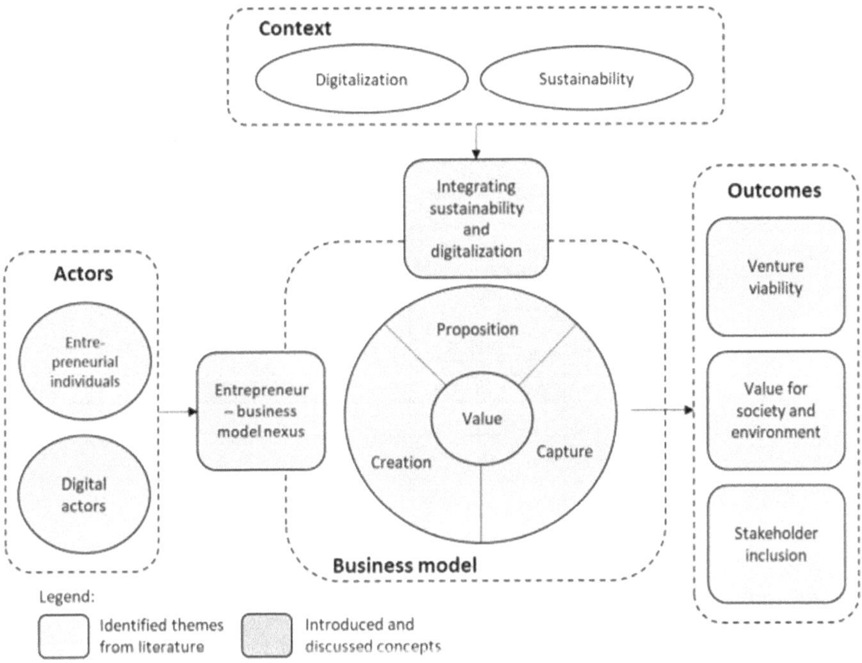

FIGURE 18.3 Integrated business model framework (Patrick Holzmann, Patrick Gregori, 2022).

- The influence of business models and their progressive development in relation to products, services, value chains, organizational structures, and more can be significant.
- As sectors undergo transformations characterized by permeable boundaries, emerging enterprises establish a world that is extraordinarily dynamic due to market forces, stakeholder engagements, and the ongoing evolution of entrepreneurial practices.
- The emphasis on sustainability pertains to outcomes or impacts that may be underestimated on various structural and contextual levels in relation to the potential of the new venture.

As a result, the modifications to the conceptual framework for evaluating new ventures are illustrated in Figure 18.4.

A sequential assessment of the organization is presented in Figure 18.4. This assessment incorporates various methodologies such as scenario analysis, environmental modeling, algorithmic models, business models, dynamic functions, and those capable of inducing changes in cognition and behavior, all of which are based on digital technologies. The effectuation-based approach to impact assessment takes into account the present state and long-term sustainability of organizations. Incorporating the CAMO logic (context, actions, mechanisms, outcomes)

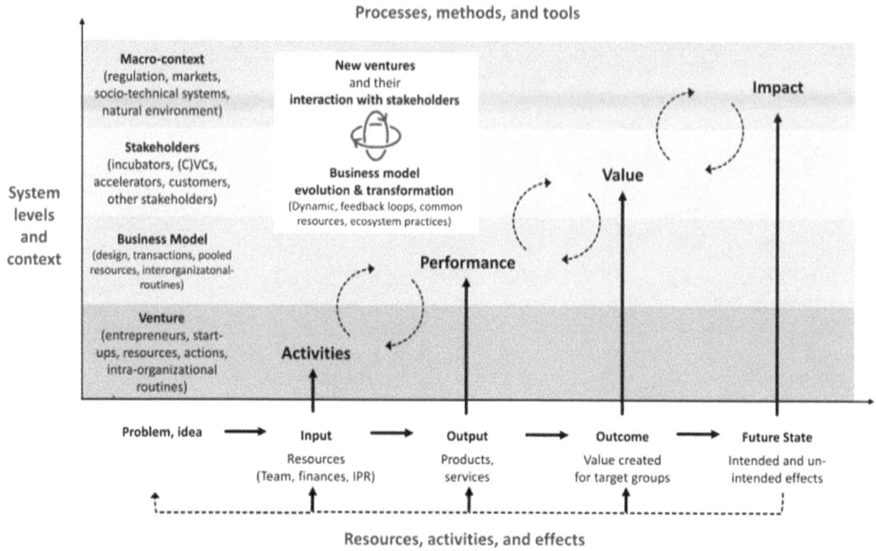

FIGURE 18.4 A conceptual framework for investigating sustainability (Klaus Fichter et al, 2023).

into the design of process-integrated assessments emphasizes co-creation, contextual awareness, and a systemic perspective. Within the realm of entrepreneurship, the "impact cycle" begins with the emergence of an innovative concept or the recognition of a perceived issue, such as climate change, which holds the capacity to provide a comprehensive or partial resolution to the said challenge (Bhambri et al., 2005). Entrepreneurial endeavors that employ resources including capital, personnel, trademarks, and intellectual property to effectuate a desired systemic transformation may generate potential sustainability impacts. These assets serve as inputs during the innovation and development of new products and services. Key performance indicators consist of the products and services that are made available to customers and other stakeholders. A macro-level effect may result from the interaction of public policies, market institutions, socio-technical systems (e.g., energy or mobility systems), and other variables with a multilevel system.

18.4 POTENTIAL DIRECTIONS FOR FUTURE RESEARCH

Potential research avenues, theoretical frameworks, and methodologies encompass the following:

- The integration of digital technologies and sustainability via business models, componential business model strategy incorporating possible research techniques such as Quantum Dot Cellular Automata (QCA), cluster analysis, and inference statistics, among others.

- (Re)conceptualizing digital technologies as actors within the business model aspects of perspective socio-materiality, a practice-oriented viewpoint, and prospective research methodologies including longitudinal qualitative analysis, netnography, and others.
- Forging the link between the entrepreneurial entity and the operational framework elements of viewpoint.

The business model places significant emphasis on a multitude of cohesive configurations of the business model, archetypes, the compatibility of digital agencies with business model configurations, the management of dynamic technologies and unanticipated changes, and the impact of these factors on the long-term trajectory of the business. Additionally, particular emphasis will be placed on the tools and techniques necessary to assist entrepreneurs.

Consideration is given to the investigation of value proposition exemplar technologies including big data and AI. Innovative forms of economic, environmental, and social value integrated with digital technologies will be emphasized. Their influence on a wide range of stakeholders, the direct or indirect channels through which any concerns can be addressed, and the process by which latent socio-environmental stakeholder requirements and emerging stakeholder segments are recognized. Social media and digital platforms may also be utilized to scale the impact of specific digital resources. A multifaceted examination of the benefits and obstacles associated with the integration of economic, environmental, and social value generation is possible. One can investigate the impact of digital connectivity that is based on shared sustainable values on the formation of stakeholder relationships and the resolution of spatial proximity issues. The mechanism by which technologies are identified and selected in accordance with business requirements is an additional obstacle for entrepreneurs.

In sustainable entrepreneurship, the function of cryptocurrencies and tokens as value capture mechanisms for novel digital payment methods can be investigated. An area of research can be devoted to analyzing the factors that influence customers' acceptability, usage, and perception of innovative digital payment alternatives when compared to predetermined forms of sustainable monetization. Sustainable entrepreneurs must devote considerable attention to the development or establishment of alternative funding sources that qualify for impact financing and quantify the value captured in the environment.

CONCLUSION

Sustainability of businesses and endeavors is a critical determinant of the human economy and way of life. This chapter elucidates the significance of sustainability in the development and expansion of the human race as a whole. It emphasizes the integration of digitization into entrepreneurial endeavors. Moreover, as understanding regarding eco-friendly products increases, so does the imperative to safeguard the environment. Sustained enterprise and individual effort could propel sustainable entrepreneurs to the forefront. The adoption of recyclable and environmentally friendly products by stakeholders and customers, in conjunction with conscientious consumption and waste reduction, could undoubtedly bolster the use and discard economy through more responsible production and consumption, as supported by digitized sustainable entrepreneurship. Engagement of various stakeholders—including intellectual and research institutions, entrepreneurial ecosystems, philanthropic organizations, consumers, governmental bodies, and industry participants—is crucial for the successful implementation of sustainability and circular economy initiatives. These factors influence the formulation of solutions to address social issues and environmental degradation. Entrepreneurs and new ventures are permitted to test out novel business models. Case study examples are presented to illustrate the concept. In addition, the factors that influence the sustainability assessment of new ventures and potential solutions to surmount them are detailed. In addition, future obstacles and strategies for their resolution are discussed. Integration with digital technologies promises exponentially expanding innovations with a customer concentration on socio-economically and environmentally friendly products and services, despite the fact that sustainable entrepreneurship has an effect on the global economy.

REFERENCES

BCG. (2021). New BCG GAMMA survey reveals that only 9% of organizations are able to measure their total greenhouse gas emissions comprehensively. Oct 13, 2021. Boston Consulting Group. https://www.bcg.com/press/13october2021-only-nine-percent-of-organizations-measure-emissions-comprehensively (Accessed 1 October 2022)

Bhambri, P. (2024). NFTs: Transforming digital ownership in the Web 3 era. In D. Darwish (Ed.), *Decentralizing the Online Experience With Web3 Technologies* (pp. 151–167). IGI Global. https://doi.org/10.4018/979-8-3693-1532-3.ch007.

Bhambri, P., & Bhandari, A. (2006). Different protocols for wireless security. In National Conference on Advancements in Modeling and Simulation at LLRIET, Moga held during January 20–21, 2006.

Bhambri, P., Singh, I., & Gupta, S. (2005). Robotics systems. In UGC Sponsored National Conference (Abstract Book) (p. 107). Presented at the "Emerging Computing Technologies" conference at HRMMV, Jalandhar held during September 29–30, 2005.

Ciriello, R. F., Richter, A., & Schwabe, G. (2018). Digital innovation. *Business & Information Systems Engineering*, 60, 563–569.

Clarke-Sather, A. R., Hutchins, M. J., Zhang, Q., Gershenson, J. K., & Sutherland, J. W. (2011). Development of social, environmental, and economic indicators for a small/medium enterprise. *International Journal of Accounting & Information Management*, 19, 247–266. https://doi.org/10.1108/18347641111169250.

Desi Hangover. (2021). Our shoes & their making: Materials. https://www.desihangover.com/making-of-ourupcycled-handmade-leather-shoes/

Fichter, K., Lüdeke-Freund, F., Schaltegger, S., & Schillebeeckx, S. J. D. (2023). Sustainability impact assessment of new ventures: An emerging field of research. *Journal of Cleaner Production*, 384, 135452. ISSN 0959-6526. https://doi.org/10.1016/j.jclepro.2022.135452.

H.G. Group. (2021). Bespoke. https://thecreativeartisans.com/bespoke/.

Holzmann, P., & Gregori, P. (2022). The promise of digital technologies for sustainable entrepreneurship: A systematic literature review and research agenda. *International Journal of Information Management*, 68, 102593. https://doi.org/10.1016/j.ijinfomgt.2022.102593.

Hübel, C., Weissbrod, I., & Schaltegger, S. (2022). Strategic alliances for corporate sustainability innovation: The 'how' and 'when' of learning processes. *Long-range Planning*, 55, 102200. https://doi.org/10.1016/j.lrp.2022.102200.

Judl, J., Mattila, T., Manninen, K., & Antikainen, R. (2015). Life cycle assessment and ecodesign in a day-Lessons learned from a series of LCA clinics for start-ups and small and medium enterprises (SMEs). Reports of the Finnish Environment Institute 18.

Lepoutre, J., & Heene, A. (2006). Investigating the impact of firm size on small business social responsibility: A critical review. *Journal of Business Ethics*, 67, 257–273. https://doi.org/10.1007/s10551-006-9183-5.

Martínez-Peláez, R., Ochoa-Brust, A., Rivera, S., Félix, V. G., Ostos, R., Brito, H., Félix, R. A., & Mena, L. J. (2023). Role of digital transformation for achieving sustainability: Mediated role of stakeholders, key capabilities, and technology. *Sustainability*, 15, 11221. https://doi.org/10.3390/su151411221.

Neumeyer, X., & Santos, S. C. (2018). Sustainable business models, venture typologies, and entrepreneurial ecosystems: A social network perspective. *Journal of Cleaner Production*, 172, 4565–4579. https://doi.org/10.1016/j.jclepro.2017.08.216.

Picken, J. C. (2017). From startup to scalable enterprise: Laying the foundation. *Business Horizons*, 60, 587–595. https://doi.org/10.1016/j.bushor.2017.05.002.

Rattan, M., Bhambri, P., & Shaifali. (2005). Institution for a sustainable civilization: Negotiating change in a technological culture. In National Conference (Abstract Book) (p. 7). Presented at the "Technical Education in Globalized Environment- Knowledge, Technology & The Teacher" conference at SBBSIET, Padhiana, Jalandhar held during February 25–26, 2005.

Shanmuga, S. M., & Bhambri, P. (2024). Pulmonary and lungs nodule classification using deep learning. In P. Bhambri, S. Rani, & M. Fahim (Eds.), *Computational Intelligence and Blockchain in Biomedical and Health Informatics* (pp. 320–332). CRC Press, Taylor & Francis Group. https://doi.org/10.1201/9781003459347.

Sharma, R., Bhambri, P., & Sohal, A. K. (2020). Energy bio-inspired for MANET. *International Journal of Recent Technology and Engineering*, 8(6), 5580–5585. https://www.ijrte.org/wp-content/uploads/papers/v8i6/F8522038620.pdf. https://doi.org/10.35940/IJRTE.F8522.038620.

Singh, G., & Bhambri, P. (2023). Simulation analysis of AODV and DSDV routing protocols for secure and reliable service in Mobile Adhoc Networks (MANETs). In *Integration of AI-Based Manufacturing and Industrial Engineering Systems with the Internet of Things* (pp. 205–216). CRC Press.

Skala, A. (2019). *Digital Startups in Transition Economies: Challenges for Management, Entrepreneurship and Education*. Springer International Publishing, Cham. https:// doi.org/10.1007/978-3-030-01500-8.

Trautwein, C., & Fichter, K. (2018). Manual for the sustainability assessment of start-ups: A practical tool for start-up teams, investors and funding organizations. Borderstep Inst. für Innovation und Nachhaltigkeit gGmbH.

Youssef, A. B., Boubaker, S., & Omri, A. (2018). Entrepreneurship and sustainability: The need for innovative and institutional solutions. *Technological Forecasting and Social Change*, 129, 232–241.

Index

Kolibri 360 GmbH
Ferdinand-Jühlke-Straße 7
99095 Erfurt, Deutschland
produktsicherheit@kolibri360.de